International Conference on Fluid and Chemical Engineering (ICFCE 2021)

Journal of Physics: Conference Series Volume 2097

Wuhan, China
29-30 October 2021

Editors:

Omid Mahian
Sébastien Déon

ISBN: 978-1-7138-5068-7
ISSN: 1742-6588

Printed from e-media with permission by:

Curran Associates, Inc.
57 Morehouse Lane
Red Hook, NY 12571

Some format issues inherent in the e-media version may also appear in this print version.

Institute of Physics
Dirac House, Temple Back
Bristol BS1 6BE UK

Phone: 44 1 17 929 7481
Fax: 44 1 17 920 0979

techtracking@iop.org

Additional copies of this publication are available from:

Curran Associates, Inc.
57 Morehouse Lane
Red Hook, NY 12571 USA
Phone: 845-758-0400
Fax: 845-758-2633
Email: curran@proceedings.com
Web: www.proceedings.com

TABLE OF CONTENTS

Journal of Physics: Conference Series **2097** (2021) 011001 doi:10.1088/1742-6596/2097/1/011001

Preface

2021 International Conference on Fluid and Chemical Engineering (ICFCE 2021) was scheduled to be held in Wuhan, China during October 29-30, 2021. Due to the continuing effect of Covid-19 epidemic, after the discussion of the conference committee, ICFCE 2021 will be postponed to be held on 19th November 2021 as an online conference. Organized by Hubei Zhongke Institute of Geology and Environment Technology, ICFCE 2021 intends to invite worldwide famous scientists, experts, scholars, and researchers for academic presentations.

ICFCE 2021 proceeding is a collection of outstanding submissions in the field of biofluid mechanics, chemically reacting fluids and combustion, computational fluid dynamics (CFD), experimental fluid mechanics, flow through porous media, fluid–solid interactions (FSI), geophysical fluid dynamics, granular/suspension flows, heat and mass transfer, hydrodynamics, physical, theoretical and computational chemistry, chemical engineering fundamentals, chemical reaction engineering, chemical engineering equipment design and process design, thermodynamics, catalysis & reaction engineering, particulate systems, crystallization, etc. 28 articles were selected from 58 submissions after peer-review, and authors participated in ICFCE 2021 with oral presentations and posters, promoting the communication among researchers and experts from institutes and universities, including Dalian University of Technology, Liaoning University of Technology School of Mechanical Engineering and Automation, University of Science and Technology of China, China North Engine Research Institute, etc.

Cordial gratitude is delivered to Technical Program Committee of ICFCE 2021 for the professional work on guarantying the high standard of conference as well as the proceedings. We believe the proceedings will enhance the exchange of ideas and state-of-art knowledge on Fluid and Chemical Engineering.

Professor Omid Mahian
Xian Jiaotong University, China

Associate Professor Sébastien Déon
University of Bourgogne-Franche-Comté, France

Technical Committee

Chair
Professor Mohamed Kheireddine Aroua, Sunway University, Malaysia

Editor
Professor Omid Mahian, Xian Jiaotong University, China.
Associate Professor Sébastien Déon, University of Bourgogne-Franche-Comté, France.

International Committee Members
Professor Mohamed Gamal El-Din, University of Alberta, Canada
Professor T V V L N Rao, Madanapalle Institute of Technology & Science, India
Professor Mohsen Sharifpur, University of Pretoria, South Africa
Professor Shengli Pu, University of Shanghai for Science and Technology, China
Professor Ilmutdin M. Abdulagatov, Dagestan State University, Russia
Professor Dr. Mohammad Mansur Rahman, Sultan Qaboos University, Sultanate of Oman
Professor Jianzhong Lin, Zhejiang University, China
Professor Alberto Brucato, University of Palermo, Italy
Professor Olivier SIMONIN, Institut de Mécanique des Fluides de Toulouse (IMFT), France
Professor P. Selvakumar, PSN College of Engineering and Technology (Autonomous), India
Professor Mohammad Mehdi Rashidi, University of Electronic Science and Technology of China, Chengdu, China
Associate Professor Mario Versaci, "Mediterranea" University, Italy
Associate Professor Ehsan Roohi, Xi'an Jiaotong University, China
Associate Professor Zhipeng Qin, Guangxi University, China
Associate Professor Dr. CHARMY S. KOTHARI, Institute of Pharmacy, Nirma University, India
Associate Professor Ramesh Kasi, University of Malaya, Malaysia
Associate Professor Renu Gupta, Dr B R Ambedkar National Institute of Technology Jalandhar
First class Research Director, Dr. Lingai LUO, Centre National de la Recherche Scientifique (CNRS), Laboratoire de Thermique et Energie de Nantes, France
Senior Lecturer (Assistant Prof.) Noor Fadiya Mohd Noor, University of Malaya, Malaysia
Senior Lecturer (Assistant Professor) Mohd Amiruddin Abd Rahman, Universiti Putra Malaysia, Malaysia
Assistant Professor Zafar Said, University of Sharjah, Sharjah, UAE
Assistant Professor Dhananjay Yadav, University of Nizwa, Sultanate of Oman
Assistant Professor Dr. Ehnber Naheed Maraj, HITEC University, Pakistan
Assistant Professor Dr. Rashid Mehmood, HITEC University Taxila, Pakistan
Assistant Professor Muhammed Abdullah Hassan Ahmed, Cairo University, Egypt
Assistant Professor Ehtsham Azhar, PMAS-Arid Agriculture University, Pakistan
Assistant Professor Md Readul Mahmud, Independent University, Bangladesh (IUB), Bangladesh

Peer review declaration

All papers published in this volume of Journal of Physics: Conference Series have been peer reviewed through processes administered by the Editors. Reviews were conducted by expert referees to the professional and scientific standards expected of a proceedings journal published by IOP Publishing.

- **Type of peer review: Single-blind / Double-blind / Triple-blind / Open / Other (please describe)**
  Double-blind: While receiving a new submission, we/our editors firstly check the scope of it whether suitable for our conference. If not, it will be rejected directly. If yes, it will be sent out to check plagiarism and then peer review. We choose 2 reviewers for each of the submission and after getting the two reviews, we conclude a final result for the paper. Once the two reviews show totally different attitudes, we will find another reviewer for it until we get the final result.
- **Conference submission management system:**
  Through our email system
  Email: academic@icfce-conf.org,
  In charge: Terasa Lei
- **Number of submissions received:**
  58
- **Number of submissions sent for review:**
  49 (papers out of topic or plagiaristic were rejected directly)
- **Number of submissions accepted:**
  28
- **Acceptance Rate (Number of Submissions Accepted / Number of Submissions Received X 100):**
  48.28%
- **Average number of reviews per paper:**
  2-3
- **Total number of reviewers involved:**
  About 35
- **Any additional info on review process:**

- **Contact person for queries:**
  Name : Terasa Lei
  Affiliation: Hubei Zhongke Institute of Geology and Environment Technology
  Email : academic@icfce-conf.org

Study on Influencing Factors and Laws of Fuel Injection Consistency of Common Rail Injector

Ziwei Zhang* and Chunlong Xu

China North Engine Research Institute, Tianjin 300400, China
Emai:751334315@qq.com

Abstract. In order to study the influence of parameters of common rail injector internal components on cycle injection consistency, its simulation model is established by AMESim, and the model is validated by the experimental injection rate data. The effects of solenoid valve spring preload, gag bit lift, fuel discharge hole diameter, fuel inlet hole diameter, needle valve lift, needle valve preload and nozzle diameter on the change of injection quantity under different operating conditions are studied by simulation method, and the impact weight of each parameter on fuel injection consistency is analyzed. The results show that the preload of solenoid valve, fuel discharge hole diameter, oil inlet hole diameter, needle valve lift and nozzle diameter are the main parameters affecting the consistency of cycle injection. The percentages of five parameters influencing on the consistency of cyclic injection are 8.68–16.84%, 11.41–23.68%, 17.2086–37.74%, 12.772–18.34% and 9.69–37.27% respectively.
Keywords. Common rail injector, Injection consistency, Operating condition, Impact weight.

1. Introduction

Due to the high injection pressure, it is very important for the common rail injection system to realize the stable and accurate control of injection quantity. The injector injects high-pressure fuel into cylinder, which is the key component of the system [1]. The solenoid valve on the injector generates electromagnetic force to control the action of ball valve and the control chamber pressure, so as to realize the needle valve movement control, and then flexibly adjust the injection duration, pulse width and injection frequency. The internal components of fuel injector are matched precisely and complexly [2]. The change of parameters of each component will cause the fluctuation of injection quantity, lead to the deterioration of injection consistency, and affect the working stability and reliability of diesel engine.

The research on the influence of structural parameters on the injection consistency can provide theoretical guidance for the design of injector, and is of great significance to improve the working stability and reliability of common rail. Through simulation analysis, this paper studies the influence of common rail injector solenoid valve spring preload, gag bit lift, fuel discharge hole diameter, fuel inlet hole diameter, needle valve lift, needle valve spring preload and nozzle diameter on the fluctuation of injection quantity, and analyzes the influence law of various parameters on injection consistency [34].

2. Simulation Model and Validation

Fuel pump, common rail pipe, injector are the main components of common rail system. On the basis of the operating principle of the system, the simulation model of the system is established by using

AMESim software, as shown in figure 1.

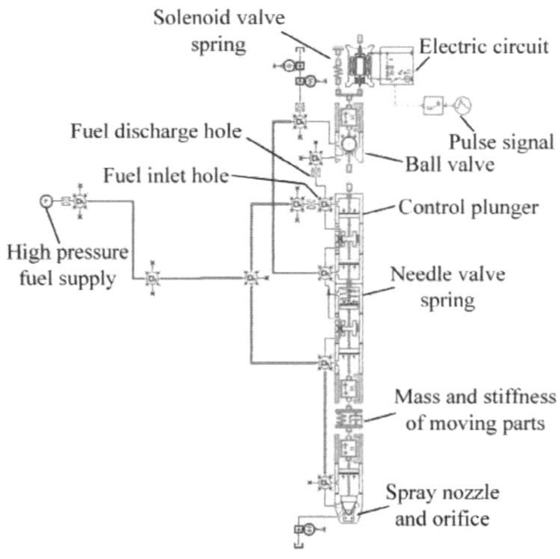

Figure 1. Simulation model.

Table 1. System parameters.

Parameter	Value
Rail pressure	180 MPa
Injection frequency	13.33 Hz
Solenoid air gap	0.065 mm
Solenoid valve coil turns	37
Inner diameter of common rail pipe	14 mm
Length of common rail pipe	720 mm

In order to validate the simulation model and enable it to reflect the injection characteristics of common rail system, it is verified by using the test data measured. Table 1 illustrates the main control and structural parameters of the system. Figure 2 shows the comparison between test values and simulation of fuel injection rate with two operating conditions, which are 180 MPa rail pressure and injection pulse of 0.7 ms and 0.8 ms respectively. From figure 2 we can know that the injection rate calculated by simulation is in high agreement with the test values, indicating that the established simulation model can accurately predict the fuel injection characteristics.

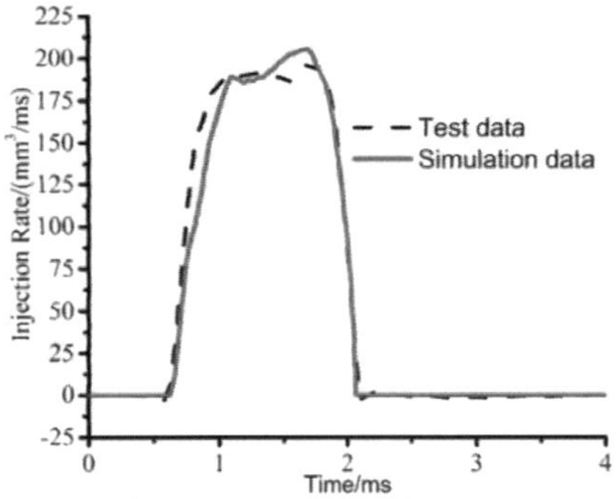

(a) Injection rate at 0.7 ms injection pulse

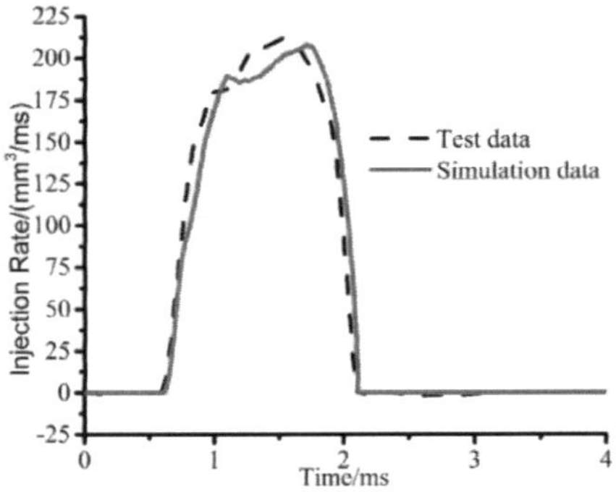

(b) Injection rate at 0.8 ms injection pulse

Figure 2. Model verification.

3. System Parameters and Test Condition

On the basis of the injector model shown in figure 1, the solenoid valve spring preload, gag bit lift, fuel discharge hole diameter, fuel inlet hole diameter, needle valve lift, needle valve spring preload and nozzle diameter are selected as influencing factors to study the injection consistency, and the fluctuation of cycle fuel injection is evaluated as an estimation index [5]. In addition, according to the injector structural parameters and their changes in the production assembly process and service cycle, the reference values and variation range of each parameter are illustrated in table 2. The injection pulse range of 0.6-1.9 ms and rail pressure range of 50-80 MPa are used for operating conditions during the research.

Table 2. Parameters of injector.

System parameter	Minimum value	Reference value	Maximum value
Spring preload of solenoid valve	62 N	64 N	66 N
Gag bit lift	0.058 mm	0.06 mm	0.062 mm
Diameter of fuel diacharge hole	0.268 mm	0.27 mm	0.272 mm
Diameter of fuel inlet hole	0.218 mm	0.22 mm	0.222 mm
Needle valve lift	0.24 mm	0.25 mm	0.26 mm
Needle valve spring preload	24 N	32 N	40 N
Nozzle diameter	0.231 mm	0.241 mm	0.251 mm

The maximum absolute value of the deviation between the injection quantity as the structure parameter is the maximum or minimum value and the quantity under the condition of reference value is taken as the fluctuation of the quantity caused by the change of parameters [6]. The calculation formula is as follows.

$$\sigma_F = \max\left(abs\left(\sigma_{\max} - \sigma_{ref}\right), abs\left(\sigma_{\min} - \sigma_{ref}\right)\right) \quad (1)$$

where σ_F is the fluctuation of injection quantity, σ_{max} is the quantity corresponding to maximum parameter value, σ_{ref} is the quantity corresponding to reference parameter value, and σ_{min} is the quantity corresponding to minimum parameter value.

4. Analysis of Parameters Effect on Fuel Injection Consistency

4.1. Spring Preload of Solenoid Valve

Figure 3 shows the fluctuation characteristics of injection quantity at three different solenoid valve spring preload levels-62 N, 64 N and 66 N. In figure 3, the injection pulse range is 0.6-1.9 ms and the rail pressure range is 50-180 MPa. The maximum injection fluctuation is 6.42707 mm^3 in figure 3, which occurs at the operating point of 180 MPa rail pressure and 0.4 ms injection pulse. When the injection pulse is more than 1.1 ms, the quantity fluctuation changes smoothly. As the rail pressure increases, the fluctuation increases. Under the same rail pressure, the change of injection pulse has little effect on the quantity fluctuation. The spring preload of solenoid valve determines its action response speed and the needle valve falling speed, then causes the injection fluctuation [7].

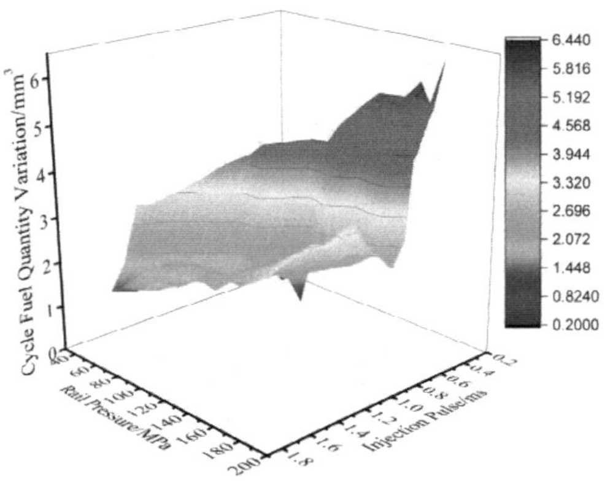

Figure 3. Spring preload of solenoid valve effects on cycle fuel injection.

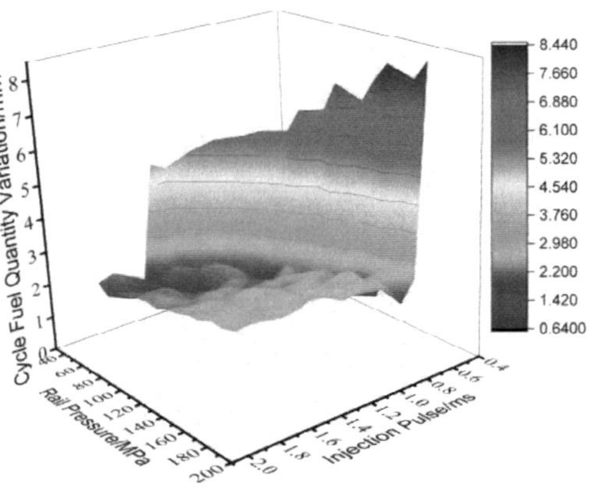

Figure 4. Gag bit lift effects on cycle fuel injection.

4.2. Gag Bit Lift

Figure 4 shows the fluctuation characteristics of injection quantity at three different gag bit lift levels – 0.058 mm, 0.06 mm and 0.062 mm. In figure 4, the injection pulse range is 0.6-1.9 ms and the rail pressure range is 50-180 MPa. The maximum injection fluctuation is 8.4341mm^3 in figure 4, which occurs at the operating point of 180 MPa rail pressure and 0.6 ms injection pulse. When the injection pulse is more than 1.6 ms, the injection fluctuation is small, and the average fluctuation is less than 2.4 mm^3. With small injector pulse, the simultaneous motion of the ball valve and control plunger causes significant fuel pressure fluctuations in the control chamber. The change of gag bit lift can affect the fuel discharge speed of control chamber [8]. In addition, it will also affect the ball valve reset time and change the control chamber fuel discharge time, thus increasing the impact on the control chamber pressure and injection duration and resulting in greater fluctuation of cycle fuel injection quantity.

4.3. Fuel Discharge Hole Diameter

Figure 5 shows the fluctuation characteristics of injection quantity at three different fuel discharge hole diameter levels-0.058 mm, 0.06 mm and 0.062 mm. In figure 5, the injection pulse range is 0.6-1.9 ms and the rail pressure range is 50-180 MPa. The maximum injection fluctuation is 6.9282 mm^3 in figure 5, which occurs at the operating point of 160 MPa rail pressure and 0.7 ms injection pulse. The injector with small injection pulse has short the fuel injection duration, and the fuel pressure setup time of control chamber is short. The injection fluctuates strongly due to the simultaneous action of the control valve and the control plunger. At this time, the change of fuel discharge hole diameter has a great impact on the control chamber pressure, resulting in dramatic injection fluctuation. With large injector pulse, the control chamber pressure has enough stability time. The change of fuel discharge diameter has little impact on the control chamber pressure before injection termination, and its impact on the injection quantity is mainly reflected in the initial stage of injection, so the fluctuation of cycle fuel injection quantity changes smoothly.

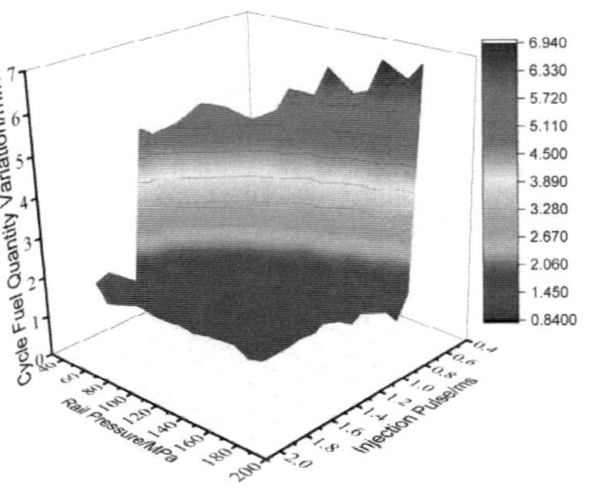

Figure 5. Fuel outlet hole diameter effects on cycle fuel injection.

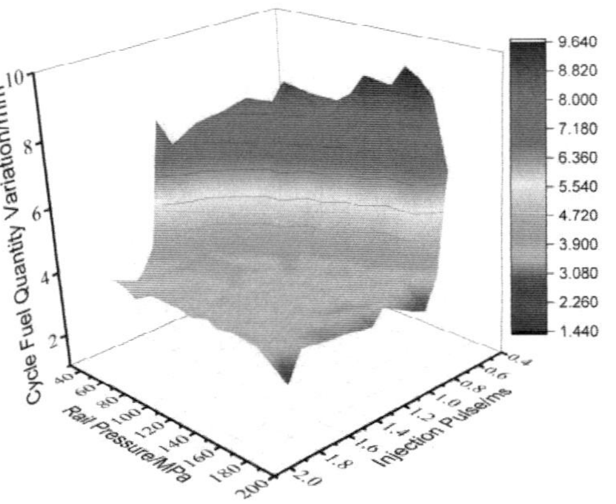

Figure 6. Fuel inlet hole diameter effects on cycle fuel injection.

4.4. Fuel Inlet Hole Diameter

The diameter of fuel inlet determines the throttling effect and flow characteristics of fuel, and its change affects the setup speed of control chamber pressure. Figure 6 shows the fluctuation characteristics of injection quantity at three different fuel inlet hole diameter levels-0.218 mm, 0.22 mm and 0.222 mm. In figure 6, the injection pulse range is 0.6-1.9 ms and the rail pressure range is 50-180 MPa. The maximum injection fluctuation is 9.6332 mm^3 in figure 6, which occurs at the operating point of 160 MPa rail pressure and 0.7 ms injection pulse. When the injection pulse is more than 1.6 ms, the injection fluctuation changes smoothly. The reason for the large injection fluctuation at small injection pulse is that the fluctuation of control chamber pressure is caused by the movement of ball valve and control plunger, and the difference of control chamber pressure due to the inlet hole diameter variation has a more obvious impact on the fuel injection duration, resulting in severe fluctuation of cycle fuel injection quantity.

4.5. Needle Valve Lift

The needle valve lift determines its motion duration, the change of which can result in the fluctuation of injection quantity. Figure 7 shows the fluctuation characteristics of injection quantity at three different needle valve lift levels-0.24mm, 0.25mm and 0.26mm. In figure 7, the injection pulse range is 0.6-1.9 ms and the rail pressure range is 50-180 MPa. The maximum injection fluctuation is 7.797mm^3 in figure 7, which occurs at the operating point of 180MPa rail pressure and 1.9ms injection pulse. Besides, the fluctuation of injection quantity caused by the change of needle valve lift is 0 on the low pressure and small pulse width area of figure 7. The reason for this is when the low-pressure pulse width is small, the injection duration is short, and the needle valve cannot reach the maximum lift. Accordingly, the change of needle valve lift has no effect on injection quantity.

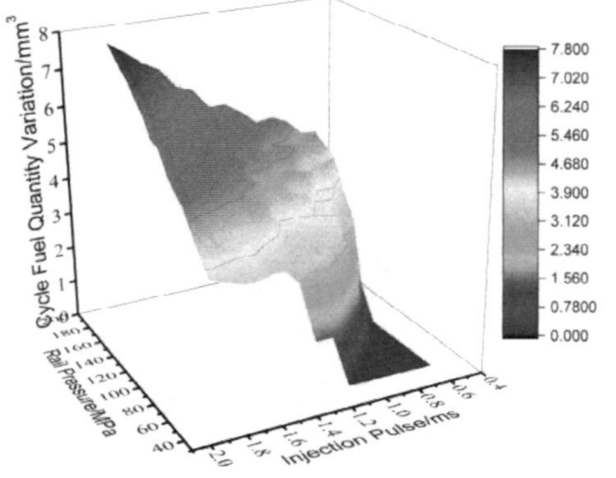

Figure 7. Needle valve lift effects on cycle fuel injection.

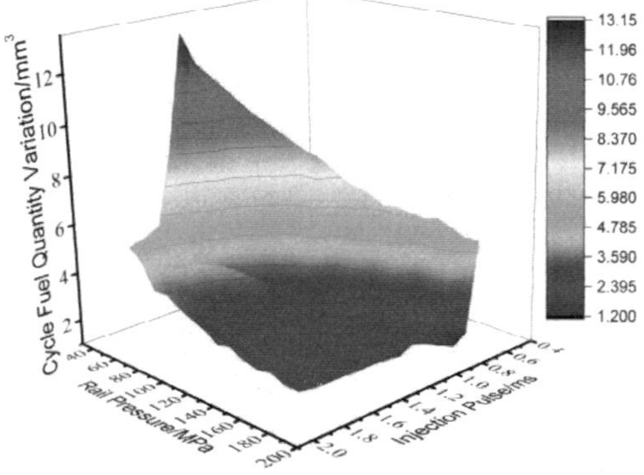

Figure 8. Needle valve spring preload effects on cycle fuel injection.

4.6. Needle Valve Spring Preload

The preload of needle valve spring affects its motion response, resulting in the fluctuation of injection quantity. Figure 8 shows the fluctuation characteristics of injection quantity at three different needle valve spring preload levels-24 N, 32 N and 40 N. In figure 8, the injection pulse range is 0.6-1.9 ms and the rail pressure range is 50-180 MPa. The maximum injection fluctuation is 13.1237mm^3 in figure 8, which occurs at the operating point of 50 MPa rail pressure and 1.5 ms injection pulse. Taking into account all operating conditions, the fluctuation of injection quantity caused by the change of needle valve preload decreases with rail pressure increasing. The fluctuation of injection quantity caused by the change of needle preload has significant effect on the small injection pulse area, and the change of pulse width has no significant effect on the injection quantity fluctuation in the large pulse width area.

4.7. Nozzle Diameter

The nozzle diameter directly determines the flow area of fuel injection into the cylinder, so it has the impact on the injection quantity. Figure 9 shows the fluctuation characteristics of the quantity at three different nozzle diameter levels-0.231mm, 0.241mm and 0.251mm. In figure 9, the injection pulse range is 0.6-1.9 ms and the rail pressure range is 50-180 MPa. The maximum injection fluctuation is 16.1671 mm^3 in figure 9, which occurs at the operating point of 180 MPa rail pressure and 1.9 ms injection pulse. The injector with large nozzle diameter has large the flow area, which makes injection quantity increase. In figure 9, the injection quantity fluctuation changes linearly with the rail pressure and injection pulse.

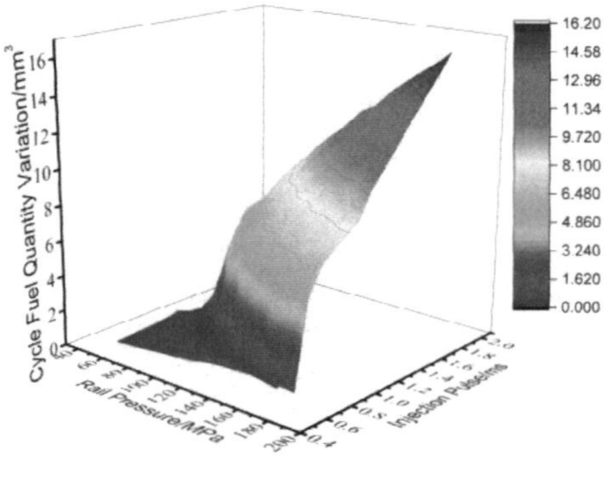

Figure 9. Nozzle diameter effects on cycle fuel injection.

5. Quantitative Analysis of Injection Fluctuation

From the above analysis, it can be known that the parameters of injector have a certain impact on the injection quantity, and the influence degree and reasons of each parameter are different due to the complex internal structure of the injector. In order to reveal the main parameters affecting the consistency of cycle injection, we quantitatively analyze the influence of each parameter on the injection quantity. We define the influence factor of cycle fuel injection quantity to eliminate the influence caused by different parameter variation ranges, and its calculation formula is as follows.

$$S_R = \frac{\Delta Q_t / (A_{\max} - A_{\min})}{Q_t / A_{ref}} \quad (2)$$

where, S_R is the influence factor of the quantity when pulse width is t, ΔQ_t is the cycle fluctuation caused by the change of parameter value under the corresponding operating conditions, Q_t is the cycle injection quantity when the parameter takes the reference value, A_{max} is the maximum value of the parameter, A_{min} is the minimum value of the parameter, and A_{ref} is the reference value.

Figure 10 shows the influence factors of injection quantity under different operating conditions. Since there are many operating points, the injection pulse of 0.6 ms, 1 ms and 1.5 ms are selected to analyze the variation law of the fluctuation percentage of injection quantity caused by the parameter variation under different rail pressure conditions.

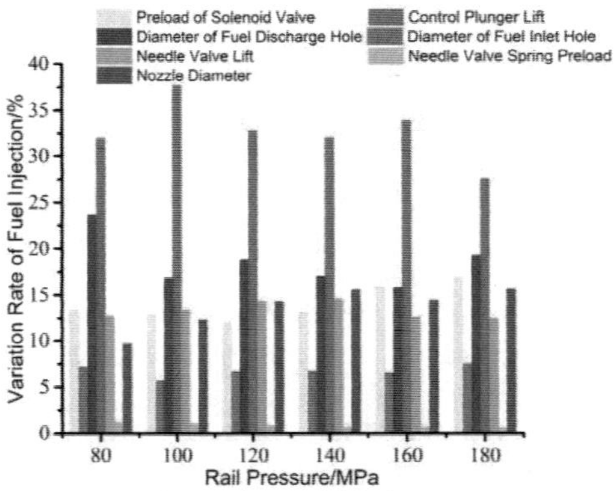

(a) Pulse Width 0.6ms

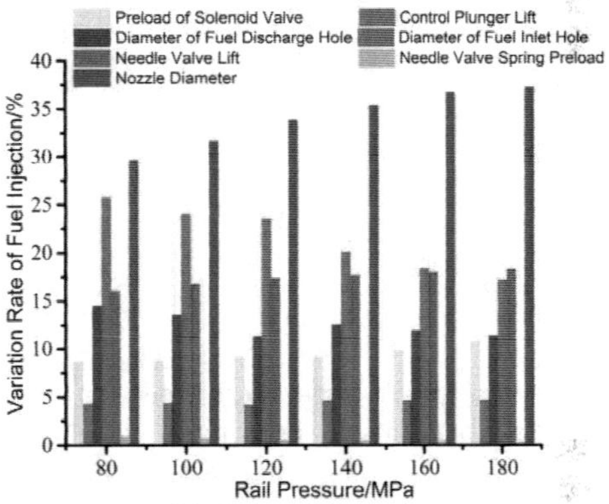

(b) Pulse Width 1ms

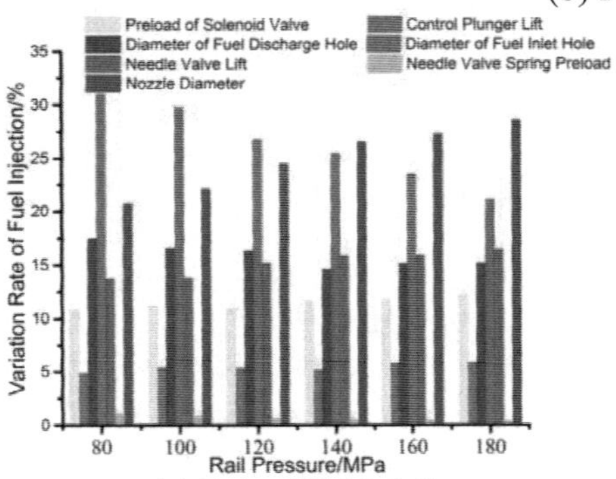

(c) Pulse Width 1.5ms

Figure 10. Variation law of injection quantity fluctuation percentage.

It can be seen from figure 10 that the quantitative indexes of the fluctuation percentage of the quantity corresponding to the change of solenoid valve spring preload and gag bit lift are 8.68–16.84% and 4.3–7.58% respectively. The quantitative index of the percentage is consistent with the change of fuel pressure, but opposite to the change of injection pulse. The quantitative indexes of the fluctuation percentage of the quantity corresponding to the changes of fuel discharge hole diameter, fuel inlet hole diameter and needle valve spring preload are 11.41-23.68%, 17.2086–37.74% and 0.319–1.23% respectively. The quantitative indexes of the percentage is opposite with the change of fuel pressure and injection pulse. The quantitative indexes of the percentage fluctuation of cycle fuel injection caused by needle valve lift and nozzle diameter are 12.772–18.34% and 9.69–37.27% respectively. The quantitative index of the percentage is consistent with the change of fuel pressure and injection pulse.

6. Conclusion

By using AMESim simulation, the influence law of parameters of injector system on injection fluctuation under all operating conditions is obtained.

The influence of parameters on injection fluctuation is more obvious in small injection pulse. Since the injector needle valve cannot reach the maximum lift in this case, the fluctuation of the quantity shows large nonlinearity and great fluctuation under the influence of parameters.

The quantitative index of the fluctuation percentage of injection caused by the change of each parameter is obtained. The analysis shows that the quantitative index caused by parameter change is mainly affected by fuel pressure. The spring preload of solenoid valve, the fuel discharge hole diameter, the oil inlet hole diameter, the needle valve lift and nozzle diameter are the most significant parameters affecting the fuel injection consistency. The quantitative indexes of the fluctuation percentage of injection caused by the changes of the above five parameters are 8.68-16.84%, 11.41-23.68%, 17.2086-37.74%, 12.772-18.34% and 9.69-37.27% respectively.

References

[1] Wang Z, Wang P, Tan X Y, et al. 2018 Effect of injector structure on fuel injection consistency *Transactions of CSICE* **36(6)** 519-523.
[2] Fan L Y, Bai Y, Jing S, et al. 2016 Fuel injection quantity fluctuation in multiple injections of high-pressure common-rail fuel injection system *J. Harbin Eng. Univ* **37(8)** 1063-1069.
[3] Han X M, Lin X D, Li D G, et al. 2017 Research on injection quantity control strategy based on the target engine speed curve during the start process *Chin. Intern. Combust. Engine Eng.* **38(3)** 46-51.
[4] Han X M, Lin X D and Li D G 2017 A study on the control strategy for the transient fuel injection quantity in starting process of a light diesel engine *Automotive Engineering* **39(1)** 23-27.
[5] Tsai W C and Zhan T S 2018 An experimental characterization for injection quantity of a high-pressure injector in GDI engines *J. Low Power Electron. Appl.* **8(36)** 1-21.
[6] Wang L, Li G X, Xu C L, et al. The investigation of geometric parameters on the injection characteristic of the high pressure common-rail injector *J. Eng. Gas Turbines Power.* **141(2)** 1116-1138.
[7] Ferrari A, Jin Z R, Vento O, et al. An injected quantity estimation technique based on time-frequency analysis *Control Engineering Practice* **116(1)** 50-62.
[8] Zhou L Y, Liang Y, Xu M F 2020 Simulation study on the impact of uneven fuel injections of each hole of the diesel engine injector on the combustion process *IOP. Ser.:Mater. Sci. Eng.* **782(1)** 1-9.

Fluid-Structure Interaction Approach to Single Particle in a Square Microchannel

Xiang Li and Ying Lin*

School of Mechanical Engineering, Shanghai Institute of Technology, Shanghai, 201418, China
Email: liny@sit.edu.cn

Abstract. Inertial microfluidic technique has been widely applied on particle/cell manipulation and detection. To understand the physical principle of this technique more detailed, the interaction of fluid and particle was studied through the Fluid-Structure Interaction (FSI) method. The equilibrium positions of finite-size particles with different diameters were simulated at moderate Reynolds numbers. The flow structure around two typical particles was analysed. The vortex in the front of the particle retards particle's translation leading to the lag velocity increasing. Finally, the rotation velocity and the rotational-induced force analysed quantitatively to demonstrate that particle's self-rotation significantly promotes its inertial migration.
Keywords. Inertial microfluidics, FSI, Rotation-induced force.

1. Introduction

In recent years, inertial microfluidic technique was widely applied on particle/cell manipulation and detection, such as cell focusing in flow cytometry [1], real-time detection of airborne bacteria [2], and monitoring chemical species in the environment [3], and so on. This technique originated from a phenomenon observed by Segre and Silberberg in 1961 [4], that particles would migrate laterally to the annulus equilibrium positions in a macroscale cylindrical pipe. To explore the mechanism of inertial migration in microscales, many researchers conducted theoretical investigations by the point-like-particle hypothesis [5-7]. They found the net lift force, $F_L \propto \rho U^2 a^4/H^2$ [5], dominated the inertial migration, where U is the flow velocity and H is the channel dimension. However, as using the F_L to focus randomly distributed particles in rectangular section channels, Di Carlo et al. [8] unexpectedly observed that particles ordered longitudinally along the channel wall, which could not be explained by theoretical results. Furthermore, Miura et al. [9] observed the particle equilibrium position moved monotonously towards the channel center with an increasing Re through experiments, which was contrary to the antecedent studies. And, Amini et al. [10] discovered the presence of particle would unavoidably disturb the fluid around it which was neglected before. These stimulated indepth researches and, more quantitively analysis need to understand the inertial effect deeply. Thus, numerical methods, such as Direct Numerical Simulation (DNS) [11], Immersed Boundary Method (IBM) [11], Lattice Boltzmann Method (LBM) [12] and Fluid-Structure Interaction methods (FSI), were applied for the available quantitative motion parameters of the fluid and particles. Among these methods, the FSI can handle boundaries of large displacements through remeshing which leads the simulation work similarly to experiment status. Meanwhile, the FSI can monitor particle's rotational velocity and calculate the force exerted on the particle [13-15], which exhibit its power to investigate the interaction of a particle and fluid field.

In this paper, the interaction of fluid and particle with different diameters was researched at moderate Reynolds numbers through the FSI method. The equilibrium positions of different particles were investigated at different Reynolds numbers in square microchannel. Then, the flow structures around two typical particles were presented and analysed to understand the particle's active property in the flow. Finally, the rotation velocity of equilibrium particles was discussed and then, preliminarily analysed the rotation-induced lift force exerted on particles.

2. Method and Model

In order to study the fluid and particle interaction and further the particle focusing characteristics, the Fluid-Structure Interaction (FSI) module in the commercial computational fluid dynamics (CFD) software COMSOL Multiphysics (version 5.5, COMSOL Inc., Stockholm, Sweden) was applied. In this module, the fluid flow is described by the Navier-Stokes equations, which provides a solution for the velocity field $\vec{u}_{fluid}$ and the total force exerted on the particle boundary:

$$\vec{f} = \vec{n} \bullet [-pI + (\mu(\nabla \vec{u}_{fluid} + (\nabla \vec{u}_{fluid})^T) - \frac{2}{3}\mu(\nabla \bullet \vec{u}_{fluid})I)] \quad (1)$$

where $\vec{n}$ is the normal vector to particle surface, p the pressure, I the identity matrix, and ρ and μ the density and viscosity of the fluid, respectively.

The 3D geometry and meshes were modeled, as shown in figure 1. The coordinate system locates at the center of channel's inlet (the left plane). In all the models, the length L was defined 12-20 times of the particle diameter to cover the flow field affected by the particle [16]. The particle was released freely in the plane of y=0 and, its initial position in the cross section was shown in figure 1(a). The residual error that controlled the internal time steps was 5×10^{-4}, which lead the time step to be about 1×10^{-6} s. And, the remeshing occurred when the mesh quality became less than the limit 0.2 (defined by the software). The mesh independence tests were conducted for all working conditions. The density of the particles was 1.05×10^3 kg·m^{-3} as the same value of general experimental particle. The density and viscosity of water was 1×10^3 kg·m^{-3} and 1×10^{-3} kg·m^{-1}·s^{-1}, respectively. The validity of the FSI method was testified through comparing the equilibrium position between present simulation and precedent experiment [17]. The maximum deviation of 8.2% was obtained, and the present method considered reliable.

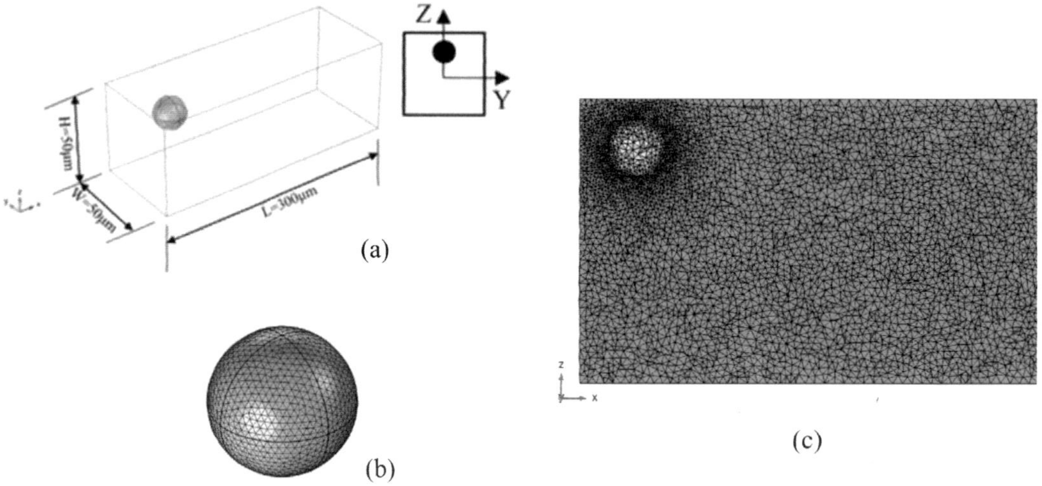

Figure 1. (a) the geometry structure and the particle initial position in the cross section, (b) the mesh of flow field and a particle on the plane of y=0, (c) the mesh of particle surface.

3. Results and Discussion

3.1. Equilibrium Positions of Different a/Hs at Different Re

The free particle would migrate to a fixed position near the center of channel face in the cross section named the channel face equilibrium (CFE) position. The equilibrium position z_{eq} was defined by the distance between the particle center and the channel axis. The nondimensional expression z_{eq}/h ($h=H/2$) was depicted as a function of *Re* (figure 2). By comparing the data at the same *Re*, it is clear that the particle with a small a/H obtains a large z_{eq}/h value. This means small particles close to channel wall and large particles to channel center relatively. As the value of a getting larger, the enlarging rate of a^6 is greater than that of a^3. The wall-induced lift force, $F_{LW}=C_{LW}\rho U^2 a^6/H^4$, which drives the particle to the channel center increases faster than the shear-induced lift force, $F_{LS}=C_{LS}\rho U^2 a^3/H$, which drives particle migrate to the channel wall, where C_{LW} is the wall-lift coefficient and C_{LS} the shear-lift coefficient.

Furthermore, the particle of $a/H=0.1$ would migrate to the channel wall slightly with *Re* risen, which is agreeable with previous investigations [17]. In Poiseuille flow, F_{LS} and F_{LW} govern the inertial migration of particles of the small a/H. As increasing *Re*, the C_{LW} keeps constant while the C_{LS} rises [18]. F_{LS} dominates the particle migration and drives particle to the channel wall. When $a/H\geq0.2$, the equilibrium position vibrates in a tiny scale as *Re* increased. That is not coincident with the trend displayed by the particle of $a/H=0.1$. The equilibrium position of the large particle would not be simply determined by Reynolds number. It reveals that the large particle does not act as a passive component in the flow only. As a positive component, the particle would disturb the flow field around it and, interfere its lateral motion. This would be discussed in following analysis.

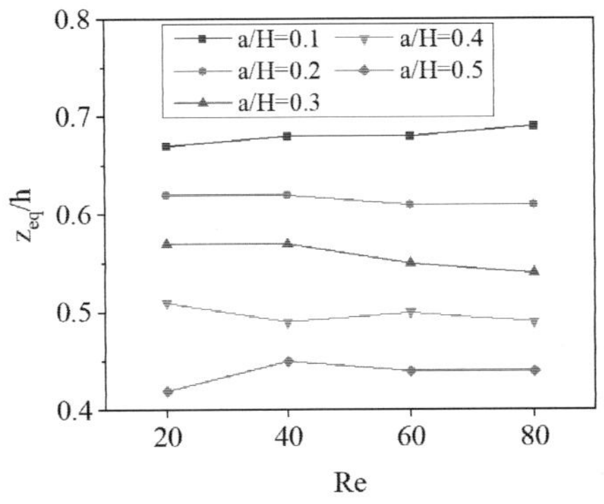

Figure 2. The nondimensional equilibrium position as a function of the Reynolds number.

3.2. Flow Structure around Particle

Two typical conditions of the migration of 5µm and 15µm particles at $Re=80$ were selected to present the flow characteristic. The absolute velocity contours of fluid around the equilibrium particles on the plane of the symmetry (y=0) were plotted in figure 3. The streamlines representing the relative velocity of the fluid to particle were illustrated on the same plane. For the 5µm particle, three vortices are formed to occupy about one eighth of the selected area as shown in the figure 3(a). The streamlines are not disturbed in a large space near the channel center (over half of the selected area). For the 15µm particle, four vortices nearly occupy half of the selected area as shown in figure 3(b).

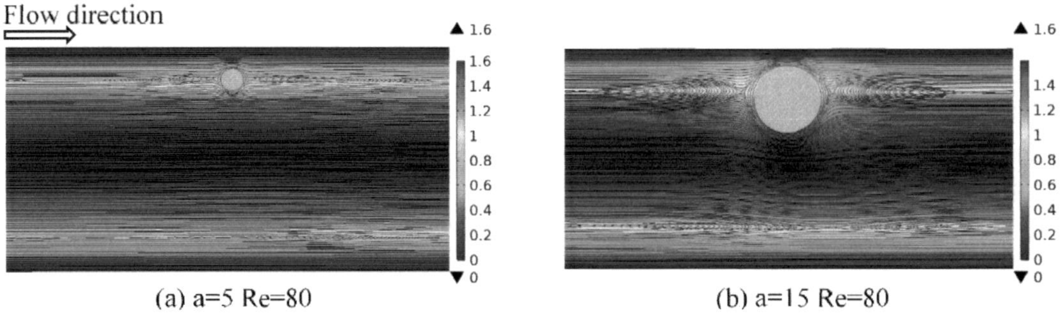

Figure 3. The absolute velocity contour as the background and the streamlines representing the relative velocity of fluid to particle of two typical conditions: (a) a=5 μm Re=80, (b) a=15 μm, Re=80.

The vortex at the downstream of the particle engenders hydrodynamic resistance force on the particle. This leads to the decrease of the actual translational velocity of the particle U_p. It was compared with the fluid velocity of Poiseuille flow at the particle center U_f to obtain the lag velocity ($\Delta U=U_f-U_p$). The lag velocity of the two selected particles was plotted in figure 4 as a function of Re. Obviously, two groups of data from a/H=0.1 and a/H=0.3 show negative and positive results, respectively. For a/H=0.1, the negative values of ΔU mean that the fluid moves slower than the particle. The traction exerted on a small particle is stronger than the blockage of the downstream fluid at moderate *Re*. With the increase of *Re*, the particle hinders the fluid flow to some extent. Then the vortex at the downstream grows up, thus the particle slowing down. For a/H=0.3, the fluid runs too fast to make the particle lag behind. At this circumstance, the vortex augments due to the constraint of the particle as shown in figure 3(b), and strongly impedes the particle move ahead. The large *Re* enhances the effect of vortices. Then it results in the increase of the lag velocity in the condition of a/H=0.3.

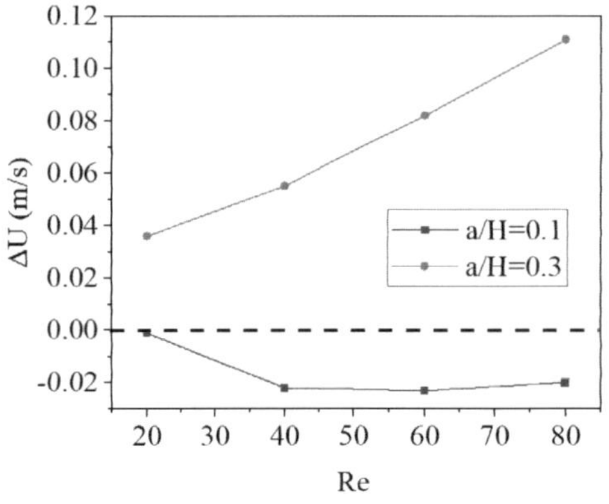

Figure 4. The lag velocity as a function of *Re*.

3.3. Particle's Rotational Velocity

The lateral migration would be promoted by the particle rotation which was ignored in most previous reports. As well known, when the velocity gradient exists between the mainstream and particle surface, the particle will be subjected by shear and then rotate. The rotation velocity Omega (Ω) was depicted as a function of a/H at different Reynolds numbers in figure 5. At all Reynolds numbers, the Ω is diminished with the a/H increasing. The trend was strongly enhanced at high Re. To further discuss the contribution of the particle rotation, the particle rotation induced lift force, $F_{LR}=\frac{1}{8}\pi a^3\rho_f(\vec{u}_f - \vec{u}_p) \times \vec{\Omega}$ [19], was introduced and calculated through all the results of simulations, where $\vec{u}_f$ is the fluid velocity and $\vec{u}_p$ the particle velocity. The net lift force F_L was also calculated, and the data of two forces were listed in table 1. The F_{LR} combines the F_{LS} and F_{LW} into the net lift force F_L which governs the lateral migration. It was found that the direction of the forces is different between *a/H*=0.1 and *a/H*≥0.2. It can be explained by the lag velocity discussed above. Moreover, the direction of F_{LR} and F_L is identical at a given *a/H*, meaning that the positive contribution of the F_{LR} to the migration. In addition, the data of the F_{LR} and F_L are both on the scale of 10^{-11}N at *a/H*=0.1. During the range of *a/H*≥0.2, the F_{LR} and the F_L amplify enormously. The F_{LR} even exceeds the F_L more than one order of magnitude. It indicates that the particle rotation has the extreme influence on the lateral migration of a large particle.

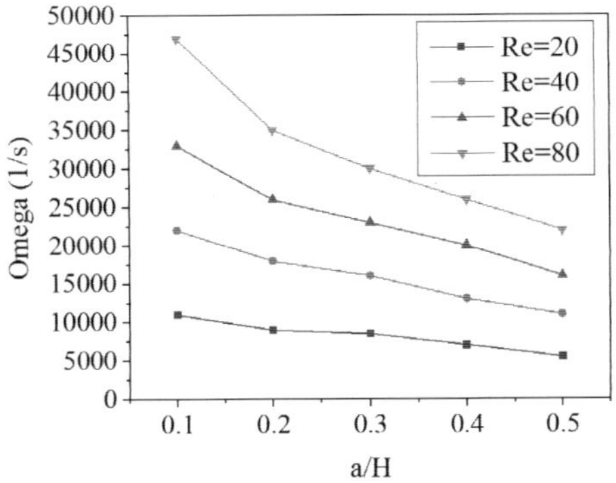

Figure 5. Rotation velocity of particles with different *a/H*.

Table 1. F_L (the z component of the surface integration of the total stress exerted on particle) and F_{LR} with different *a/Hs* at *Re*=80. Negative sign denotes the force directs to the channel center.

a/H=0.1		*a/H*=0.2		*a/H*=0.3		*a/H*=0.4		*a/H*=0.5	
F_L (×10^{-11}N)	F_{LR}	F_L (×10^{-11}N)	F_{LR}	F_L (×10^{-10}N)	F_{LR}	F_L (×10^{-10}N)	F_{LR}	F_L (×10^{-9}N)	F_{LR}
-3.6	-2.1	3.89	57.6	1.99	46.1	6.3	51.0	2.3	39.2

4. Conclusion

The interaction of fluid and particle was studied at moderate Reynolds numbers through the FSI method in this paper. The equilibrium positions of different particles (0.1≤*a/H*≤0.5) were investigated at different Reynolds numbers (20≤*Re*≤80). Then, flow structure around two typical particles (*a/H*=0.1 and *a/H*=0.3) was analysed. The vortex in the front of the particle retards particle's translation leading to the lag velocity increasing. Finally, the particle rotation and the rotation-induced force were discussed quantitatively. The rotation-induced force tremendously promotes large particle's inertial migration. In the present research, the particle in a fixed position was selected to analyse which was limited by the computer resources. In the future, more attention needs to be paid to the process of the overall inertial migration to explore the kinetics of inertial microfluidics.

Acknowledgments

This study was financially supported by Foundation of Scientific and Technological Talents Development for Young and Middle-aged Teacher of Shanghai Institute of Technology (No. ZQ2019-9).

References

[1] Hur S C, Tse H T and Di Carlo D 2010 *Lab Chip* **10** 274-80
[2] Choi J, Hong S C, Kim W and Jung J H 2017 *ACS Sens* **2** 513-21
[3] Marle L and Greenway G M 2005 *TrAC Trends in Analytical Chemistry* **24** 795-802
[4] Segre G and Silberberg A 1961 *Nature* **189** 209
[5] Asmolov E S 1999 *Journal Of Fluid Mechanics* **381** 63-87
[6] Matas J P, Morris J F and Guazzelli É 2004 *Journal of Fluid Mechanics* **515** 171
[7] Schonberg J A and Hinch E 1989 *Journal of Fluid Mechanics* **203** 517-24
[8] Di Carlo D, Irimia D, Tompkins R G and Toner M 2007 *Proc. Natl. Acad. Sci. U. S. A.* **104** 18892-7
[9] Miura K, Itano T and Sugihara-Seki M 2014 *Journal of Fluid Mechanics* **749** 320-30
[10] Amini H, Sollier E, Weaver W M and Di Carlo D 2012 *Proc. Natl. Acad. Sci. U. S. A.* **109** 11593-8
[11] Esipov D V, Chirkov D V, Kuranakov D S and Lapin V N 2020 *Journal of Fluids Engineering* **142**
[12] Jiang D, Tang W, Xiang N and Ni Z 2016 *RSC Advances* **6** 57647-57
[13] Tian F B, Dai H, Luo H, Doyle J F and Rousseau B 2014 *Journal of Computational Physics* **258** 451-69
[14] Pedrol E, Massons J, Díaz F and Aguiló M 2018 *Fluids* **3**
[15] He Y, Bayly A E and Hassanpour A 2018 *Powder Technology* **325** 620-31
[16] Yuan C, Pan Z and Wu H 2018 *Microfluidics and Nanofluidics* **22** 102
[17] Di Carlo D, Edd J F, Humphry K J, Stone H A and Toner M 2009 *Phys. Rev. Lett.* **102** 094503
[18] Shi P Y and Rzehak R. 2020. *Chemical Engineering Science* **211**
[19] Zhang J, Yan S, Yuan D, Alici G, Nguyen NT, Ebrahimi W M and Li W 2016 *Lab Chip* **16** 10-34

Numerical Simulation on Gas-Solid Two-Phase Flow in Horizontal Pneumatic Conveying Pipe Based on DPM Model

Zihan Guo[1, 2], Jun Zhang[1, 2*] and Jinman Huang[3]

[1]School of Marine Equipment and Mechanical Engineering, Jimei University, Xiamen 361021, China
[2]Fujian Province Key Laboratory of Energy Cleaning Utilization and Development, Xiamen 361021, Fujian Province, China
[3]Xiamen Anmaixin Automation Technology Co., LT, Xiamen 361021, Fujian, China
Email: bull0202@sina.com

Abstract. The determination of pipe deposition and optimum conveying velocity in pneumatic conveying has an important impact on conveying efficiency. The Euler-Lagrange method DPM model is used to analyse five different particle sizes and densities of small particles, and the flow pattern in the horizontal pipeline at different particle sizes and densities is derived from the graphs of the maximum discrete phase concentration, particle trajectory and discrete phase concentration distribution for each working condition. The simulation results show that the deposition increases with particle size and density, the optimum conveying speed increases with particle size and density, the larger the deposition, the larger the required conveying velocity. The velocity of 2 m/s can make the particles below 20μm suspended transport, the velocity of 4 m/s allows particles with a particle size of 30μm and a density of 1000 kg/m^3 or less to be transported in suspension and 6 m/s allows particles with a density of 2000 kg/m^3 or less to be transported in suspension. The aim is to provide a reference for the design of pneumatic conveying systems and the selection of the optimum conveying velocity.
Keywords. Pneumatic conveying, DPM model, gas-solid two-phase flow, numerical simulation.

1. Introduction

Pneumatic conveying is a method used to transport loose particles, powders etc., from one location to another via closed pipes [1]. Compared to traditional material conveying methods, pneumatic conveying has the advantages of simple structure, small footprint, clean and environmental protection, and flexible arrangement. Pneumatic conveying has been widely used in food, chemical and energy [2-4]. Pneumatic conveying systems consist mainly of horizontal, vertical and bent pipes. Due to the limitations of the measuring equipment, numerical simulations are often used as a necessary complement to the flow analysis. Pneumatic conveying is a gas-solid two-phase flow category, and computational fluid dynamics (CFD) provides a more convenient and efficient method for complex models [5], and Fluent is used as reliable simulation software for pipeline science research. In the operation of pneumatic conveying systems, pipe deposition is an important factor affecting the conveying efficiency, and determining the optimum conveying velocity is its main problem. In recent years, scholars have done a lot of researches on pneumatic conveying pipeline deposition. Li et al. [6] established a pneumatic conveying experimental bench and used Eulerian method to analyze the flow characteristics of sand in the horizontal pipeline of pneumatic conveying, and obtained the characteristics of pressure drop and particle concentration in the horizontal pipeline, and agreed with the experimental results, which laid a foundation for the engineering design of pneumatic conveying system. Miao Z et al. [7] carried out numerical simulation of powder dense phase pneumatic conveying by CFD to obtain the influencing factors of different flow pattern conversion in horizontal pipeline. Qin et al. [8] studied the flow law of micron particles in horizontal pipeline under the influence of different particle size, shape, lift force and other factors by DPM model, analyzed the number of particle deposition under different conditions, and obtained the main factors affecting horizontal pipeline pneumatic conveying. However, the current studies on deposition using CFD methods have mainly focused on Eulerian methods, and relatively few studies of deposition using the Fluent sub-model Lagrange discrete phase model (DPM), which provides a model basis for gas-solid two-phase flows with solid phase volume fraction less than 10 % and without consideration of collisions, and which provides a visual observation of the particle trajectory motion compared to Eulerian methods, and is important for the optimal running velocity. The model can be used to visualise the particle trajectory motion compared to the Eulerian method.

In this paper, we start from the study of pneumatic conveying horizontal pipe, and use DPM model to analyse the maximum discrete phase concentration, particle trajectory, outlet discrete phase concentration distribution to analyse the flow state and optimum conveying velocity under different particle sizes and densities. Fluent software is used to simulate and analyse the flow phenomena in the horizontal pipeline under different conditions, with the aim of providing a reference for the selection of the optimum conveying velocity in the horizontal pipeline for pneumatic conveying.

2. Mathematical Calculation Equation

2.1. Gas Phase Control Equation [9]

In this paper, the volume fraction of solid phase is less than 10%, so the DPM model can be used. It is assumed that both gas-solid phase temperatures are the same as the atmospheric temperature, so the energy equation is not considered. In the DPM model, the gas phase is treated as the continuous phase, the continuous phase and momentum equation are as follows.

$$\frac{\partial \rho}{\partial t}+\frac{\partial}{\partial x}(\rho u)+\frac{\partial}{\partial y}(\rho v)+\frac{\partial}{\partial z}(\rho w)=0 \tag{1}$$

$$\begin{cases} \frac{\partial}{\partial t}(\rho u)+\frac{\partial}{\partial x}(\rho uu)+\frac{\partial}{\partial y}(\rho uv)+\frac{\partial}{\partial z}(\rho uw)=\frac{\partial \tau_{xx}}{\partial x}+\frac{\partial \tau_{yx}}{\partial y}+\frac{\partial \tau_{zx}}{\partial z}-\frac{\partial p}{\partial x}+F_x \\ \frac{\partial}{\partial t}(\rho v)+\frac{\partial}{\partial x}(\rho vu)+\frac{\partial}{\partial y}(\rho vv)+\frac{\partial}{\partial z}(\rho vw)=\frac{\partial \tau_{xy}}{\partial x}+\frac{\partial \tau_{yy}}{\partial y}+\frac{\partial \tau_{zy}}{\partial z}-\frac{\partial p}{\partial x}+F_y \\ \frac{\partial}{\partial t}(\rho w)+\frac{\partial}{\partial x}(\rho wu)+\frac{\partial}{\partial y}(\rho wv)+\frac{\partial}{\partial z}(\rho ww)=\frac{\partial xz}{\partial x}+\frac{\partial \tau_{yz}}{\partial y}+\frac{\partial \tau_{zz}}{\partial z}-\frac{\partial p}{\partial x}+F_z \end{cases} \tag{2}$$

where ρ is the density of the fluid, p is pressure, F_x, F_y, F_z are volume forces acting on the micro-element, respectively, u, v, w are velocity components along x, y, z axes, respectively.

2.2. Solid Phase Control Equations

Considering the solid phase as a discrete phase and using the Lagrangian method for particle tracking, the controlling equation for the solid phase according to Newton's second law can be obtained as [9-10]

$$\frac{du_p}{dt}=F_D(u-u_p)+\frac{g_x(\rho_p-\rho)}{\rho_p}+F_x \tag{3}$$

$$F_D=\frac{18\mu}{\rho_p {d_p}^2}\frac{C_D Re}{24} \tag{4}$$

$$Re=\frac{\rho d_p|u-u_p|}{\mu} \tag{5}$$

$$C_D = \frac{24}{Re} f(Re) \tag{6}$$

where ρ_p is the density of the particle, F_D is fluid drag forces on particles, μ is fluid dynamic viscosity for gas phase, d_p is particle diameter, u,u_p are fluid and particle velocities, respectively. In different flow regions, there are different equations for f(Re). f(Re)=1 in the Stokes zone, f(Re)=1+3/16 Re in the Olsen zone and f(Re)=1+1/6 Re^(2/3) for the standard resistance.

2.3. Turbulence Models

The two-equation k-ε model is widely used in the calculation of turbulence parameters, which includes the standard k-ε, RNG k-ε, and Realizable k-ε models. In this paper, the flow in the tube is a typical turbulent flow, so the standard k-ε model is used. The standard k-ε model equation is:

$$\frac{\partial(\rho k)}{\partial t} + \frac{\partial}{\partial x_i}(\rho k u_i) = \frac{\partial}{\partial x_j}\left[\left(\mu + \frac{\mu_t}{\sigma_k}\right)\frac{\partial k}{\partial x_j}\right] + G_k + G_b - \rho\varepsilon - Y_m + S_k \tag{7}$$

$$\frac{\partial(\rho \varepsilon)}{\partial t} + \frac{\partial}{\partial x_i}(\rho \varepsilon u_i) = \frac{\partial}{\partial x_j}\left[\left(\mu + \frac{\mu_t}{\sigma_\varepsilon}\right)\frac{\partial k}{\partial x_j}\right] + C_{1\varepsilon}\frac{\varepsilon}{k}(G_k + C_{3\varepsilon}G_b) - C_{2\varepsilon}\rho\frac{\varepsilon^2}{k} + S_\varepsilon \tag{8}$$

where μ_t is viscosity factor, Y_m is contribution of pulsational expansion in compressible turbulence, σ_k, σ_ε are prandtl number against turbulent energy and turbulent dissipation rate, respectively, G_k, G_b are turbulent kinetic energy due to mean velocity gradient and buoyancy, respectively, S_k, Sε are user-defined source term, respectively. C_μ=0.09, $C_{1\varepsilon}$=1.44, $C_{2\varepsilon}$=1.92, $C_{3\varepsilon}$=0.09 are standard empirical values and the turbulent kinetic energy k and turbulent dissipation rate ε are defaulted to 1.0 and 1.3 in the Fluent software.

3. Horizontal Pipeline Gas-Solid Two-Phase Flow Model

3.1. Model Assumptions

In this paper, the object of study are micron particles, due to the small size of the particles, so the following assumptions: (1) the gas flow itself can be compressed, but the gas flow velocity is low, Mach number below 0.3, so the gas is assumed to be incompressible. (2) The velocity of the gas is equal to the velocity of the particles. (3) The particles are all spherical and the effect of shape on flow is ignored. (4) The particles are uniformly incident on the vertical inlet surface into the calculation domain. (5) The force F_x acting on the particles by the fluid is ignored and only the effect of the field force on the particles is considered.

3.2. Pre-Processing Geometry Model

The horizontal pipe is selected as the object of study, due to the long pipe in pneumatic conveying, the calculation volume is large, so the full development of 2 m horizontal pipe is selected for analysis. The geometric model of pipe diameter 50 mm is shown in figure 1. In order to improve the mesh quality and the accuracy of the solution, ICEM-CFD was used to carry out hexahedral structured o-grid of the horizontal pipe, the geometric model of the pipe diameter 50 mm inlet mesh division is shown in figure 2, the final mesh cell number is 436872, the mesh quality is greater than 0.7.

3.3. Solver Parameter Settings

A pressure-based solver is used for the continuous phase, discrete phase steady-state simulations, and a gravity term is added. The inlet and outlet boundary conditions are velocity inlet and pressure outlet boundary conditions. In the discrete phase setup, the particles are set to inlet surface injection with a mass flow rate of 0.16 kg/s and a spherical single particle size is chosen for the particle discrete phase model. The discrete phase increases the effect of turbulence in the random wander model (RWM). The particle boundary conditions are escape conditions for inlet and outlet and reflect conditions for the wall.

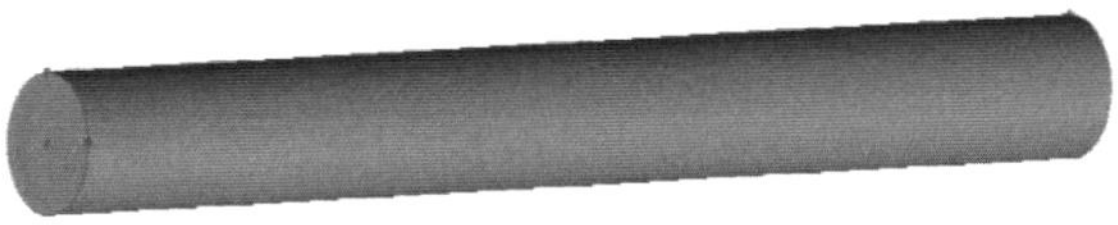

Figure 1. Geometric model of horizontal pipe.

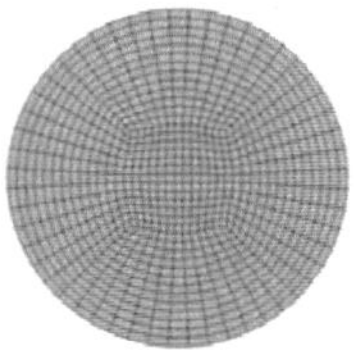

Figure 2. Grid division of inlet.

4. Simulation Results and Analysis

It is essential to determine the validity of a simulation before it is carried out for a gas-solid two-phase flow, and the reliability of the simulation has been verified by a combination of simulation and experiment in references [11-12]. In a horizontal pipe with gas-solid flow, the maximum discrete phase concentration is to a certain extent representative of the deposition in the pipe. In the following, the flow pattern will be analysed by the maximum discrete phase concentration for different particle sizes and densities, and the particle trajectory diagram and the discrete phase concentration distribution at the outlet will be used to visualise the particle flow and provide reference for determining the optimum velocity under different operating conditions.

4.1. Effect of Particle Size on Flow

The maximum discrete phase concentrations were quantified for five different particle sizes (10, 20, 30, 40 and 50μm) at different velocities when the density was 1120 kg/m^3, and the results are shown in figure 3. As can be seen from figure 3, the maximum discrete phase concentration gradually decreases with increasing velocity and particle deposition gradually decreases. At the same velocity, the larger the particle size, the larger the maximum discrete phase concentration and the more serious the deposition, which changes more obviously with particle size at low velocity. This is because the greater the velocity, the greater the kinetic energy of the air, the easier it is for the particles to be suspended, and as the particle size increases, the more pronounced the effect of gravity, the easier it is for the particles to be deposited towards the bottom of the tube, and it is more difficult for particles of larger size to obtain larger velocities under the same conditions. In addition, smaller particles are more likely to be suspended by gas turbulence.

Due to the large number of working conditions, the trajectory diagram for five different particle sizes of 10, 20, 30, 40 and 50μm at 500 mm at the end of the horizontal pipe and the discrete phase concentration distribution at the outlet are visualised in figure 4, using a velocity of 2 m/s as an example. As can be seen from the graph, at a particle size of 10μm, the discrete phase concentration is evenly distributed and in full suspension flow, but the energy consumption is greater and the wear is more serious in the pneumatic conveying bend. When the particle size is 20μm, most of the particles are in suspension, only a small number of particles sink under the action of gravity, but can maintain the normal flow state. In industrial production, taking into account energy consumption and wear needs, often to maintain the state of the velocity as the conveying velocity. Particle size from 30μm onwards, the interaction between particles and particles, particles and the wall surface to enhance the momentum loss increased, until 50μm, the particles are completely deposited and the deposition area is concentrated at the bottom of the pipe, which no longer meets the industrial conveying requirements and requires a greater conveying velocity. Therefore, when the velocity is 2 m/s, it is possible to maintain the industrial conveying requirements for particles below 20μm.

Journal of Physics: Conference Series **2097** (2021) 012003 doi:10.1088/1742-6596/2097/1/012003

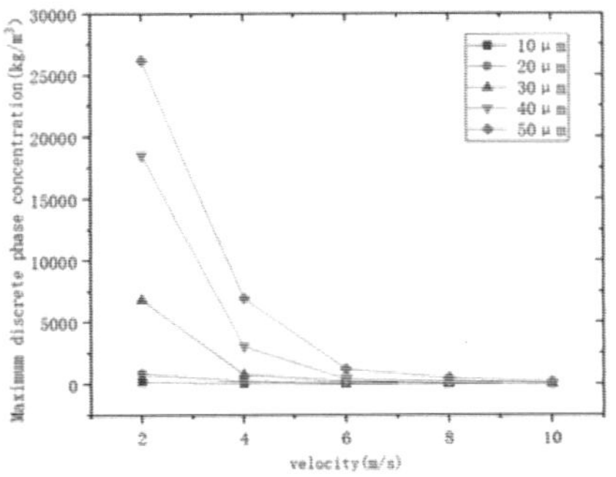

Figure 3. Maximum discrete phase concentration under different velocity particle sizes.

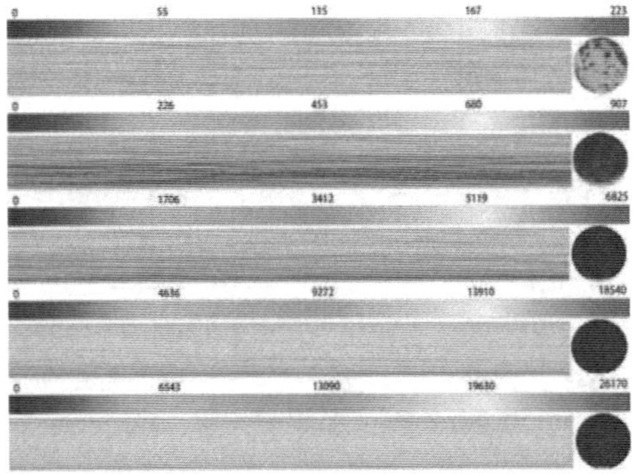

Figure 4. Trajectory diagram of 2m/s particles with different particle sizes and concentration of discrete phase at outlet.

4.2. Influence of Particle Density on Flow

Due to the large number of working conditions, the maximum discrete phase concentrations are now quantified for different particle sizes of 10, 20, 30, 40 and 50 μm at different densities of 1000, 1500, 2000, 2500 and 3000 kg/m^3 at a velocity of 2 m/s and the results are shown in figure 5. It can be seen from the figure that as the density increases, the maximum discrete phase concentration of the particles increases and the particles are more likely to be deposited, and the larger the particle size, the more serious the deposition. This is because as the particle density increases, the influence of gravity increases and the particles are more likely to be deposited towards the bottom of the pipe. When the particle size is 10μm, the maximum discrete phase concentration changes very little, combined with the conclusion of 3.1, it can be assumed that 2 m/s velocity can make 10μm, 3000 kg/m^3 below the particles remain suspended, so that 50μm, 1120 kg/m^3 density particles suspended need more than 10m/s speed. As there are many different particle size, the optimum flow velocity at different densities is now chosen for particles with a particle size of 30μm.

Figure 6 quantifies the maximum discrete phase concentrations of five different densities at different velocities, which shows that it is more difficult to obtain large velocities for denser particles under the same conditions. Velocity of 4m/s must enable 1000 kg/m^3 of particles to be transported in suspension. Therefore, to determine the optimum conveying velocity for different densities at 30μm, the simulation only needs to start with a velocity of 4 m/s and a density of 1500 kg/m^3.

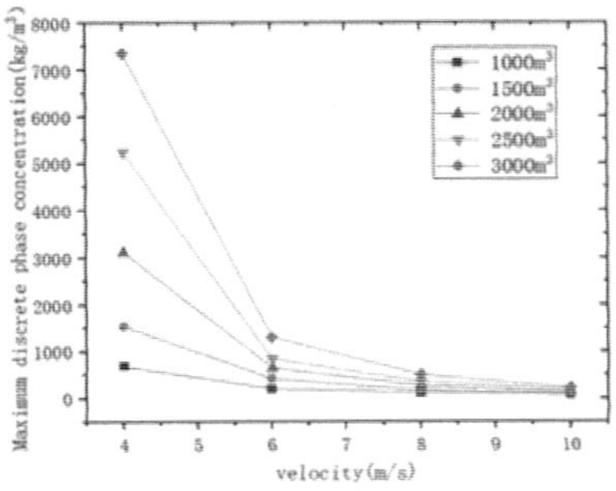

Figure 5. Maximum discrete phase concentration at different velocities.

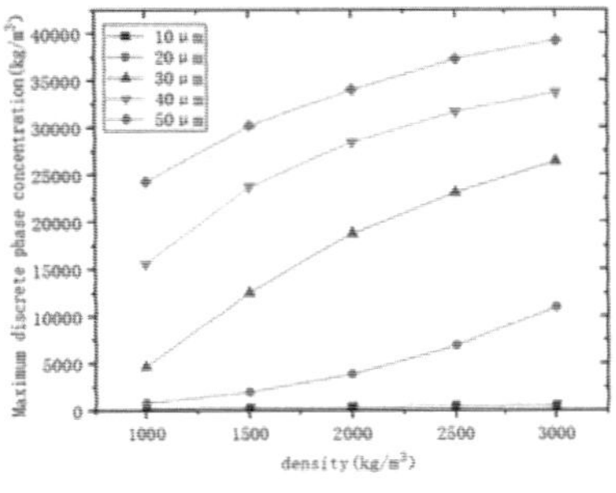

Figure 6. Maximum discrete phase concentration at different densities.

Figure 7 shows the 500mm trajectory at the end of the pipe for 4m/s, 30μm particles at densities of 1500, 2000, 2500 and 3000kg/m^3 and the discrete phase concentration distribution at the outlet, yielding that most of the particles are still deposited at a density of 1500kg/m^3 and as the density increases the deposition becomes more severe and higher velocities are required to maintain normal transport.

Figure 8 shows the 500 mm trajectory at the end of the pipe and the discrete phase concentration distribution at the outlet for the same conditions at 6 m/s. From 2500 kg/m^3 onwards, the particles sink and a higher velocity is required to maintain normal conveying. Therefore, a velocity of 6 m/s is sufficient for industrial conveying of particles with a density of 2000 kg/m^3 or less.

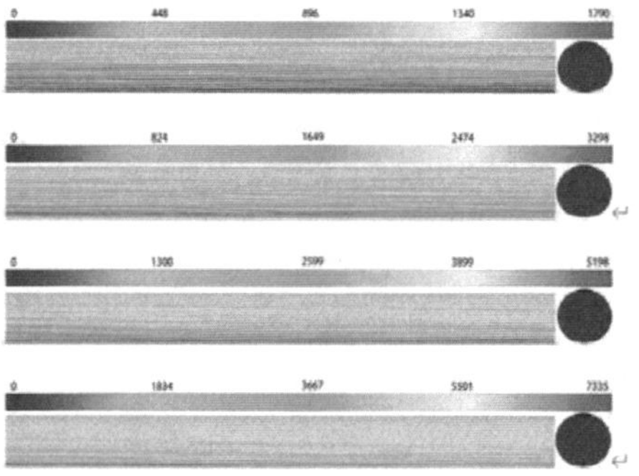

Figure 7. 4m/s particle trajectories of different densities and discrete phase concentration distribution at outlet.

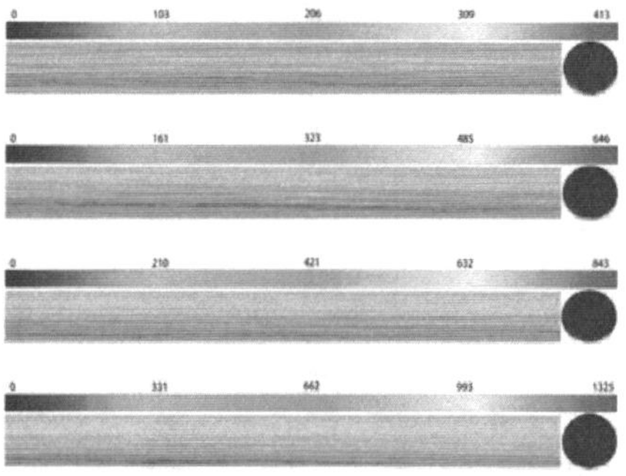

Figure 8. 6 m/s particle trajectories of different densities and discrete phase concentration distribution at outlet.

5. Conclusions

The determination of pipe deposition and optimum conveying velocity is an important issue in engineering design, so this paper analyses the flow characteristics of different particle sizes and densities based on the DPM model, and draws the following conclusions.

(1) Compared with the Eulerian method, the model provides an intuitive and clear observation of the particle trajectory, which provides a reference for the determination of the optimal conveying speed.

(2) The larger the particle size, the greater the effect of gravity, the more likely it is to be deposited, the less likely it is to be suspended by gas turbulence, the lower the kinetic energy of mutual collision, and the more difficult to obtain larger velocities. 2m/s velocity can make the particles below 20 μm suspended transport.

(3) The greater the density, the greater the pipeline deposition, and the effect of density changes on deposition is more pronounced for particles of larger particle size. When the particle size is 30 μm, 4m/s velocity can make the mainstream suspension transport of particles below 1000 kg/m^3, and 6 m/s velocity can make the mainstream suspension transport of particles below 2000 kg/m^3.

References

[1] Alkassar Y, Agarwal V K, Pandey R K and Behera N 2021 Analysis of dense phase pneumatic conveying of fly ash using CFD including particle size distribution *Particulate Science and Technology* **39**(3) 322-337

[2] Fitzpatrick J J 2007 Particle Properties and the Design of Solid Food Particle Processing Operations *Food & Bioproducts Processing* **85**(4) 308-314.

[3] Takabatake K, Mori Y, Khinast J G and Sakai M 2018 Numerical investigation of a coarse-grain discrete element method in solid mixing in a spouted bed *Chemical Engineering Journal* **346** 416-426

[4] Zhou J W, Xu L G and Du C L2018 Prediction of lump coal particle pickup velocity in pneumatic conveying *Powder Technology* **343** 599-606

[5] Faris A F A 2021 Propeller design and performance evaluation by using computational fluid dynamics (CFD): a review *Journal of Aeronautics, Astronautics and Aviation* 263-274
[6] Li D Y, Liu S and Wang X N 2016 Research on the pneumatic conveying of the sand in the horizontal pipe based on fluent *Chemical Product and Process Modeling* **11**(2) 159-165
[7] Miao Z, Kuang S and Zughbi H 2020Numerical simulation of dense-phase pneumatic transport of powder in horizontal pipes *Powder Technology* **361** 62-73
[8] Qin Si 2021 Numerical simulation of pneumatic conveying characteristics of micron particles in horizontal pipe *Vibroengineering PROCEDIA* **38** 135-141
[9] Zhang R, Li L and Xing X 2021 Study of the hydraulic performance of PID bits based on a discrete phase model *Arabian Journal of Geosciences* **14**(9) 1-9
[10] Du X C, Liu S and Liu Peng 2021 Numerical simulation of pipeline erosion of particulate matter in LBE based on DPM model *Nuclear Power Engineering* **42**(01) 48-53
[11] Behera N, Agarwal V K, Jones M G and Williams K C 2013 CFD modeling and analysis of dense phase pneumatic conveying of fine particles including particle size distribution *Powder Technology* **244**(4) 30-37
[12] Yang D and Li G 2020 Prediction of horizontal pneumatic conveying of large coal particles using discrete phase model *Advances in Materials Science and Engineering* 1-15

The Flotation Modification Test of Chrysocolla Research on RSM

Rui Guo*, Siwei Liu,Zhenhong Liao and Ren Liu

Changsha Research Institute of Mining and Metallurgy Co. Ltd., Changsha, China
Email: 424481397@qq.com

Abstract. This study is focused on the flotation of a cooper mineral.Chrysocolla is poor flotability, surface porous, high porosity, nonuniform property, so it has strong hydrophilic and difficult dissolution. XRD and SEM were used to detect the properties and surface morphology of chrysocolla. The paper make an experiment, it contain modified polymer adsorption - intermediate metal copper ion connection - collector adsorption testing program. The experiment can exchange mineral surface property which enhancing mineral flotation and hydrophobicity. With the conclusion, the results have a trend that increasing the agents can increase mineral recovery, then mineral recovery reach the stable trend. In the simulation of RSM, mineral recovery is based on 3 factors ammonium, xanthate and agent, those factors interact with each other, simulation find the main factor is agent. RSM response surface method has the function of optimizing test results, improving test efficiency, inputting test influence factors and results, and getting the best test factors and results through test simulation.

Keywords. Chrysocolla; flotation; RSM.

1. Introduction

Copper is an important production raw material in the national economy main products , with good electrical conductivity, ductility and easy to form alloys, so it is widely used in electronic, national defense, construction, machinery, light industry and other industries. With the continuous exploitation of copper resources, relatively easy beneficiation has been reduced year by year and the shortage of resources has worsened. Therefore, great attention has been paid to the application research and development of low-grade oxidized copper ores [1]. This type of minerals generally have the characteristics of high oxidation rate, high binding rate, fine dissemination particle size, large mud content, polymetallic symbiosis, etc., so it is difficult to obtain better technical and economic indexes with conventional processing and smelting technology. Study on beneficiation of oxidized copper ore in China has a history of more than 50 years, and its processing technology has made great progress [2-5]. The copper mines of the world are mined for a long time and grade decreased.

However, the copper demand is increasing, and the mining cost is getting higher and higher. The copper in the refractory complex copper oxide ore can be deeply developed and utilized, and it is significant to study the complex refractory copper oxide ore [6]. The malachite studied in this paper generally occurs in the copper-lead-zinc ore, which belongs to the refractory, complex and high oxidation rate copper, it is difficult to be floatated by conventional collector. On the one hand, chrysocolla surface porous and high porosity, the nature of the change is uneven, belongs to the typical strong hydrophilicity, refractory, difficult to float mineral [7]. The hydrophobic differences of various minerals are enhanced by the selective adsorption of flotation reagents according to the physical and chemical properties of the mineral surfaces, so that the target minerals are selectively separated from the gangue minerals [8]. Oxidized ore due to the strong hydrophilicity, specific surface area (128 m^2/g), soluble and surface copper ion diffusivity is poor, collector is hard to copper oxide ore recovery, on the other hand because of chrysocolla predecessors research content is less. It is a kind of copper oxide ore with scientific research significance. In order to overcome these challenges, it is necessary to further understand the activation properties of the activator adsorbed on the mineral and to design more efficient activators for chrysocolla. Found that ammonium sulfate and ethylenediamine phosphate had certain activation effect on malachite, and sodium sulfide [9-11] would inhibit the floatability of malachite. In this study, an appropriate amount of copper sulfate was added to activate malachite, and then the polymer agent reacted with the surface of malachite through surface organic modification. After the surface modification of malachite, the added copper ions are combined with the outer layer of the modified polymer agent, and finally the amyl xanthate is added to combine with the copper ions outside the modified polymer agent.

2. Mineral Experiment Research

2.1. Experiment Raw Mineral

The composition of malachite is $CuSiO_3$ $2H_2O$. The test material was taken from the ore of malachite in Wuhan, Hubei Province, and the purity was over 90% after manual selection. The manually selected pure mineral of malachite is ground to less than 200 mesh through a three-headed agate grinder. Under the grinding of agate grinding machine, pure minerals keep their original composition, and no impurities will be entrapped after grinding.

Table 1. Multi-element analysis results of ore sample/%.

Element	Cu	SiO_2
Content	35.82	34.25

As shown in the table 1, the Cu and SiO_2 content is 35.82% and 34.25% in the pure mineral sample by multi-element analysis.

Figure 1 shows the SEM spectrum. (a), (b), (c) and (d) respectively show the surface morphology of ore under the detection distance of 1 mm, 200 um, 100 um and 20 um by SEM. $CuSiO_3$ is the main component of malachite. There are some white points raised on the surface of malachite. The ore is fragile and fragmented after crushing grinding. The atomic contents of O, Cu and Si were 40.90%, 8.74% and 10.89%, respectively, and the ratio of atomic content of Cu: Si: O was approximately 1:1:4.

Due to the strong hydrophilicity and poor floatability of malachite, it is of research significance to understand the modification mechanism of activator to improve the flotation efficiency of valuable minerals recovered from ores, and to develop more powerful activator. The combined application of SEM, XRD, ATR-FTIR, X-ray photoelectron spectroscopy (XPS), DFT calculation and solution chemistry can accurately clarify the activation properties of the activator on the surface of malachite, and provide a new way to understand the modification configuration and hydrophobicity mechanism of the activator on mineral surface.

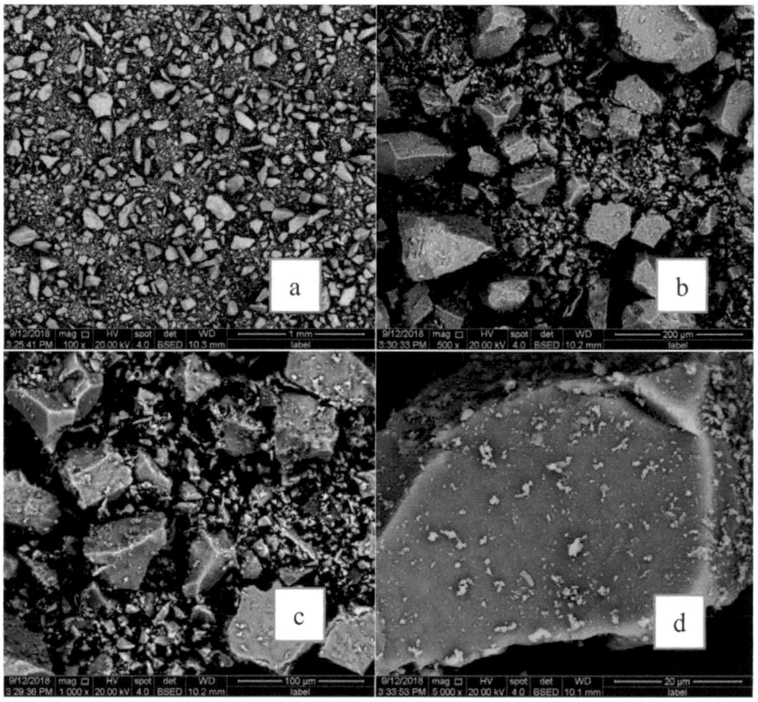

Figure 1. Raw ore SEM picture-(a); (b); (c); (d).

2.2. Experiment Agent

The main agents used in the test are anhydrous copper sulfate, sodium hydroxide, ammonium sulfate, surface-modified polymer agents, ethylenediamine phosphate, amyl xanthate, No.2 oil. Polyacrylamide is a broad spectrum of flocculant, for the majority of the microgranular mineral grain has the very good flocculation effect, to transform, to remove oxygen atoms with strong hydrophilic, join the sulfur atom of copper ion selective adsorption ability, can obtain the surface with a surface modification effect of chrysocolla more atoms polymer reagents. The modified polyacrylamide is an organic modified polymer agent, and the modified polymer agent in this paper uniformly uses the GFZ code.

2.3. Experiment Method

The test was carried out in XFGC laboratory aerated floatation tank with a volume of 60ml. First, 2g of 200 mesh pure mineral silachite was weighted, and 40ml of deionized water was injected into the floatation tank. 2g of pure mineral silachite was added. The spindle speed of flotation machine is set at 1800r/min. The flotation reagent amyl xanthate concentration test, the activator ammonium sulfate concentration test, the activator ethylenediamine phosphate and ammonium sulfate concentration test and the ammonium sulfate GFZ reagent test were carried out successively. After flotation, each product will be placed in a conical funnel for natural filtration, drying, weighing, sample preparation and testing to obtain the corresponding weight and finally calculate the recovery rate.

3. Experiment Project and Results Analysis

3.1. GFZ Agent Experiment Theory

As shown in the equations (1), (2), (3), $CuSiO_3$ in the water is dissolving to Cu and SiO_3^{2-}, then SiO_3^{2-} react with H_2O to $HSiO^{3-}$ and OH^-, finally $HSiO_3^{2-}$ react with H2O to H_2SiO_3 and OH^-.

$$CuSiO_3 \rightarrow Cu^{2+} + SiO_3{}^{2-} \tag{1}$$

$$SiO_3{}^{2-} + H_2O \rightarrow HSiO_3{}^{-} + OH^{-} \quad (2)$$

$$HSiO_3{}^{-} + H_2O \rightarrow H_2SiO_3 + OH^{-} \quad (3)$$

3.2. RSM Analysis Experiment Results

Design-expert is the world's top experimental Design software.

As shown in the figure 2 RSM recovery of ammonium and xanthate diagram can intuitively see the influence of two factors on the dependent variable, and can intuitively find out the optimal range. With the increase of ammonium and addition amount of xanthate, the recovery will correspondingly increase, and finally the recovery reaches the maximum value of 81.5%, then the recovery tends to be stable [12-14].

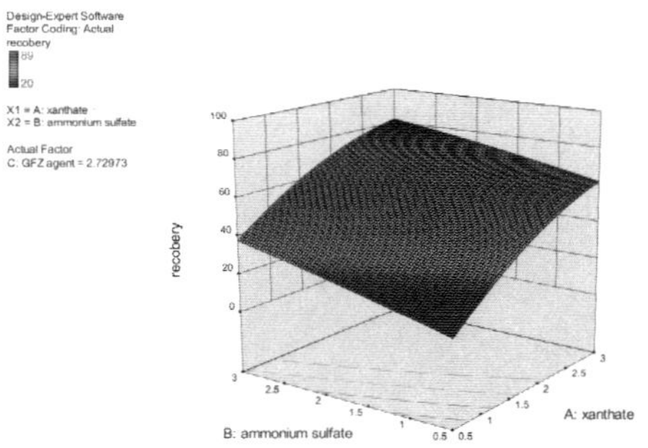

Figure 2. RSM recovery of ammonium and xanthate.

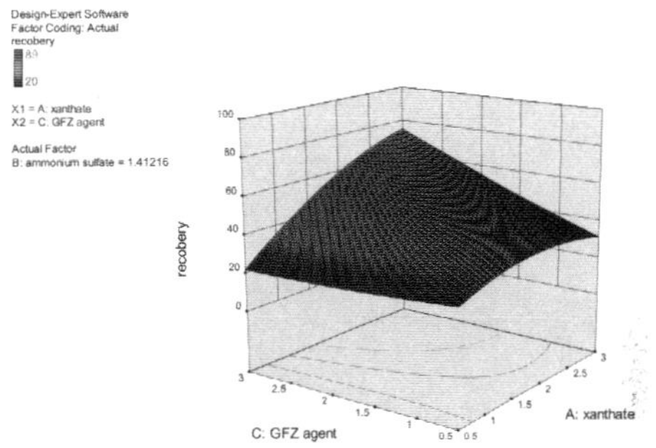

Figure 3. RSM recovery of GFZ agent and xanthate.

As shown in the figure 3 RSM recovery of GFZ agent and xanthate, with the increase of GFZ agent and xanthate addition, the recovery gradually increases to the maximum value.

Table 2. RSM model results analysis.

Source	Sum of squares	Freedom	F value	p-value prob>F	Significance
Model	4.52	9	45.34	0.0023	Significant
A-xanthate	3.62	1	72.5	0.0031	
B-ammonium sulfate	2.41	1	8.50	0.0012	
C-GFZ agent	2.63	1	81.98	0.0014	
AB	0.76	1	1.29	0.2941	
AC	0.21	1	13.75	0.0076	
BC	0.74	1	0.043	0.8411	
A^2	0.72	1	0.43	0.5309	
B^2	0.43	1	0.025	0.8788	
C^2	0.67	1	1.27	0.2973	
Residual	0.35	7			

In this example, as shown in the table 2 above, the p values of recovery is 0.0001 respectively, and the significance test of the model was $p < 0.05$, indicating that the model had statistical significance. If the regression equation is favorable to the model, there is no defect, so the regression equation can be used to replace the real point of the experiment to analyze the experimental results.

Final Equation in Terms of Coded Factors:

$$Recovery=43.26+18.64*A+3.82*B+12.47*C+2.20*AB+7.20*AC+0.46*BC+1.46*A^2+0.39*B^2+2.62*C^2 \quad (4)$$

As shown the equation (4), multiple quadratic response surface regression models were obtained through quadratic response surface regression analysis with the design-expert software.

As shown in the table 3 RSM prediction of concentration recovery, quadratic response surface regression analysis was carried out by design-expert software, and the predicted value calculated based on the obtained multivariate quadratic response surface regression model was compared with the actual value. The prediction effect is good, which indicates that the modeling is of good quality.

Table 3. RSM prediction of concentration recovery.

Actual value	Predicted value	Error	Residual
30.00	28.08	1.92	0.869
38.00	39.84	-1.84	-0.504
87.00	85.66	1.34	0.449
35.00	37.16	-2.16	-0.946
40.00	43.26	-3.26	-1.213
75.00	69.77	5.23	2.446
35.00	30.44	4.56	2.395
56.00	57.73	-1.73	-0.733
62.00	58.36	3.64	1.139
45.00	46.31	-1.31	-0.580
26.00	24.85	1.15	0.406
25.00	23.43	1.57	0.587
32.00	33.97	-1.97	-0.876
55.00	54.47	0.53	0.206
46.00	47.47	-1.47	-0.416
20.00	22.66	-2.66	-0.947
89.00	92.53	-3.53	-1.707

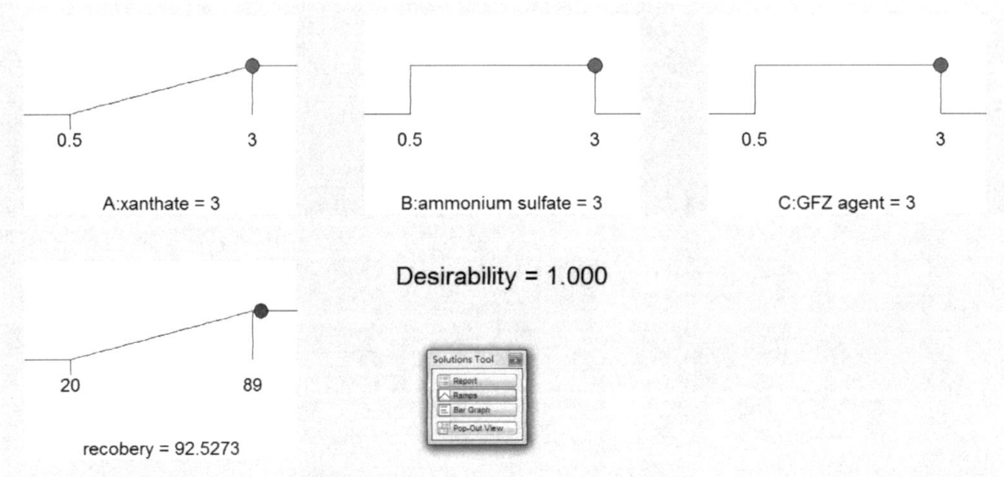

Figure 4. The best fotation agent dosge.

As shown in the figure 4, the best flotation agent dosage, based on the response surface design experiment and the central combination test design principle of box-benhnken, three factors that have significant influence on the grade and recovery of copper concentrate were selected: xanthate A, ammonium sulfate B and GFZ agent C. Three factors and three levels of response surface analysis test were conducted. Finally, the best process was obtained: xanthate 3, ammonium sulfate3, GFZ agent C 3, concentrate recovery rate 92.52 [15-17].

4. Conclusion

In this paper, the flotation mechanism and the flotation results optimization are studied.

(1) Due to the experiment results, with GFZ agent and xanthate increase, recovery also increase, so GFZ agent has a certain activation effect on malachite.

(2) RSM is a great experiment tool, it contains data analysis function, and the RSM model p value is reliable.

(3) RSM surface response method has the function of optimizing the test results, improving the test work efficiency, and inputting the influence factors and results of the test, the optimized experimental results are improved to a certain extent compared with the actual test results.

Reference

[1] Brest K K, et al. 2021 Statistical investigation of flotation parameters for copper recovery from sulfide flotation tailings *Results in Engineering* **9**.

[2] Lotter N O, Bradshaw D J and Barnes A R 2016 Classification of the Major Copper Sulphides into semiconductor types, and associated flotation characteristics *Minerals Engineering* **96-97** 177-184.

[3] Yin W Z, et al. 2019 Mechanism and application on sulphidizing flotation of copper oxide with combined collectors *Transactions of Nonferrous Metals Society of China* **29**(1) 178-185.

[4] Moimane T, Plackowski C and Peng Y 2020 The critical degree of mineral surface oxidation in copper sulphide flotation *Minerals Engineering* **145** 106075.

[5] Feng Q, Zhao W and Wen S 2018 Ammonia modification for enhancing adsorption of sulfide species onto malachite surfaces and implications for flotation *Journal of Alloys & Compounds* **744**.

[6] Tian M J, et al. 2018 Activation role of lead ions in benzohydroxamic acid flotation of oxide minerals: New perspective and new practice *Journal of Colloid And Interface Science* **529** 150-160.

[7] Yin W Z, Sun Q Y, Dong L I, et al. 2019 Mechanism and application on sulphidizing flotation of copper oxide with combined collectors *Transactions of Nonferrous Metals Society of China* **29**(1) 178-185.

[8] Xu X H, Shen P L, Zhang X L, et al. 2018 Effects of ethylene diamine phosphate on the flotation of chrysocolla *Bulletin of the Chinese Ceramic Society* **37**(07) 2291-2296. .

[9] Feng Q C, Zhao W J and Wen S M 2018 Surface modification of malachite with ethanediamine and its effect on sulfidization flotation *Appl. Surf. Sci.* **436** 823–831.

[10] Wu D, Ma W, Wen S, et al. 2017 Enhancing the sulfidation of smithsonite by superficial dissolution with a novel complexing agent *Minerals Engineering* **114** 1-7.

[11] Wei S N, Cooper L, Connal L A, et al. 2018 Tuneable collector/depressant behaviour of xanthate-functional temperature-responsive polymers in the flotation of copper sulfide: Effect of shear and temperature *Minerals Engineering* **117** 91-99.

[12] Chahal M, Singh V and Garg R 2017 Optimum surface roughness evaluation of dies steel H-11 with CNC milling using RSM with desirability function *International Journal of System Assurance Engineering and Management* **8**(2) 432-444.

[13] Shirzad H, Niknam V, Taheri M, et al. 2017 Ultrasound-assisted extraction process of phenolic antioxidants from Olive leaves: a nutraceutical study using RSM and LC–ESI–DAD–MS *Journal of Food Science and Technology* **54**(8) 1-11.

[14] Khatti T, Naderi-Manesh H and Kalantar S M 2017 Prediction of diameter in blended nanofibers of polycaprolactone-gelatin using ANN and RSM *Fibers and Polymers* **18**(12) 2368-2378.

[15] Torabi S H R, Alibabaei S, Bonab B B, et al. 2017 Design and optimization of turbine blade preform forging using RSM and NSGA II *Journal of Intelligent Manufacturing* **28**(6) 1409-1419.
[16] Peishu Z, Jianping T, Shuyu W, et al. 2016 Dynamical downscaling of regional climate over eastern China using RSM with multiple physics scheme ensembles *Theoretical and Applied Climatology* **129**(3-4) 1263-1277.
[17] Tavakoli F, et al. 2017 Optimization of influential factors on the photocatalytic performance of TiO2–graphene composite in degradation of an organic dye by RSM methodology *Journal of Cluster Science* **28**(5) 2979-2995.

Optimization and Application of D850 Dredge Pump Blades under Medium Coarse Sand Condition of Cutter Suction Dredger

Haifei Zhuang*, Mingming Liu and Yongding Wu

CCCC National Engineer Research Center of Dredging Technology and Equipment CO., TD, room 2902, 1088 Yangshupu Road, Shanghai, China
Email: zhuanghaifei@cccc-drc.com

Abstract. Regarding wear issues of a dredge pump's impeller as a cutter suction dredger transports medium coarse sand slurry, blades of the D850 dredge pump are modified and optimized, which extends the distance from the blade inlet root to the impeller suction and avoids damages of the impeller suction anti-wear ring. Analyses via computational fluid dynamics (CFD) simulation show that the head and the efficiency after blade modification have little changes compared with before optimization in the construction flow range of 10000-12000 m^3/h under coarse sand condition. While it improves the flow field of impeller's channels, decreases the vortex at the inlet root of the blades, ensures more uniform distribution of the solid particles. Meanwhile, this is beneficial to reductions of the channels' wears. Applications from constructions show some improvements in the wears of the blade root. Through the construction data comparison, after replacing the modified impeller, dredging productivity will be increased by 15.1% and the fuel consumption per 10000 m^3 will be then reduced by 11.5%.
Keywords. Dredge pump, worn, CFD, blade modification, application.

1. Introduction

As a key equipment of cutter suction dredger, its performance of dredge pump has directly affects on efficiency of construction. There exist great differences in the characteristics of dredged soil during dredger construction [1-2], such that the dredge pump performs variously with different dredged soil. At present, although the design of dredge pump considers the influence of a certain transmission medium, the performance of dredge pump can not meet the construction requirements of dredger in the face of more and more complex dredged soil. This will result in low efficiency of dredge pump and serious wear of overflow parts. For example, in the land reclamation project in Xiamen, the D850 dredge pump of 3500 m^3/h series cutter suction dredger was seriously damaged due to coarse sand wear and cavitation.

In order to improve the dredger construction efficiency under special conditions, the dredger shall be equipped with targeted special equipment. Many foreign R & D institutions carry out targeted research. Cutter suction dredgers are equipped with different mud pumps to meet the requirements of construction conditions. Bugdayci, H.H et al. [3] in the Netherlands, the large channel special pump,was specially used for heavy rock operations that is equipped with a curve blade impeller; Ukhin BV(Russia) [4], experimentally studied the effect of the number and shape of the blades on the performance of dredge pump. Domestic scholars have studied the influence of slurry on the performance of mud pump: e.g. Zhang Xiaosong et al. [5] have conducted a steady numerical study on the cavitation performance of a six blade mud pump. The numerical simulation analysis shows that the greater the mud concentration, the more likely the cavitation occurs in the mud pump. With the same mud concentration, it is more likely that the cavitation happens at the mud pimp with the larger diameter of solid particles in the mud. Cheng Zejiu et al. [6] studied the critical sediment concentration in cavitation and abrasion under experiment tests. When the sediment concentration is lower than the critical value, it promotes cavitation. And when the sediment concentration is greater than the critical value, it inhibits cavitation. Li Longhua et al. [7] found that the cavitation wear occur at the same time for the impurity pump, which is often difficult to be strictly distinguished. By the performance from the centrifugal pump test, it can be found that the critical NPSH has little changes with increaseing of solid content. That is, it has little impacts on its cavitation performance. However, the performance of the pump head and efficiency as well as the anti cavitation performance decreases rapidly with the increase of particle size.

Scholars from many domestic universities and scientific research institutes have studied on the blade design of the centrifugal pump: e.g. the author keeps the overall size of the impeller unchanged, developed a twisted blade impeller with suction port diameter of 900 mm the pump for the cutter suction dredge, which improves the construction efficiency [8]. Optimal the designs of the dredge pump is implemented by combining numerical simulation and model test method, which improves the efficiency of dredge pump [9-10]; Zhang Renhui et al. [11] improved the cavitation resistance under high flow conditions by increasing the blade inlet placement angle; The analysis shows that the inlet flow passage area and the projection surface of the impeller can significantly improve the pump performance [12]. By comparing the numerical simulation results of five impellers, Luo Xianwu et al. [13] analyzed the influence of impeller inlet geometric parameters on the cavitation performance of centrifugal pump. The results showed that properly extending the blade inlet position and increasing the blade inlet placement angle can improve the cavitation performance of centrifugal pump, and pointed out that homogenizing the flow at the impeller inlet is a necessary measure to improve the cavitation performance of centrifugal pump. Based on the numerical simulation and external characteristic test, Wang Yanyan et al. [14] improved the design of the impeller by optimizing the blade wrap angle and outlet placement angle. After this improvement, the head at the design working point was increased by 2.74m and the efficiency was increased by 5.77%. Li Jing et al. [15] studied the influence of blade profile on the hydraulic performance and impeller wear characteristics of slurry pump. Under the condition that the parameters such as impeller axial surface and blade inlet and outlet placement angle remain unchanged, logarithmic helix is used to control the blade profile. The numerical results show that under different blade profile conditions, the hydraulic performance of slurry pump and its impeller wear characteristics restrict each other; The blade with small wrap angle leads to the decline of the hydraulic performance of the pump, but the wear strength of the impeller is relatively low, and the wear strength of the impeller is positively correlated with the solid concentration. Zhao Wanyong et al. [16] numerically simulated and analyzed the impact of solid particles at the inlet of centrifugal pump on the inlet wear of centrifugal pump blades. Under the investigation, it was shown that reducing the inlet placement angle at the middle streamline of impeller can improve the wear resistance of impeller within a certain range.

In view of the damage of the dredge pump blades of the cutter suction dredger under the coarse sand mud condition, optimizing the pump blade enable a realization of the reasonable mud flow a reduction of the sediment impact on the blade, wear and cavitation.These improve the efficiency of the dredge pump and improve the dredging performance of the cutter suction dredger under the medium coarse sand condition.

2. Impeller Damage and Cause Analysis

One month after the construction progress of a project in Xiamen, it was found that the dredge pump was seriously worn. According to the engineering geological survey report, the soil in the construction area is mainly medium coarse sand which containing a large amount of loose quartz sand particles and shell debris. Samples are taken at three locations on the construction site, with average particle sizes of

0.58 mm, 0.36 mm and 0.49 mm. During the construction of cutter suction dredger, the main operating parameters of dredge pump are: flow rate of 5-6 m/s, concentration of 20-30%, and the diameter of delivery pipe is 850 mm.

The impeller suction port of D850 dredger dredged in Xiamen was seriously damaged (figure 1). The damage was mainly located near leading edge of the blade and had nearly worn through impeller body. The contact position between the impeller anti-wear ring and the impeller was the most seriously damaged. The perforation of impeller anti-wear ring causes leakage and backflow at the suction port of the dredger pump, which reduces the performance of the dredge pump and then reduces the construction efficiency of cutter suction dredger. The D850 dredge pump dredged in Quanzhou (D_{50}=0.35 mm) was used for one month, and the same damage occurred at the impeller suction position.

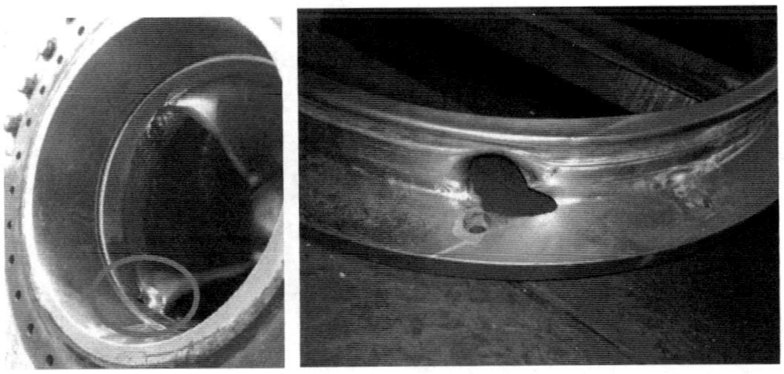

Figure 1. Damage position: impeller suction blade root and impeller wear ring.

By investigating the impeller damage of D850 dredge pump in different soil construction, only when the construction soil is medium coarse sand, the impeller suction is seriously damaged. Therefore, it is considered that medium coarse sand is the main factor causing serious damage to the impeller suction. Relevant studies show that with the increase of particle size, the movement speed of particles to the leading edge of blade working face increases and forms point impact wear [17]; The motion track of large particles deflects greatly towards the blade working face, is easy to collide with the blade head, and there are multiple impact processes, which has a large degree of wear on the blade [18].

The main reasons for the increased damage of the suction port of the dredge pump impeller under coarse sand conditions are as follows: 1) The starting position of the impeller blade is very close to the suction port of the impeller, that is, it is very close to the installation position of the impeller and the impeller wear ring. The blade root and the clearance between the impeller and the impeller wear ring are all factors inducing vortex, which affect each other and further local wear and cavitation [19]. 2) The flow pattern at the impeller suction port is complex. When the mud enters the impeller suction port, the fluid pressure near the front cover plate is low and the flow rate is high. The blade root blocks the mud flow, changes the mud flow direction and forms a local vortex. Large particle sand can promote cavitation and produce local wear. Cavitation damage is induced by pits after wear, which will further aggravate cavitation and wear [19].

3. Blade Leading Edge Modification Optimization

The damage of the root of the suction port of the blade shortens the service life of the impeller, and the anti-wear ring of the suction port is perforated, which reduces the construction efficiency of the dredger. According to the medium coarse sand soil and the damaged position of the blade, the blade is modified.The modification includes increasing the distance between the root of the blade suction and the anti-wear ring of the impeller suction, and reducing the blade wrap angle. So as to improve the flow field at the inlet of the impeller and prolong the service life of the impeller.

As shown in the figure 2, the distance between the blade root and the suction port increases to 130mm after blade modification, while the distance between the blade suction port root and the suction port before modification is only 20-30mm. At the same time, the blade wrap angle is correspondingly reduced (about 30° less than that before modification).The projection of impeller shaft surface and blade outlet angle remain unchanged.

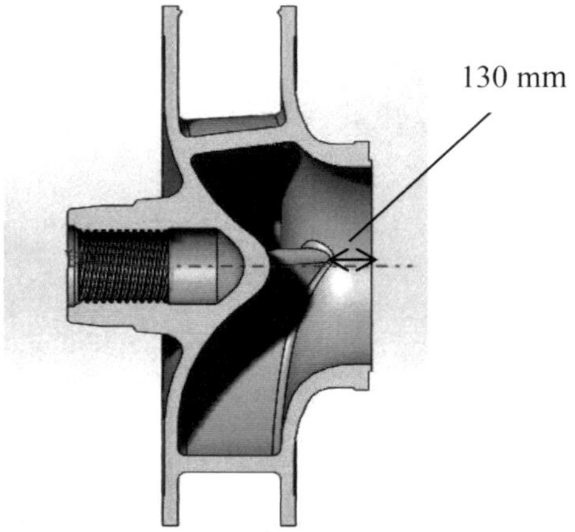

Figure 2. Schematic diagram of suction port root position after blade modification.

4. Performance Prediction and Flow Field Analysis before and after Blade Modification

4.1. Performance Curve Comparison
Computational fluid dynamics (CFD) technology has been successfully applied to the flow field analysis and external characteristic prediction of dredging pump. It has the advantages of low cost and short cycle. Especially when the laboratory test conditions of large dredge pump are limited, simulation can replace the test under certain conditions. Applying CFD technology to simulate and analyze the internal flow field of pump can not only save experimental resources, but also display many internal characteristic of flow field such as pressure distribution and velocity distribution [20].

Table 1. Comparison of performance data before and after import modification.

	Before modification		After modification		Contrast	
Flow m^3/h	Head m	Efficiency %	Head m	Efficiency %	Head difference m	Efficiency difference %
10000	80.5	83.8	82.2	82.8	1.7	-1.0
12000	77.6	85.9	77.9	85.1	0.3	-0.8

As shown in table 1, after the blade modification, the average change in clean water head and efficiency is very small under the condition of construction flow $10000m^3/h$ and $12000m^3/h$. That is , the external characteristics of the pump change little after the blade modification.

4.2. Blade Inlet Flow Field Analysis
Figure 3 shows the location of the velocity cloud. When the mud enters the impeller suction port, the blade root blocks the mud flow and changes the mud flow direction. The flow field at the impeller inlet forms turbulence (figure 4 (a)). After the blade is modified, the flow field at the same position is

improved (figure 4 (b)), decreadsing the possibility of local vortex formed by turbulence and cavitation damage induced by coarse sand wear, and prolonging the service life of the impeller.

(a) Schematic diagram of section. (b) Before blade modification. (c) After blade modification.

Figure 3. Relative position between section (red surface) and blade at 200mm from impeller suction.

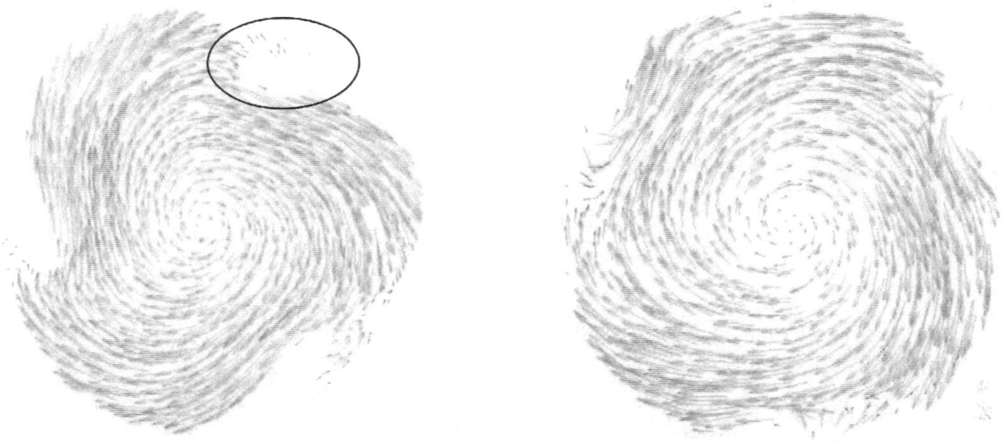

(a) Before blade suction modification. (b) After blade suction modification.

Figure 4. Section relative velocity at 200mm from impeller suction surface.

4.3. Analysis of Solid-Liquid Two-Phase Flow Field

The Euler-Euler model is used for the numerical unsteady calculation of two-phase flow. Taking the impeller rotation of 1.5 degrees as a time step, five cycles are calculated, and the data of the last calculation cycle is used as the basis for analysis. The boundary conditions of the dredge pump are velocity inlet and free outlet. It is assumed that the inlet velocity has no tangential and radial components and the outlet flow is fully developed. The wall surface is non slip wall surface and treated by standard wall function. The turbulence model adopts standard k-ε turbulence model. The second-order upwind scheme is discretized in space to solve the turbulence equation.

Figure 5. Particle concentration distribution on the surface of the front cover plate of the impeller before (left) and after (right) improvement of blades.

Figure 5 shows that the particle concentration distribution of the front cover plate of the impeller before and after the improvement of the blade suction. It can be seen from the figure that the surface concentration of the front cover plate before the improvement is high and the high concentration area is large. After the improvement of the blade, the particle concentration distribution on the surface of the front cover plate is significantly reduced, meanwhile reducing the wear on the surface of the impeller channel. At this time, the high concentration area is mainly close to the blade suction. After the improvement of the suction port, the blade suction port is far away from the impeller suction port, which reduces the wear of the particles on the suction anti-wear ring, reduces the replacement frequency of the anti-wear ring during construction, and improves the dredging efficiency.

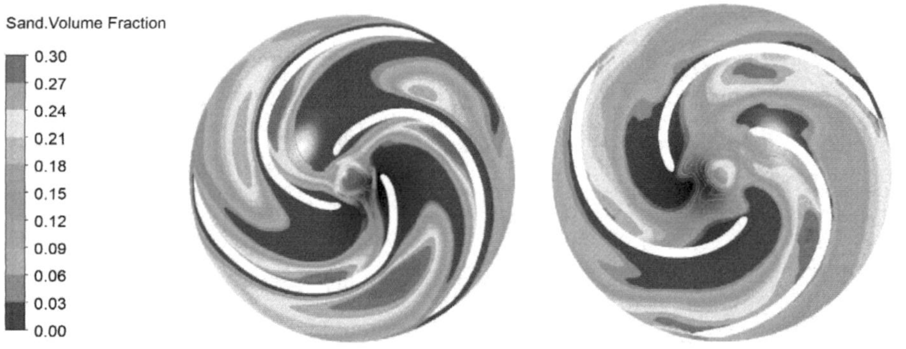

Figure 6. Particle concentration distribution on the surface of impeller rear cover plate before (left) and after (right) improvement of blades.

Figure 6 shows the particle concentration distribution shown on the rear cover plate of the impeller before and after the improvement of the suction port of the blade. It can be seen from the figure that before the improvement of the suction port, the particles are mainly concentrated near the back of the blade, and the mud concentration distribution is uneven. After the improvement of the impeller, the particle distribution uniformity on the surface of the rear cover plate is enhanced, the range of high concentration area is reduced, and local wear is reduced.

5. Tracking of Use of Mud Pump Impeller

Track and investigate the service condition of the improved impeller. The wear of the impeller is as follows: after the mud pump is replaced with the improved impeller, the intersection of the impeller blade inlet and the front cover plate (the root of the blade suction) is greatly worn. The wear starting surface is relatively smooth, and fish scale pits gradually appear with the aggravation of wear. The damaged position gradually extends from the root of the blade inlet to the front cover plate, and the wear thickness reaches 40mm. The damaged pits of the impeller after 148 hours and 640 hours are shown in figure 7.

The suction port improvement impeller shortens blade suction port. Before the improvement, the damaged position of the root was changed from the position where the original suction anti-wear ring meets the impeller. After the improvement, there was still pit damage at the root of the blade, but the damaged position is moved back to avoid the damage of the impeller anti-wear ring. The impeller improves the blade wear to a certain extent, prolongs the service life of the impeller and reduces the repair frequency of the dredge pump, that ensures the continuity of dredging construction.

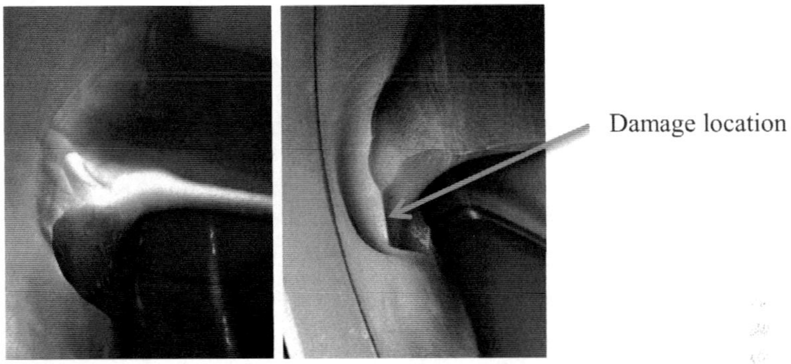

Figure 7. Modification impeller blade root (148 h and 640 h).

Comparison of construction data, "Xin Hai E" dredger pump impeller was constructed in a project before and after the suction port modification, with conveying distance of about 2.17km. The construction data of impeller under similar working conditions before and after the suction port modification were counted, and the productivity and energy consumption per 10000 m^3 before and after the impeller modification were analyzed. The comparison results are shown in table 2. After the impeller modification, the productivity is increased by 15.1% and the unit fuel consumption is reduced by 11.5%, indicating that the dredger construction efficiency is significantly improved and the energy-saving effect is obvious under this working condition.

Table 2. Comparison of construction efficiency of "Xin Hai E".

	Before modification (397h)	After modification (439h)	Contrast %
Productivity(m^3/h)	1351	1555	15.1
Unit fuel consumption (ton/10^4 m^3)	11.3	10.0	-11.5

6. Conclusion

- During the dredging construction of medium coarse sand soil, the erosion damage of the impeller is serious which is supposed to be repaired in time.In addition, the local pits shall be fixed to prevent local slurry backflow, induced vortex, cavitation and wear;
- After the blade modification, the distance from the blade inlet root of the dredge pump in to the impeller suction port is increased to 130mm. This is used to avoid the damage of the impeller suction anti-wear ring due to the damage of the blade root;
- After the modification, it brings little changes for the external characteristics of the dredge pump. According to the flow field analysis, the vortex at the root of the blade inlet is reduced after the modification of the blade. And a more uniform distribution of the solid particle is obtained, which is good for reducing the wear of the channel;

- The application shows that the damage of suction port of dredge pump impeller is improved to a certain extent after blade modification, which provides a reference for further modification and optimization.

References

[1] Xu Y H and Yue W F 2016 Analysis of transport process of cutter-suction dredger *China Rural Water and Hydropower* (9) 62-65
[2] Li M Z, He Y P, Zhuge W and Huang C 2015 Working range and operation optimal of slurry transportation system in dredger *Port Waterway Engineering* (4) 201-204, 213
[3] Bugdayci H H, Munts E and Grinwis H 2013 Latest developments in dredge pump technology *20th World Dredging Congress and Exhibition (WODCON XX)* (Brussels, Belgium: Central Dredging Association) p 260-277
[4] Ukhin B V 2002 Effect of the quantity and shape of impeller blades on dredge pump parameters *Power Technology & Engineering* **36**(3) 168-172.
[5] Zhang X S 2010 *Cavitation Research of High-Power Dredge Pump* (Harbin: Harbin Engineering University).
[6] Cheng Z J 1990 Experimental study on critical sediment concentration incavitation and erosion *Water Conservancy and Hydropower Technology* (2) 57-63
[7] Li L H, Miao Y and Gong Y F 2006 Discussion on cavitation in centrifugal impurity pump *Water Pump Technology* (6) 8-11
[8] Zhuang H F, Liu M M and Jin B X 2020 Research and application of centrifugal pump impeller modification for cutter-suction dredger *Fluid Machinery* **48**(4) 42-45
[9] Zhuang H F, Wang W K, Zheng L Z, Yang C and Hu J Z 2015 Optimization design of dredge pump based on CFD and model experiment *China Rural Water and Hydropower* (10) 130-132
[10] Zhuang H F, Yang C and Hu J Z 2017 Research and application of dredge Pump for 4500m^3 trailing suction hopper dredger *China Harbour Engineering* (1) 63-66
[11] Zhang R H, Zhang X J and Yang J H 2003 Flow field around intake of impeller under off-design conditions and its control *Journal of Gansu University of Technology* **29**(4) 64-66
[12] Wu R R.1995 Influence of impeller structure on cavitation performance of centrifugal pump *Mechanical and Electrical Equipment* (6) 18-21
[13] Luo X W, Zhang Y, Peng J Q and Xu H Y 2008 Effect of impeller inlet geometry on centrifugal pump cavitation performance *Journal of Tsinghua University (Science and Technology)* **48**(05) 836-839
[14] Wang Y Y, Zhao W G, Han X D and Zheng Y J 2019 Improved centrifugal pump impeller design in terms of blade wrap and exit angles *Journal of Drainage and Irrigation Machinery Engineering* **37**(7) 574-579.
[15] Li J, Zhang R H, Guo R and Li R N 2020 Influence of blade camber profile on hydraulicperformance of slurry pump and impeller wear characteristic *Journal of Drainage and Irrigation Machinery Engineering* **38**(1) 21-27.
[16] Zhao W Y, Guo Q, Song Q B and Hu X N 2018 Effects of blade entrance geometrical parameter on centrifugal pump wear rule *Journal of Gansu Sciences* **30**(4) 126-131
[17] Liu J, Xu H Y, Tang S and Lu L 2008 Numerical simulation of erosion and particle motion trajectory in centrifugal pump *Transactions of the Chinese Society for Agricultural Machinery* **39**(6) 54-59
[18] Huang X B, Yang S, Liu Z Q, Yang W and Li Y J 2016 Numerical simulation of erosion prediction in centrifugal pump based on particle track model *Transactions of the Chinese Society for Agricultural Machinery* **47**(8) 35-41
[19] Guan X F 2011 *Modern Pumps Theory and Design* (Beijing: China Aerospace Press) chapter 11 p 387–388

[20] Wang F J 2004 *Computational Fluid Dynamics Analysis* (Beijing: Tsinghua University Press) chapter 1 p 2-3

Experimental Study on the Effect of Locally Convergent Configurations on the Flow Pattern in PMMA Microchips

Jinsong Zhang[1], Zhongxin Liu[1], Chao Wang[1] and Zhiliang Wang[2*]

[1] School of Mechatronic Engineering and Automation, Shanghai University, Shanghai, 200444, China
[2] School of Mechanics and Engineering Science, Shanghai University, Shanghai, 200444, China
Email: wng_zh@shu.edu.cn

Abstract: The geometries of micro-channel play a key role in forming of digital droplets, and can be real-time or effective controlling methodologies. Local convergence regions are designed in the rectangular cross-section channels on PMMA microchips, in which two-phase coaxial jets are introduced by inserting a syringe needle. The two-phase flow (lubricating oil (continuous phase, flow rate Q_c) / deionized water (dispersed phase, flow rate Q_d)) is considered. Two geometric control variables, the relative position (needle displacement x) and tapering characteristics (convergence angle α), are naturally adopted to discribe such geometry configurations. The micro-flow under the change of these two parameters is mainly studied in this paper. Four kinds of characteristic flow patterns, namely, sausages, slug, dripping and jetting, are found in the experiment, and their occurring parameters and developing dynamic characteristics are discussed. The experiment shows that the increase of inner needle displacement x can produce higher frequency and finer droplets, which is in consistent with our previous results obtained in round tube experiments and simulations. While increasing the convergence angle α, contrarily, takes opposite effects.
Keywords. Two-phase flow, Convergent coaxial flow, Flow pattern.

1. Introduction

Microfluidic equipment can produce monodisperse sub-micron droplets, bubbles, emulsions and capsules in a continuous way by deforming, stretching and splitting certain substances in the fluids, which can be precipitated and solidified if necessary.

Microfluidic technology prepares droplets with controllable volume, controllable frequency, controllable composition or controllable movement, and large specific surface area with high consistency, and also realizes large-scale droplet productions. Moreover, it can contribute to wide areas such as rapid chemical reaction [1], drug delivery [2], cell capsule, digital PCR (polymerase chain reaction) system [3], protein crystallization [4], polymer microcapsule [5], or microreactor [6], and so on.

Geometrically, there are three main types of micro-channels: cross flow (T-shaped and Y-shaped) [7–10], flow focusing [11–13], and coaxial type [14–16]. The two-phase flow patterns of laminar, slug/plug, drilling, jetting, and annular have been observed in these channels. Moreover, when the viscosity ratio and interfacial tension of the dispersed phase fluid and the continuous phase fluid are small, the sausage flow pattern will appear [17].

2. Experimental Set-ups and Materials

The co-flowing configuration used and experimental apparatus in this research are shown in figure 1. The materials used are 5W-20 lubricating oil (continuous phase, density ρ_c=826 *kg/m*^{3}, viscosity μ_c=40.96 *mPa·s*, interfacial tension σ=20.08 *mN/m*) and deionized water (dispersed phase, density ρ_d=986.2 *kg/ m*^{3}, viscosity μ_d=1.23 *mPa·s*, interfacial tension σ=20.08 *mN/m*) separately. The devices include microchip, micro-injection-pump (LSP02-2A, Lange), highspeed camera (Phantom V611-16G-M, AMETEK), microlens (AT-X M100 PRO-D, Tokina), light source and computer, as shown in figure 1(a). The material of microfluidic chip is polymethylmethacrylate (PMMA), and its total dimensions are 30 *mm**30 *mm**4 *mm*.

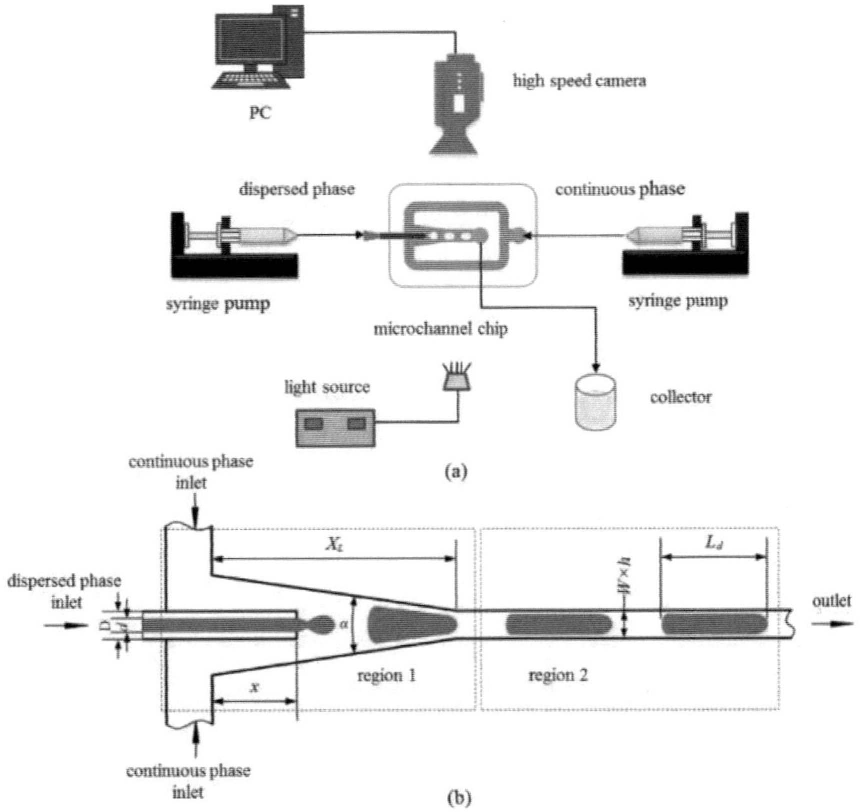

Figure 1. Schematic diagram of experiments. (a) The apparatus set-ups, (b) the co-flowing in the tapering micro-channel geometry.

Figure 1(b) shows the coaxial flow configuration and geometric parameters of convergent microchannel on a PMMA microchip. The dispersed phase and continuous phase flow in from the left side, and flow out toward the right side. Both phases mix at the convergence area, whose length is X_L = 2680 μm and convergence angle α. The syringe needle, whose inner diameter is d = 200 μm and outer diameter D = 400 μm, of dispersed phase is inserted into this tapering area to form two-phase flow (region 1). The downstream micro-channel is a straight tunnel (region 2), with rectangular cross-section of width W_{out} = 400 μm and height h = 600 μm. L_d is the axial size (length) of the droplet generated.

The two immiscible liquids, driven by a high-precision two-channel injection pump97 (LSP02-2A, Lange), are injected into a nested coaxial outer-inner annular micro-channel. The dispersed phase (deionized water) flow rate Q_d ranges from 8 to 152 *mL/h*, whereas the continuous phase (5W-20 lubricating oil) flow rate Q_c varies from 8 to 152*mL/h*, too.

Many samples of PMMA microchips are prepared. The combinations of the two geometrical parameters, convergence angle α and needle displacement x, are shown in table 1, labeled as sample A3, B3, C3, D3, C1, C2, C4, C5 and E. The microchips of certain convergence angle α (0°, 5°, 11°, 17° and 29°) and needle displacement x (0, 620, 1170, 1620 and 2160 μm) are designed to avoid redundant experimental work. Note 'A, B, C, and D' are symbols for convergence angle α at values of 5°, 11°, 17°, and 29° respectively. The numbers '1, 2, 3, 4 and 5' represent the needle displacement x at relative locations 0, 620, 1170, 1620, and 2160μm. The sample E, which is used to check the consistency and effectiveness of our experiments with the work of others, is a straight micro-channel without tapering configuration, which is a commonly used and well studied microchip structure.

Table 1. Two parameters to define tapering geometries of PMMA microchip samples.

Samples	Convergence Angle α (°)	Nozzle Stretch x (μm)
A3	5	1170
B3	11	1170
C3	17	1170
D3	29	1170
C1	17	0
C2	17	620
C4	17	1620
C5	17	2160
E	0	-

3. Result and Discussion

3.1. Flow Patterns

Many flow patterns can be obtained. Their descriptions are often based on qualitative and sometimes subjective visual discriminations. For instance, in the sample C4 (α = 17°, and x = 1620 μm), four main flow patterns are observed: slug, dripping, jetting and sausages, seen in figure 2. The slug pinches off at the tip of the inner needle tip, and produces a large droplet whose length L_d far exceeds the size of the channel width. The dripping droplets are gained the similar size to the channel width. These two flow patterns are periodic for generating monodisperse droplet. For the type of sausages, it seems a widening jet with bamboo waves at the interface, which never breaks up and brings any droplets. While the jetting droplets form faraway from the inner needle tip and at the downward inner jet end. The jetting mode may cause very tiny satellite droplets.

In such small scales of micro-channels, the gravity force is usually ignored. The countable transition of two-phase flow pattern is mainly determined by the competitions among three forces, namely, the viscous force of continuous phase, the interfacial tension of two phases and the inertial force of dispersed phase, and we define two non-dimensional parameters to depict their competitions. Those are the capillary number of the continuous phase, $Ca_c = u_c\mu_c/\sigma$, representing the ratio of the viscous force of the continuous phase to the interfacial tension force; and the Weber number of the dispersed phase, $We_d = \rho_d u_d^2 d_h/\sigma$, representing the ratio of the inertial force of the dispersed phase to the interfacial tension force. Here, the superficial velocities $u_c = Q_c/(W_{out}\ h)$ and $u_d = Q_d/(W_{out}\ h)$ are defined on the downward area of rectangular cross-section, and $d_h = 2W_{out}\ h/(W_{out} + h)$. Note that the subscript 'c' represents the continuous phase and 'd' represents the dispersed phase on no special cases. We draw a typical flow chart to explain the occurrence of flow patterns in $Ca_c \sim We_d$ space, as shown in figure 3 for sample C4.

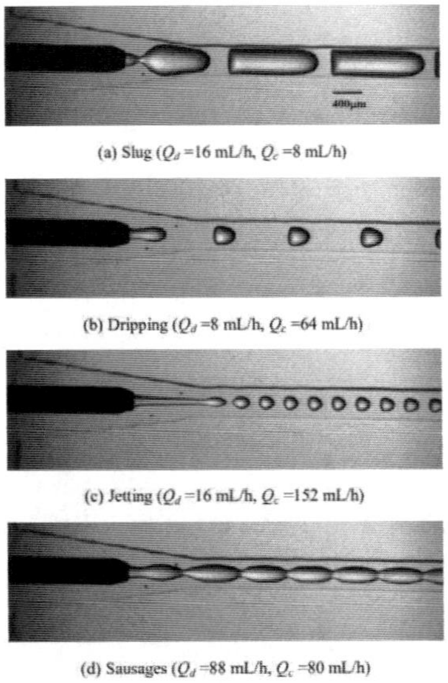

Figure 2. Four typical flow patterns vary with flow rates for sample C4: (a) Slug, (b) dripping, (c) jetting and (d) sausages.

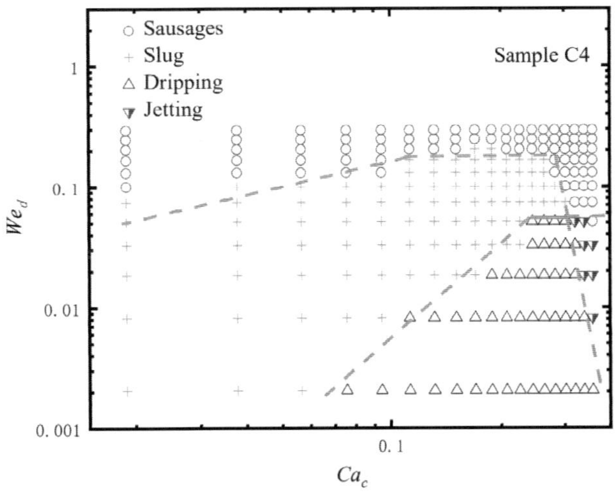

Figure 3. Flow pattern map in Ca_c–We_d space for sample C4 ($\alpha = 17^{\circ}$, and $x = 1620 \mu m$).

The four flow regimes, namely, slug, dripping, jetting and sausage as pictures represented in figure 2, are identified in figure 3. Dashed lines for pattern transition separation are used to divide these flow regimes. As a general, the slug regime occurs at low Ca_c with low We_d region, the dripping regime occurs at higher Ca_c with low We_d region, the jetting regime occurs at highest Ca_c with low We_d region, and the sausages regime occurs at high We_d region. The transitions among these regimes are highly correlated with the relative power superiority of interfacial tension force, inertial force and shear stress force. Apparently, the interfacial tension dominates the proceeding of slug forming, the inner inertial force maintains the wavy interface of sausages, the moderate outer shear force and low inner inertial force produce dripping process, and the high outer shear force causes thin jetting.

3.2. The Variations of Convergence Angle α and Needle Displacement x on Flow Charts

When the needle displacement x increases, the flow pattern map varies correspondingly. We draw collectively the flow pattern maps for samples C1, C2, C3, C4 and C5 in figure 4. Apparently the convergence angle α is a constant of 17°, and the needle displacement x increases from 0, 620, 1170, 1620 to 2160 μm.

As can be seen from figure 4, with the increase of x, the flow pattern transition boundaries between sausages and slug move downwards, and the area of slug shrinks. The flow pattern transition boundary between slug and dripping moves left-upwards and also suppresses the slug region, and the area of dripping increases. Jetting begins to appear into the scope of our experimental setting parameters from the right side only for samples C4 and C5.

The influence of the convergence angle α on flow patterns is also investigated for samples A3, B3, C3 and D3, as shown in figure 5 for the fixed needle displacement x = 1170 μm. The corresponding convergence angle α of each sample increases from 5°, 11°, 17° to 29° respectively.

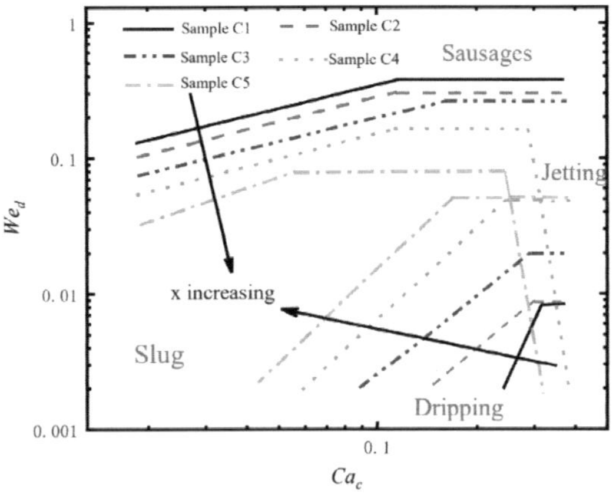

Figure 4. Four typical flow patterns vary with flow rates for sample C4. (a) Slug, (b) dripping, (c) jetting and (d) sausages.

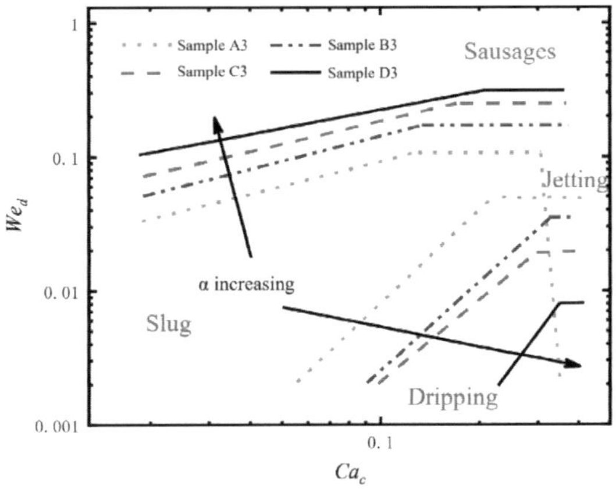

Figure 5. Flow pattern transitions in Ca_c–We_d space for samples A3, B3, C3 and D3 as the convergence angle α increases from 5°, 11°, 17° and 29°.

It can be seen from figure 5 that with the decrease of convergence angle α, the flow pattern transition lines between slug and sausages move downwards and compress the area of slug regime. The slug-dripping transition lines move left-upwards and also suppress the slug regime, and the regime of dripping increases with α getting smaller. Jetting begins to appear from the right side only for sample A3 for the smallest α in the present parameter scope.

By comparing figure 5 with figure 4, it is intuitive to see a correlation between the effects of x and α on the occurrence of flow patterns. The increase of the needle displacement x and decrease of the convergence angle α almost have the same influence on flow pattern transitions.

3.3. Scaling up Droplet Formation in Slug and Dripping Regimes

Slug and dripping are idealized flow patterns in heat and mass transfer processes or bio-testing. The droplets formed have high consistency in size and specific surface area, and monodisperse droplets can be obtained efficiently. The size of droplet and the frequency generated are the focus for applications.

The droplet formation at the inner needle tip can be divided into two processes: growth and necking off. Figure 6 shows photographs of droplet growth stages (a)-(c) and necking-shedding stages (d)-(f) in the slug flow pattern, and photographs of droplet growth stages (g)-(i) and necking-shedding stages (j)-(l) in the dripping flow pattern.

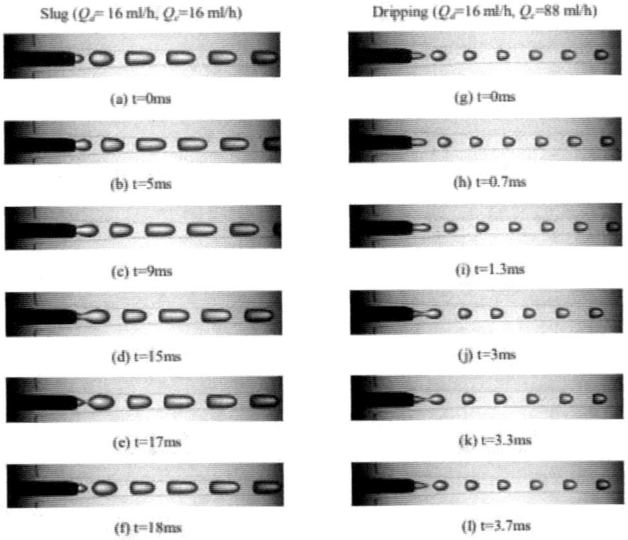

Figure 6. Optical micrographs of the evolution of the periodic droplet formation sequences for sample A3. Slug ((a)–(c) for growing and (d)–(f) for necking-pinchoff), and Dripping pattern((g)–(i) for growing and (j)–(l) for necking-pinchoff).

4. Conclusion

The PMMA microchips are designed to take into account the convergence configurations on behaviors of oil/water two-phase coaxial flows in rectangular cross-section micro-channels. The influence of two geometric parameters, needle displacement x and convergence angle α on droplet forming characters is mainly investigated.

Slug, dripping, jetting and sausage modes occur in the present experimental observations. Although all the modes are periodical, the sausage waves never pinch off, and the jetting mode has very tiny satellite droplets evolved, only slug and dripping modes can produce perfect uniform droplets. The decrease of needle displacement x and increase of convergence angle α can both expand the slug and dripping regimes. It is proved that the needle displacement x and convergence angle α are essentially correlated variables, and can be degenerated into a single geometrical variable, which means the great improvement on controlling costs.

Acknowledgements

This article is funded by the National Key Research and Development Program "8.5 Generation Printing OLED Display Industrial Demonstration Project" (2017YFB0404503).

References

[1] Teh S Y, Lin R, Hung L H and Lee A P 2008 Droplet microfluidics *Lab on a Chip* **8** 198–220.
[2] Kleinstreuer C, Li J and Koo J 2008 Microfluidics of nano-drug delivery *Int. J. of Heat and Mass Transfer* **51** 5590–97.
[3] Demello A J 2006 Control and detection of chemical reactions in microfluidic systems *Nature* **442** 394-402.
[4] Zheng B, Tice J D and Ismagilov R F 2004 Formation of arrayed droplets by soft lithography and twophase fluid flow, and application in protein crystallization *Advanced Materials* **16** 1365–68

[5] Wang Q, Liu S, Wang H, Zhu J and Yang Y 2015 Alginate droplets pre-crosslinked in micro-channels to prepare monodispersed spherical microgels *Colloids and Surfaces A: Physicochemical and Engineering Aspects* **482** 371–377.

[6] Cygan Z T, Cabral J T, Beers K L and Amis E J 2005 Microfluidic platform for the generation of organic-phase microreactors *Langmuir* **21** 3629–34.

[7] Yao C, Dong Z, Zhao Y and Chen G 2014 The effect of system pressure on gasliquid slug flow in a microchannel *AIChE Journal* **60** 1132–42.

[8] Pan D, Liu M, Li F, Chen Q, Liu X, Liu Y, Zhang Z, Huang W and Li B 2018 Formation mechanisms of solid in water in oil compound droplets in a horizontal t-junction device *Chemical Engineering Science* **176** 254–263.

[9] Yin Y, Zhu C, Guo R, Fu T and Ma Y 2018 Gas-liquid two-phase flow in a square micro-channel with chemical mass transfer: Flow pattern, void fraction and frictional pressure drop *Int.J. of Heat and Mass Transfer* **127** 484–496.

[10] Zhang Q, Li H, Zhu C, Fu T, Ma Y and Li H Z 2018 Micro-magnetofluidics of ferrofluid droplet ormation in a t-junction *Colloids and Surfaces A: Physicochemical and Engineering Aspects* **537** 572–579.

[11] Cramer C, Fischer P and Windhab E J 2004 Drop formation in a co-flowing ambient fluid *Chemical Engineering Science* **59** 3045–58.

[12] Kashid M N, Kowaliski W, Renken A, Baldyga J and Kiwi-Minsker L 2012 Analytical method to predict two-phase flow pattern in horizontal micro-capillaries Chemical Engineering Science 74 219-232.

[13] Wang K, Xie L, Lu Y and Luo G 2013 Generating microbubbles in a co-flowing microfluidic device *Chemical Engineering Science* **100** 486–495.

[14] Wu L, Liu X, Zhao Y and Chen Y 2017 Role of local geometry on droplet formation in axisymmetric microfluidics *Chemical Engineering Science* **163** 56–67.

[15] Rahimi M, Shams Khorrami A and Rezai P 2019 Effect of device geometry on droplet size in co-axial flow-focusing microfluidic droplet generation devices *Colloids and Surfaces A: Physicochemical and Engineering Aspects* **570** 510–517.

[16] Svetlov S D and Abiev R S 2018 Formation mechanisms and lengths of the bubbles and liquid slugs in a coaxial-spherical micro mixer in taylor flow regime *Chemical Engineering Journal* **354** 269–284.

[17] Harvie D J E, Davidson M R, Cooper-White J J and Rudman M 2006 A parametric study of droplet deformation through a microfluidic contraction: Low viscosity newtonian droplets *Chemical Engineering Science* **61** 5149–58.

Singularly Perturbed Solutions for a Class of Thermoelastic Weakly Coupled Problems

Liping Bao*, Xin Shuai and Liqun Wu

School of science, Hangzhou Dianzi University, Hangzhou, China
Email: baolp@hdu.edu.cn

Abstract. Based on the basic equation of Green Lindsay (G-L) theory, the thermoelastic weak coupling problem under the basic equation is discussed, that is, two thermal relaxation parameters are added to the constitutive equation, the influence of the coupling term on the temperature field and elastic field is considered, and the asymptotic solution of the governing equation is constructed. Firstly, in order to obtain the asymptotic solution, the singularly perturbed expansion method is used.Then,combined with the corresponding boundary conditions, the partial differential equation method is used to solve the external solution and the boundary layer correction term. Secondly, in the case of weak coupling, the uniformly efficient estimation of the remainder of the asymptotic solution is obtained by using Gronwall inequality, so as to obtain the uniformly efficient of the formal asymptotic solution. Finally, the first term of the asymptotic solution is numerically analyzed by using the singularly perturbed numerical method. The present work will be conducive to the analysis of thermoelastic processes and numerical simulation of different materials in the case of weak coupling.
Keywords. Thermoelasti ccoupling, singular perturbation, numerical solution, uniform valid estimate.

1. Introduction

In thermoelastic theory, with the emergence of ultrashort laser pulses, non Fourier law has a significant effect on the generation of heat flow, and then on the elastic field, and no longer meets the conditions of classical thermoelastic theory. Based on the green Naghdi (G-N) theory of type III, Othman et al. [1] studied the effect of laser pulse on the heat load of three-dimensional generalized micropolar thermoelastic homogeneous isotropic medium, solved the dimensionless equation of the problem by using normal analysis technology. Rajneesh et al.[2] studied the homogeneous isotope micropolar porous thermoelastic circular plate by using the eigenvalue method in the thermoelastic three-phase lag theory caused by thermomechanical source. The expressions of each component were obtained in the transformation domain by using Laplace transform and Hankel transform. Paul et al. [3] studied the thick plate with diffusion of two-dimensional magnet thermoelastic problem, the analytical solution was deduced in Laplace Fourier transform domain , and obtained the numerical inversion of stress, displacement and chemical potential in space-time domain. Shakeriaski et al. [4] studied the nonlinear thermoelastic response of elastic media heated by short laser pulses, established the nonlinear coupled L-S-type generalized thermoelastic equation by using the finite strain theory (FST), and solved the time-dependent nonlinear equations by using Newmark's numerical time integration technique and an updated finite element algorithm.

In recent years, many scholars have modified and extended the classical thermoelastic coupling theory by adding thermal relaxation parameters in combination with practical problems. Tang et al. [5]

obtained the combined form of the basic equations of L-S theory, extended L-S theory and G-L theory. On this basis, the thermoelastic coupling problem of semi infinite medium surface after instantaneous heating is considered, and the equations of motion and energy are solved by numerical inverse Laplace transform. El-Maghraby et al. [6] obtained the state space equation of one-dimensional generalized thermoelastic problem with a thermal relaxation parameter based on L-S theory.Youssef et al. [7, 8] established a new generalized thermoelastic theory with two thermal relaxation parameters, obtained the uniqueness theorem of the two temperature generalized linear thermoelastic equation of a homogeneous isotropic object, and used this theory to solve the thermoelastic interaction problem of a cylindrical cavity subjected to thermal shock on the boundary surface and a moving heat source with constant velocity in an elastic infinite medium. Kumar et al. [9-11] studied the thermoelastic interaction under the dual temperature G-L theory, deeply analyzed the plane harmonic in the dual temperature thermoelastic medium with two relaxation parameters, obtained the dispersion relation solution of its propagating in the medium. Shivay et al. [12] established the temperature rate dependent two temperature thermoelasticity theory which based on the generalized thermodynamic law, obtained a new and more general two temperature relationship, and proved the uniqueness of the solution of the general mixed initial boundary value problem for anisotropic media under that theory. Jangid [13] considered homogeneous and anisotropic materials under the background of this model, based on the generalized thermodynamic law, a modified trdtt theory with a new two temperature relationship is established, and an alternative equation for the mixed boundary value initial value problem is proposed. Using this equation, the convolution variational principle is derived, and the reciprocal principle of this theory is derived. Zhang et al. [14] proposed a thermal mechanical coupling propagation model based on non Fourier heat conduction, established a dimensionless three-dimensional generalized thermal mechanical coupling equation of equal diameter two spherical particles and the corresponding initial boundary value conditions.

The exact analytical solution of ultrasonic wave excited by pulsed laser in materials is very difficult, because this physical process includes not only transient thermal diffusion, but also transient elastic wave excitation and propagation in finite space. The singular perturbation method is mainly used to solve the problem with small parameters in the high-order term. When the thermal relaxation parameter is added to the thermoelastic coupling problem, the singular perturbation method can be used to analyze this kind of problem. Therefore, it is very necessary to study it from the perspective of theoretical research and practical application.

In this paper we describe the temperature field generated by pulsed laser acting on the material surface by using the non Fourier heat conduction law, and the thermoelastic coupling model is established in conjunction with the motion equation in G-L theory. It is considered that the coupling coefficient has a positive linear relationship with the time of heat flow relaxation. Using the singularly perturbed asymptotic expansion method, the internal and external solutions of the equation are constructed. The solution of the coupled equation is expanded into parabolic equation and hyperbolic equation. The formal asymptotic solutions are solved one by one through the corresponding boundary conditions. The uniform validity of the remainder of the asymptotic solution is obtained by Gronwall inequality and L^2 estimation, so as to obtain the distribution of temperature field and elastic field in the bounded region. Then the numerical analysis method is used to approximate the asymptotic solution equation, and the effectiveness of the numerical method is verified by a numerical example.

2. Governing Equations

In this paper, we consider a class of the Green-Lindsay models with weak coupling, so as to study the properties of temperature field and elastic field of thermoelastic plane wave from uniform isotropic heat conduction in finite solid space.According to reference [5], the basic governing equations of the homogeneous isotropic thermal conductive elastic medium model (when the density and thermal conductivity remain unchanged) are as follows (in general Cartesian coordinates system):

The strain-displacement relation:

Journal of Physics: Conference Series **2097** (2021) 012007 doi:10.1088/1742-6596/2097/1/012007

$$e_{i,j} = \frac{1}{2}(u_{i,j} + u_{j,i}) \tag{1}$$

The constitutive relations:

$$\sigma_{ij} = \lambda e_{kk}\delta_{ij} + 2\mu e_{ij} - \gamma(T - T_0 + \tau_1 \dot{T})\delta_{ij} \tag{2}$$

The equation of motion:

$$\sigma_{ji,j} = \rho \ddot{u} \tag{3}$$

Cattaneo-Vernotte(C-V) non-Fourier heat flow equation:

$$q_i + \tau_0 \dot{q}_i = -kT_{,i} \tag{4}$$

Combining equation (2) and (3), we get the movement displacement equation is as follows:

$$\rho \ddot{u}_i = (\lambda + \mu)u_{j,ij} + \mu u_{i,jj} - (3\lambda + 2\mu)\alpha_T(T + \tau_1 \dot{T})_{,i} \tag{5}$$

The energy balance equation which is derived from Cattaneo-Vernotte(C-V) non-Fourier heat flow equation :

$$kT_{,ii} = \rho C_E(\dot{T} + \tau_0 \ddot{T}) + \gamma T_0 \dot{\varepsilon} \tag{6}$$

In the above equations, u_i are the displacement components, e_{ij} are the components of the elasticstrain tensor, σ_{ij} are the components of the elastic stress tensor, λ and μ are Lamé's constants, α_T is the coefficient of linear thermal expansion, $\gamma=(3\lambda+2\mu)\alpha_T$ is the parameter of thermoelastic coupling,, ρ is the constant mass density, C_E is the specific heat at constant strain, q_i are the components of the heat flux vector, k is the thermal conductivity of the materia, τ_0 and τ_1 are the thermal relaxation parameters.

3. Theoretical Analysis

One dimensional temperature and deformation fields are considered, for simplifications we will use the following non-dimensional variables:

$$x' = \left(\frac{\lambda+2\mu}{\rho}\right)^{\frac{1}{2}} \frac{\rho C_E}{k} x \quad t' = \left(\frac{\lambda+2\mu}{\rho}\right) \frac{\rho c_E}{\kappa} t \quad U' = \left[\rho\left(\frac{\lambda+2\mu}{\rho}\right)^{\frac{3}{2}} \frac{\rho c_E}{(3\lambda+2\mu)\alpha T_0 \kappa}\right] u$$

$$\sum = \frac{\sigma}{\gamma T_0} \quad \theta = \frac{T - T_0}{T_0}.$$

Equations (5)-(6) in the non-dimensional forms (after suppressing the primes) reduce to:

$$\begin{aligned} U'' - \ddot{U} &= \theta' + \varepsilon_1 \dot{\theta}' \\ \theta'' &= \dot{\theta} + \varepsilon_0 \ddot{\theta} + \delta \dot{U}' \\ \sum &= U' - (\theta + \varepsilon \dot{\theta}) \end{aligned} \tag{7}$$

where $\varepsilon_i = (\frac{\lambda + 2\mu}{\rho})\frac{\rho C_E \tau_i}{k}, \delta = \frac{(3\lambda + 2\mu)^2 \alpha_T{}^2 T_0}{(\lambda + 2\mu)\rho C_E}$, we order $\varepsilon_0 = \varepsilon_1 = \varepsilon$, $\delta = \beta\varepsilon$, then considering an initial values and boundary conditions, thus, we can rewrite the equation (7) as:

$$
\begin{cases}
U_{xx} = U_{tt}\theta_x + \varepsilon\theta_{xt} \\
\theta_{xx} = \theta_t + \varepsilon\theta_{tt} + \beta\varepsilon U_{xt} \\
\theta(x,t)|_{t=0} = \varphi_1(x) \\
\theta_t(x,t)|_{t=0} = \varphi_2(x) \\
\theta(x,t)|_{x=0} = \varphi_3(t) \\
\theta(x,t)|_{x=L} = \varphi_4(t) \\
U(x,t)|_{t=0} = \psi_1(x) \\
U_t(x,t)|_{t=0} = \psi_2(x) \\
U(x,t)|_{x=0} = \psi_3(t) \\
U(x,t)|_{x=L} = \psi_4(t)
\end{cases}
\tag{8}
$$

where $x \in [0, \mathrm{L}], t \in (0, \mathrm{T})$, We make the following assumptions:

$[H_1]$ $(x,t) \in \Omega_T, \Omega_T = \bar{\Omega} \times (0,\mathrm{T}), \Omega = (0,\mathrm{L})$, L and T are finite values

$[H_2]$ $\varphi_1(x), \varphi_2(x), \psi_1(x), \psi_2(x)$ are some known arbitrary functions

$[H_3]$ $\varphi_3(t), \varphi_4(t), \psi_3(t), \psi_4(t)$ are any known function of at least first order differentiable

By canonical expansion of the solution of equation (7), we obtain

$$
U(x,t,\varepsilon) = \sum_{i=0}^{N} \varepsilon^i U_i(x,t) \qquad \theta(x,t,\varepsilon) = \sum_{i=0}^{N} \varepsilon^i \theta_i(x,t)
\tag{9}
$$

Comparing the same power of ε, we have:

$$
\begin{cases}
U_{0,xx} = U_{0,tt} + \theta_{0,x}, \\
\theta_{0,xx} = \theta_{0,t}, \\
\theta_0(x,t)|_{t=0} = \varphi_1(x) \\
\theta_0(x,t)|_{x=0} = \varphi_3(t) \\
\theta_0(x,t)|_{x=L} = \varphi_4(t) \\
U_0(x,t)|_{t=0} = \psi_1(x) \\
U_{0,t}(x,t)|_{t=0} = \psi_2(x) \\
U_0(x,t)|_{x=0} = \psi_3(t) \\
U_0(x,t)|_{x=L} = \psi_4(t)
\end{cases}
\tag{10}
$$

$$
\begin{cases}
U_{i+1,xx} = U_{i+1,tt} + \theta_{i+1,x} + \theta_{i,xt}, \\
\theta_{i+1,xx} = \theta_{i,tt} + \theta_{i+1,t} + \beta U_{i,xt}, \\
\theta_{i+1}(x,t)|_{t=0} = -Q_i(x,0) \\
\theta_{i+1}(x,t)|_{x=0} = 0 \\
\theta_{i+1}(x,t)|_{x=L} = 0 \\
U_{i+1}(x,t)|_{t=0} = -P_i(x,0) \\
U_{i+1,t}(x,t)|_{t=0} = P_{i+1}(\mathrm{x},0) - P_{i,t}(\mathrm{x},0) \\
U_{i+1}(x,t)|_{x=0} = -P_i(0,t)\mathrm{e}^{-\varepsilon^{-1}t} \\
U_{i+1}(x,t)|_{x=L} = -P_i(\mathrm{L},t)\mathrm{e}^{-\varepsilon^{-1}t}
\end{cases}
\tag{11}
$$

By separating variables to solve the equation (10), we can obtain:

$$
\theta_0(x,t) = V(x,t) + \sum_{n=1}^{\infty} A_n \mathrm{e}^{-\frac{n^2\pi^2 t}{\mathrm{L}^2}} \sin\frac{n\pi x}{\mathrm{L}} - \sum_{n=1}^{\infty} \int_0^t \bar{V}_n(\tau) \mathrm{e}^{-\frac{n^2\pi^2(\tau - t)}{\mathrm{L}^2}} d\tau \sin\frac{n\pi x}{\mathrm{L}}
\tag{12}
$$

$$
U_0(x,t) = W(x,t) + \sum_{n=1}^{\infty}\left(B_n \cos\frac{n\pi x}{\mathrm{L}} + C_n \sin\frac{n\pi x}{\mathrm{L}} \right) \sin\frac{n\pi x}{\mathrm{L}} + \sum_{n=1}^{\infty} \frac{\mathrm{L}}{\mathrm{n}\pi} \int_0^t \bar{W}_n(\tau) \sin\frac{n\pi(\tau - t)}{\mathrm{L}} d\tau \sin\frac{n\pi x}{\mathrm{L}}
\tag{13}
$$

where

$$V(x,t)=\frac{\varphi_4(t)-\varphi_3(t)}{\mathrm{L}}x+\varphi_3(t) \quad A_n=\frac{2}{\mathrm{L}}\int_0^{\mathrm{L}}(\varphi_1(x)-V(x,0))\sin\frac{n\pi x}{\mathrm{L}}\mathrm{d}x$$

$$\bar{V}(x,t)=V_t(x,t)=\sum_{n=1}^{\infty}\bar{V}_n(t)\sin\frac{n\pi x}{\mathrm{L}} \quad \bar{V}_n(t)=\frac{2}{\mathrm{L}}\int_0^{\mathrm{L}}\bar{V}(x,t)\sin\frac{n\pi x}{\mathrm{L}}\mathrm{d}x$$

$$W(x,t)=\frac{\psi_4(t)-\psi_3(t)}{\mathrm{L}}x+\psi_3(t) \quad B_n=\frac{2}{\mathrm{L}}\int_0^{\mathrm{L}}(\psi_1(x)-W(x,0))\sin\frac{n\pi x}{\mathrm{L}}\mathrm{d}x \quad C_n=\frac{2}{\mathrm{L}}\int_0^{\mathrm{L}}(\psi_2(x)-W_t(x,0))\sin\frac{n\pi x}{\mathrm{L}}\mathrm{d}x$$

$$\bar{W}(x,t)=W_{tt}(x,t)+\theta_{0,x}=\sum_{n=1}^{\infty}\bar{W}_n(t)\sin\frac{n\pi x}{\mathrm{L}}$$

Now give the composite expansion of the solution:

$$U^*=\sum_{i=0}^{N}\varepsilon^i U_i(x,t)+\sum_{i=0}^{N-1}\varepsilon^{j+1}P_j(x,t)\mathrm{e}^{-\varepsilon^{-1}t} \quad \theta^*=\sum_{i=0}^{N}\varepsilon^i\theta_i(x,t)+\sum_{i=0}^{N-1}\varepsilon^{j+1}Q_j(x,t)\mathrm{e}^{-\varepsilon^{-1}t} \tag{14}$$

Combining initial values and boundary conditions, we can be concluded that $P_0(x,t)=P_1(x,t)=P_2(x,t)=0$, then, comparing the same power of ε, we have:

$$\begin{cases} -Q_{0,t}=Q_{0,xx} \\ Q_0(x,\mathrm{t})|_{t=0}=\theta_{0,t}(x,0)-\varphi_2(x) \\ Q_0(x,t)|_{x=0}=0 \\ Q_0(x,t)|_{x=L}=0 \end{cases} \tag{15}$$

$$\begin{cases} Q_{j+1,xx}=Q_{j,tt}-Q_{j+1,t}+\beta P_{i+1,xt}-\beta P_{i+1,x} \\ Q_{j+1}(x,t)|_{t=0}=\theta_{j+1,t}(x,0)+Q_{j,t}(x,0) \\ Q_{j+1}(x,t)|_{x=0}=0 \\ Q_{j+1}(x,t)|_{t=L}=0. \end{cases} \tag{16}$$

$$P_{j+3}=P_{j+1,xx}-P_{j+1,tt}+2P_{j+2,t}-Q \tag{17}$$

.

By separating variables, we can obtain

$$Q_0(x,t)=\sum_{n=1}^{\infty}D_n\mathrm{e}^{\frac{n^2\pi^2 t}{\mathrm{L}^2}}\sin\frac{n\pi x}{\mathrm{L}} \tag{18}$$

where $D_n=\frac{2}{\mathrm{L}}\int_0^{\mathrm{L}}(\theta_{0,t}(x,0)-\varphi_2(x))\sin\frac{n\pi x}{\mathrm{L}}\mathrm{d}x$, according to reference [15], on the bound domain, the existence and uniqueness of equation (11) and equation (16) can be obtained.

4. Remainder Estimation

Lemma 1. The remainder of solution of equation (8)

$$\begin{aligned} U(x,t,\varepsilon)&=\sum_{i=0}^{N}\varepsilon^i U_i(x,t)+\sum_{i=0}^{N-1}\varepsilon^{j+1}P_j(x,t)\mathrm{e}^{-t/\varepsilon}+\varepsilon^{N+1}R \\ \theta(x,t,\varepsilon)&=\sum_{i=0}^{N}\varepsilon^i\theta_i(x,t)+\sum_{i=0}^{N-1}\varepsilon^{j+1}Q_j(x,t)\mathrm{e}^{-t/\varepsilon}+\varepsilon^{N+1}S \end{aligned} \tag{19}$$

Satisfies

$$\begin{aligned} &\|R\|_{L^2(\Omega)} + \|R_x\|_{L^2(\Omega)} + \|R_t\|_{L^2(\Omega)} \le \\ &\|S\|_{L^2(\Omega)} + \|S_x\|_{L^2(\Omega)} + \|S_t\|_{L^2(\Omega)} \le \end{aligned} \tag{20}$$

Then, the formal asymptotic solution of equation (8) is uniformly valid.

Proof. Replace equation (19) with equation (8), we have

$$\begin{cases} R_{xx} = R_{tt} + S_x + \varepsilon S_{xt} + F_1 \\ S_{xx} = \varepsilon S_{tt} + S_t + \beta\varepsilon R_{xt} + F_2 \\ S(x,t)|_{t=0} = \varphi_1(x) \\ S_t(x,t)|_{t=0} = \varphi_2(x) \\ S(x,t)|_{x=0} = \varphi_3(t) \\ S(x,t)|_{x=L} = \varphi_4(t) \\ R(x,t)|_{t=0} = \psi_1(x) \\ R_t(x,t)|_{t=0} = \psi_2(x) \\ R(x,t)|_{x=0} = \psi_3(t) \\ R(x,t)|_{x=L} = \psi_4(t) \end{cases} \tag{21}$$

where

$$F_1 = \theta_{N,xt} + Q_{N-1,xt}\mathrm{e}^{-t/\varepsilon} \quad F_2 = \theta_{N,tt} + Q_{N-1,tt}\mathrm{e}^{-\varepsilon^{-1}t} + \beta U_{N,xt} + \beta P_{N-1,xt}\mathrm{e}^{-\varepsilon^{-1}t}$$

Similar to reference [16], equation (21) multiplies R_t on both sides of the first equation, multiplies S_t on both sides of the second equation, then integrate on Ω_T and use the corresponding boundary conditions to obtain

$$\begin{aligned} &\frac{1}{2}\int_0^L R_x^2\mathrm{d}x + \frac{1}{2}\int_0^L R_t^2\mathrm{d}x + \int_0^t\int_0^L S_x R_t\mathrm{d}x\mathrm{d}t + \varepsilon\int_0^t\int_0^L S_{xt}R_t\mathrm{d}x\mathrm{d}t + \int_0^t\int_0^L R_t F_1\mathrm{d}x\mathrm{d}t = 0 \\ &\frac{1}{2}\int_0^L S_x^2\mathrm{d}x + \frac{1}{2}\varepsilon\int_0^L S_t^2\mathrm{d}x + \int_0^t\int_0^L S_t^2\mathrm{d}x\mathrm{d}t + \beta\varepsilon\int_0^t\int_0^L S_t R_{xt}\mathrm{d}x\mathrm{d}t + \int_0^t\int_0^L F_2 S_t\mathrm{d}x\mathrm{d}t = 0 \end{aligned} \tag{22}$$

According to equation (22) , we have:

$$\begin{aligned} &\frac{\beta}{2}\int_0^L R_x^2\mathrm{d}x + \frac{\beta}{2}\int_0^L R_t^2\mathrm{d}x + \frac{1}{2}\int_0^L S_x^2\mathrm{d}x + \frac{1}{2}\varepsilon\int_0^L S_t^2\mathrm{d}x + \frac{1}{2}\int_0^t\int_0^L S_t^2\mathrm{d}x\mathrm{d}t \\ &\le \frac{\beta}{2}\int_0^t\int_0^L S_x^2\mathrm{d}x\mathrm{d}t + \beta\int_0^t\int_0^L R_t^2\mathrm{d}x\mathrm{d}t + \frac{1}{2}\int_0^t\int_0^L \beta F_1^2 + F_2^2\mathrm{d}x\mathrm{d}t. \end{aligned} \tag{23}$$

Now we considering two cases of β

(i) When $0 < \beta < 1$, we have

$$\begin{aligned} &\frac{1}{2}\int_0^T\int_0^L \beta F_1^2 + F_2^2\mathrm{d}x\mathrm{d}t = \mathrm{M}_1 \\ &\frac{1}{2}\int_0^L (S_x^2 + \beta R_t^2)\mathrm{d}x \le \mathrm{M}_1 + \beta\int_0^t\int_0^L (S_x^2 + R_t^2)\mathrm{d}x\mathrm{d}t \le \mathrm{M}_1 + \int_0^t\int_0^L (S_x^2 + \beta R_t^2)\mathrm{d}x\mathrm{d}t \end{aligned} \tag{24}$$

According to the Gronwall inequality, we have

$$\int_0^L (S_x^2 + \beta R_t^2)\mathrm{d}x \le 2\mathrm{M}_1\mathrm{e}^{2t} \tag{25}$$

Thus

$$\int_0^t\int_0^L (S_x^2+\beta R_t^2)\mathrm{d}x\mathrm{d}t \le \int_0^t 2\mathrm{M}_1\mathrm{e}^{2t}\mathrm{d}t \le \mathrm{M}_1\mathrm{M}_2 \tag{26}$$

where $\mathrm{M}_2=\mathrm{e}^{2\mathrm{T}}-1$.

(ii) When $1\le\beta<\varepsilon^{-1}$, we have

$$\frac{1}{2}\int_0^L (S_x^2+\beta R_t^2)\mathrm{d}x \le M_1+\beta\int_0^t\int_0^L (S_x^2+R_t^2)\mathrm{d}x\mathrm{d}t \le \mathrm{M}_1+\beta\int_0^t\int_0^L (S_x^2+\beta R_t^2)\mathrm{d}x\mathrm{d}t \tag{27}$$

Similarly, it can be obtained

$$\int_0^t\int_0^L (S_x^2+\beta R_t^2)\mathrm{d}x\mathrm{d}t \le \int_0^t 2\mathrm{M}_1\mathrm{e}^{2\beta t}\mathrm{d}t \le \beta\mathrm{M}_1\mathrm{M}_3 \tag{28}$$

where $\mathrm{M}_3=\mathrm{e}^{2\beta\mathrm{T}}-1$.From equation (23), we can obtain

$$\int_0^t\int_0^L S_t^2\mathrm{d}x\mathrm{d}t \le 2\mathrm{M}_1(\mathrm{M}_4+1) \quad \int_0^t\int_0^L R_x^2\mathrm{d}x\mathrm{d}t \le \frac{2\mathrm{TM}_1}{\beta}(\mathrm{M}_4+2) \tag{29}$$

where $\mathrm{M}_4=\mathrm{M}_2$ or $\mathrm{M}_4=\beta\mathrm{M}_3$, because

$$\begin{gathered}\int_0^t\int_0^L (R^2)_t\mathrm{d}x\mathrm{d}t = 2\int_0^t\int_0^L R_tR\mathrm{d}x\mathrm{d}t \le \int_0^t\int_0^L R_t^2\mathrm{d}x\mathrm{d}t + \int_0^t\int_0^L R^2\mathrm{d}x\mathrm{d}t \\ \int_0^t\int_0^L (R^2)_t\mathrm{d}x\mathrm{d}t \le \int_0^L R^2\mathrm{d}x\end{gathered} \tag{30}$$

Combining equation (26), equation (28) and equation (30), according to Gronwall inequality, we have:

$$\int_0^t\int_0^L R^2\mathrm{d}x\mathrm{d}t \le \mathrm{M}_5 \quad \int_0^t\int_0^L S^2\mathrm{d}x\mathrm{d}t \le \mathrm{M}_6 \tag{31}$$

where

$$\mathrm{M}_5=\frac{\mathrm{M}_1\mathrm{M}_4}{\beta}(\mathrm{e}^{\mathrm{T}}-1), \mathrm{M}_6=2\mathrm{M}_1(\mathrm{M}_4+1)(\mathrm{e}^{\mathrm{T}}-1)$$
$$\mathrm{M}=3\max[\mathrm{M}_1\mathrm{M}_4, \frac{\mathrm{M}_1\mathrm{M}_4}{\beta}, 2\mathrm{M}_1(\mathrm{M}_4+1), \frac{2\mathrm{TM}_1(\mathrm{M}_4+2)}{\beta}, \mathrm{M}_5, \mathrm{M}_6]$$

According to reference [16], we obtain that the remainder terms R and S uniformly establish the estimation equation (20) on the regional Ω_T. Therefore, R and S are bounded, that is, equation (14) is uniformly valid.

5. Numerical Approximation

Define grid on area $[0,\mathrm{L}]\times[0,\mathrm{T}]$:

$$\begin{aligned} x_i &= ih \qquad i=0,1,2,\cdots \mathrm{m} \qquad \mathrm{m}=\mathrm{L}/h \\ t_j &= j\tau \qquad j=0,1,2,\cdots \mathrm{n} \qquad \mathrm{n}=\mathrm{T}/\tau \end{aligned}$$

Then, we shall use further the following notations:

$$
\begin{aligned}
&U_{tt,i,j}=\frac{U_{t,i,j}-U_{t,i,j-1}}{\tau^2}\quad U_{xx,i,j}=\frac{U_{x,i,j}-U_{x,i,j-1}}{h^2}\quad \theta_{t,i,j}=\frac{\theta_i^{j+1}-\theta_i^j}{\tau}\quad \theta_{x,i,j}=\frac{\theta_{i+1}^j-\theta_{i-1}^j}{2h}\\
&\theta_{xx,i,j}=\frac{\theta_{i+1}^j-2\theta_i^j+\theta_{i-1}^j}{h^2}\quad Q_{t,i,j}=\frac{Q_i^{j+1}-Q_i^j}{\tau}\quad Q_{xx,i,j}=\frac{Q_{i+1}^j-2Q_i^j+Q_{i-1}^j}{h^2}
\end{aligned}
\tag{32}
$$

According to references [17] and [18], we order $r_0=\tau^2h^{-2}, r_1=\tau^2h^{-1}, r_2=\tau h^{-2}$, for the heat conduction equation, the Grank Nicolson scheme is adopted, and the second-order part of the non-homogeneous hyperbolic equation adopts the central difference scheme. Equation (10) is transformed into:

$$
\begin{aligned}
&-\frac{r_2}{2}\theta_{0,i-1}^{j+1}+(1+r_2)\theta_{0,i}^{j+1}-\frac{r_2}{2}\theta_{0,i+1}^{j+1}=\frac{r_2}{2}\theta_{0,i-1}^{j}+(1-r_2)\theta_{0,i}^{j}+\frac{r_2}{2}\theta_{0,i+1}^{j}\\
&\theta_{0,i}^0=\varphi_1(x_i)\\
&\theta_{0,0}^j=\varphi_3(t_j)\\
&\theta_{0,N}^j=\varphi_4(t_j)
\end{aligned}
\tag{33}
$$

$$
\begin{aligned}
&U_{0,i}^{j+1}=r_0U_{0,i+1}^{j}+2(1-r_0)U_{0,i}^{j}+r_0U_{0,i-1}^{j}-U_{0,i}^{j-1}+\frac{r_1}{2}(\theta_{0,i+1}^{j}-\theta_{0,i-1}^{j}),\\
&U_{0,i}^0=\psi_1(x_i)\\
&U_{0t,i}^0=\psi_2(x_i)\\
&U_0^j=\psi_3(t_j)\\
&U_{0,N}^j=\psi_4(t_j)
\end{aligned}
\tag{34}
$$

Using the equation (32), the equation (15) is transformed into:

$$
\begin{aligned}
&Q_{0,i}^{j+1}=-r_2Q_{0,i}^{j}+(1-r_2)Q_{2,i}^{j}+r_0Q_{0,i+2}^{j}\\
&Q_{0,i}^0=\frac{\theta_i^0-\theta_{i-1}^0}{\tau}-\varphi_2(x_i)\\
&Q_{0,0}^j=0\\
&Q_{0,N}^j=0
\end{aligned}
\tag{35}
$$

According to references [17] and [18], the truncation errors of equations (33), equations (34) and equations (35) are $O(h^2+\tau), O(h^2+\tau^2),\ O(h^2+\tau)$.

6. **Numerical Example**

We order $\mathrm{L}=\pi, \mathrm{T}=1, \beta_1=[-\frac{1}{2}+(\frac{23}{108})^{\frac{1}{2}}]^{\frac{1}{3}}+[-\frac{1}{2}-(\frac{23}{108})^{\frac{1}{2}}]^{\frac{1}{3}}, \beta_2=\beta_1^2$,the given equation and definite solution conditions are as follows:

$$
\begin{aligned}
&p_{xx}=p_t\\
&p(x,t)|_{t=0}=\mathrm{e}^{\beta_1x}\\
&p(x,t)|_{x=0}=\mathrm{e}^{\beta_2t}\\
&p(x,t)|_{x=\pi}=\mathrm{e}^{(\beta_1\pi+\beta_2t)}
\end{aligned}
\tag{36}
$$

Journal of Physics: Conference Series **2097** (2021) 012007 doi:10.1088/1742-6596/2097/1/012007

$$\begin{aligned} &q_{xx} = q_{tt} + p_x \\ &q(x,t)|_{t=0} = \mathrm{e}^{\beta_1 x} \\ &q_t(x,t)|_{xt=0} = \beta_2 \mathrm{e}^{\beta_1 x} \\ &q(x,t)|_{x=0} = \mathrm{e}^{\beta_2 t} \\ &q(x,t)|_{x=\pi} = \mathrm{e}^{(\beta_1 \pi + \beta_2 t)} \end{aligned} \tag{37}$$

$$\begin{aligned} &r_{xx} = -r_t \\ &r(x,t)|_{t=0} = \sin(x) \\ &r(x,t)|_{x=0} = 0 \\ &r(\mathrm{x},t)|_{x=\pi} = 0 \end{aligned} \tag{38}$$

The exact solution of equation equation (36), equation (37) and equation (38) are

$$p = \mathrm{e}^{(\beta_1 x + \beta_2 t)} \quad q = \mathrm{e}^{(\beta_1 x + \beta_2 t)} \quad r = \mathrm{e}^t \sin x$$

For equations (36) and (37), $h = 0.01 \times \pi, \tau{=}0.01$, the numerical error diagram is as follows:

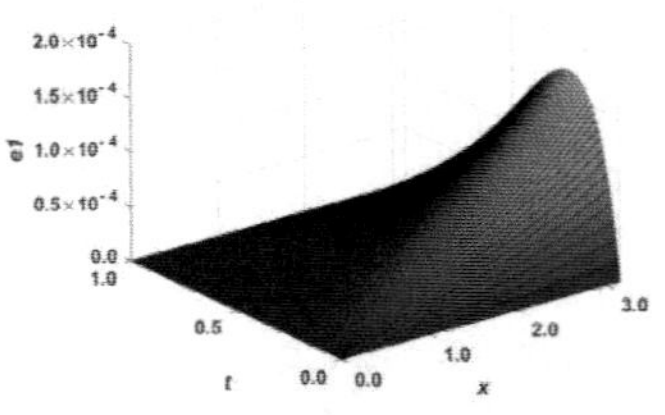

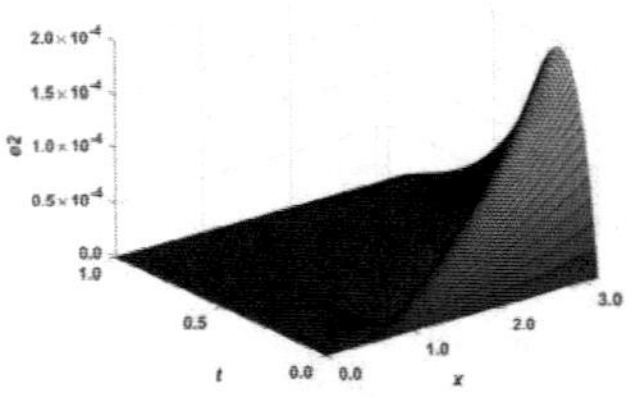

Figure 1 Error surface graph of p. **Figure 2** Error surface graph of q.

For equation (38), we select $h = 0.1 \times \pi, \tau{=}0.01$, then the numerical error diagram is as follows:

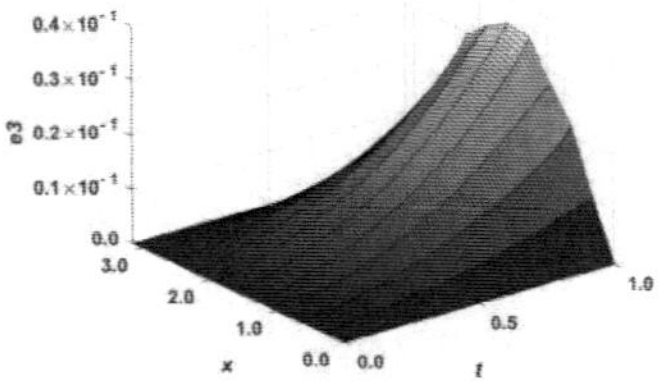

Figure 3 Error surface graph of r.

7. Conclusions

By adjusting the coupling coefficient and small parameter scale coefficient, the method can be applied to G-L thermoelastic coupling problems with different materials and corresponding different coupling coefficients. The original G-L thermoelastic coupling model is reduced by the singular perturbation method, and the asymptotic solutions of singularly perturbed hyperbolic and parabolic equations in bounded domain are obtained by using the matched asymptotic expansion method. The L^2 estimate of the remainder of the asymptotic solution is given. By numerical method, the numerical solution of formal asymptotic solution can be obtained in turn. Thus, the specific forms of temperature field and stress field in bounded domain are described.

Acknowledgement

The research was supported by the Nation Natural Science Foundation of China (China general prohect No.51775154).

References

[1] Othman M I A 2017 Effect of thermal loading due to laser pulse on 3-D problem of micropolar thermoelastic solid *Mechanics and Mechanical Engineering* **21**(3) 679-701.

[2] Rajneesh K, Aseem M and Rekha R 2020 Response of thermoelastic interactions in micropolar porous circular plate with three phase lag model *Mechanics and Mechanical Engineering* **22**(4) 999-1014

[3] Paul K and Mukhopadhyay B 2019 Two-dimensional generalized magnetothermoelastic diffusion problem for a thick plate under laser pulse heating with three-phase lag effect *Journal of Engineering Physics and Thermophysics* **9**(26) 516–527.

[4] Shakeriaski F and Ghodrat M 2020 The nonlinear response of cattaneo-type thermal loading of a laser pulse on a medium using the generalized thermoelastic model *Theoretical and Applied Mechanics Letters* **10**(4) 286-297.

[5] Tang D W and Huangmu X X 2003 Non-fourier heat conduction behavior considering thermoelastic coupling *The Chinese Society of Engineering Thermophysicson Heat and Mass Transfer Conference* 58-61

[6] El-Maghraby N M and Yossef H M 2004 State space approach to generalized thermoelastic problem with thermomechanical shock *Applied Mathematics and Computation* **156(2)** 577-586

[7] Youssef H M 2006 Theory of two-temperature-generalized thermoelasticity *IMA Journal of Applied Mathematics* **71**(3) 383-390

[8] Youssef H M 2010 Two-temperature generalized thermoelastic infinite medium with cylindrical cavity subjected to moving heat source *Archive of Applied Mechanics* **80**(11) 1213-1224

[9] Kumar R, Kumar A and Mukhopadhyay S 2016 An investigation on thermoelastic interactions under two-temperature thermoelasticity with two relaxation parameters *Mathematics and Mechanics of Solids* **21**(6) 725-736.

[10] Kumar A, Kant S and Mukhopadhyay S 2017 An in-depth investigation on plane harmonic waves under two-temperature thermoelasticity with two relaxation parameters *Mathematics and Mechanics of Solids* **22**(2) 191-209

[11] Kumar A, Shivay O N and Mukhopadhyay S 2019 Infinite speed behavior of two-temperature Green–Lindsay thermoelasticity theory under temperature-dependent thermal conductivity *Zeitschrift Für Angewandte Mathematik Und Physik* **71**(1) 1-16

[12] Shivay O N and Mukhopadhyay S 2020 On the Temperature-Rate dependent Two-Temperature thermoelasticity theory *J. Journal of Heat Transfer.* **142**(2) 1-10

[13] Jangid K and Mukhopadhyay S 2020 Variational and reciprocal principles on the temperature-rate dependent two-temperature thermoelasticity theory *Journal of Thermal Stresses* **43**(17) 1-13

[14] Zhang L, Zhang X M and Chu Z X 2014 Study on three-dimensional generalized thermoelastic analysis of Al_2O_3 powder sintering based on non-fourier heat conduction *Proceedings of China Computational Mechanics Conference 2014 and the Third Qian Lingxi Computational Mechanics Award Presentation Conference* p 1139-1148

[15] Wu Z Q, Yin J X and Wang C M 2003 *Introduction to Elliptic and Parabolic Equations* (Beijing: Science Press) 197

[16] Bao L P, Li W Y and Wu LQ 2019 Adaptive mesh method for solving a second-order hyperbolic equation *Applied Mathematics and Mechanics* **40**(5) 536-545

[17] Johnson O B and Oluwaseun L I 2020 Crank-Nicolson and modified Crank-Nicolson scheme for one dimensional parabolic equation *International Journal of Applied Mathematics and Theoretical Physics* **6**(3) 35-40

[18] Zhou Q and Yang Y 2020 Singularly perturbed solutions of non-fourier temperature field distribution in single-layer Materials *Acta Mathematica Scientia* **39**(4) 240-248

Review of Calibration and Improvement Methods of Light-Scattering Airborne Particle Concentration

Z L Gao[1,2*], Q D Cheng[1,2], G L Zeng[1,2], Y Wen[2], G F Li[1,2], J Chen[3], Y B Dong[2] and Q Z Ji[2]

[1] First Branch Company of Beijing Hangtianhe Science Technology Development Co., Ltd., Changping District, Beijing, 102205, China.
[2] Beijing Orient Institute for Measurement and Test, Haidian District, Beijing, 100094, China.
[3] Suzhou Sujing Automation Equipment Corporation, Suzhou, 215000, Jiangsu, China.
Email: gzl514cast@126.com

Abstract. Clean environment and its internal airborne particle concentration have been paid more and more attention, the demand for use and measurement of light-scattering airborne particle counter, as the main instrument for measuring airborne particle concentration, has increased synchronously. This paper untangles the worldwide standards and specifications for calibration of light-scattering airborne particle counter, analyses the shortcomings of traditional comparative calibration method, introduces the research progress of non-traditional calibration method based on statistical analysis of membrane and scanning electron microscope, then based on the theory of discrete phase model and gas-solid fluid dynamics, puts forward two improved calibration methods to obtain more reliable "true value" of the number of the standard particles passing through the calibrated OPC, to provide an innovative idea for improving the measurement accuracy of airborne particle concentration worldwide.
Keywords. Airborne particle concentration, comparative calibration, aerodynamics, measurement error.

1. Introduction
Particulate pollution in the air is the main source of pollution in the production and processes of medical, microelectronics, precision machinery, aerospace and other industries, accurate evaluation of air cleanliness has become an urgent requirement for the production and product testing [1-6], such as large-scale integrated circuit manufacturing, spacecraft assembly, pharmaceutical production, precision instrument manufacturing, emission of ultrafine particles from motor vehicles, virus aerosol transmission, negative pressure ward operation, etc. Light-scattering airborne particle counter (also known as Optical Particle Counter, OPC) has been widely used to measure the airborne particle concentration in these clean environment and confined space [7-9].

At present, the comparative test method is widely used to calibrate OPC airborne particle concentration in the world, that is, a "precise particle counter" is used to calibrate the "calibrated particle counter". Generally, the "precise particle counter" is another OPC corrected by capability comparison, or a Condensation Particle Counter (CPC) is directly selected, so that a comparative calibration traceability path of "Optical Particle Counter (OPC) – (Condensation Particle Counter) CPC–Aerosol Electrometer (AE)" is established. However, due to the differences in the particle size range, working flow and measured particles of the instruments at the three levels, the comparative calibration method needs to and improve the accuracy of calibration result continuously, combined with the theory of measurement error and the characteristics of the calibrated OPC itself.

2. Traditional Comparative Calibration Method for Airborne Particle Concentration

2.1. Standards and Specifications for the Comparative Calibration Worldwide

Airborne particle concentration calibration has been highly concerned by worldwide metrology institutes and standardization organizations, there are many standards of OPC airborne particle concentration calibration method. ASTM International implemented ASTM F328-1998 and ASTM F649-2001, which tried to make the particle counting results could be traceable to the voltage pulse signal generated by the instrument itself, but the counting loss caused by instrument design, parameter setting and laser power attenuation couldn't be determined [10].

At present, the OPC comparative calibration method has been widely used for many years worldwide, the "precise particle counter" comes from two sources: one is another OPC after capacity comparison, relevant standards and specifications such as IEST-RP-CC014.1-2006, JIS B9921-1997/2010, JJF1190-2008, JJF 1562-2016, JJF 1864-2020, GB/T 6167-2007, GB/T 29024.4-2017 (equal utilization of ISO 21504-4-2007) and GJB/J 5416-2005, the other is a CPC, which way has been gradually applied worldwide in recent years, relevant standards and specifications such as ISO 21501-4-2007/2018, ISO 15900-2009/2020 and ISO/DIS 27891-2013, to establish the traceability path of "OPC - CPC - AE" [11-14].

2.2. Shortcomings of the Traditional Comparative Calibration Method

Some Shortcomings are found in the method of step-by-step comparative calibration and traceability path of "OPC - CPC - AE". Firstly, based on the measurement principle, the lower limit of particle size measured by CPC is smaller, and the light-scattering measurement is carried out after particle size amplification through the working solution, so that a CPC is essentially an OPC (the physical object of the light scattering cavity is shown in figure 1, and the working principle of OPC is shown in Figure 2). OPC and CPC slightly overlap in the range of particle size, so their coverage is incomplete, what's more, the sampling working flow is also inconsistent between them (see table 1).

Table 1. Comparison of OPC and CPC technical indicators.

Category/ Specifications	OPC	CPC
Particle size range	0.1μm~25μm	5nm~3 000nm
Sampling flow	2.83L/min or 28.3L/min	0.30L/min~0.60L/min
Maximum particle concentration	100 000/28.3L(≈3.53/cm^3)	7 000/cm^3
Working fluid	—	Butanol, isopropanol, water or other

Secondly, about the standard particle size for calibration, JJF 1562 gives the average particle size range (80~120) nm for CPC calibration, and the standard requires aerosol generator to produce polystyrene standard particle size range of (50~150) nm [15], but JJF 1190 requires that the size of monodisperse particles for calibration is (0.1~10) μm, 0.4μm and 0.6μm are used in the calibration of particle size distribution error, 0.5μm is used in the calibration of particle concentration indication error.

Thirdly, about the working flow setting during calibration, aerosol dispenser and aerosol diluter need to be used for flow matching in the process of calibrating OPC with CPC. For one channel, it is necessary to increase the clean gas flow through a series of clean air injection, mixing and other steps

[16-19]. But the calibration process does not estimate the particle loss in the pipeline or evaluate the uncertainty of additional air replenishment operation, including the 600mm long aerosol diluter pipe and the adsorbed particles at exhaust port outside the dilutor.

Finally, OPC comparative calibration methods, generally used or recognized worldwide, the calibration process is that the precision particle counter and the calibrated particle counter simultaneously sample and extract gas from a sampling chamber containing stable aerosol composed of standard particles [20-24]. The sampling chamber device is generally a buffer (JJF 1190), a mixing box (GB/T 29024.4, ISO 21501-4), a distributor (GB/T 6167) or a sampler (JIS B 9921), etc. (see figure 1).

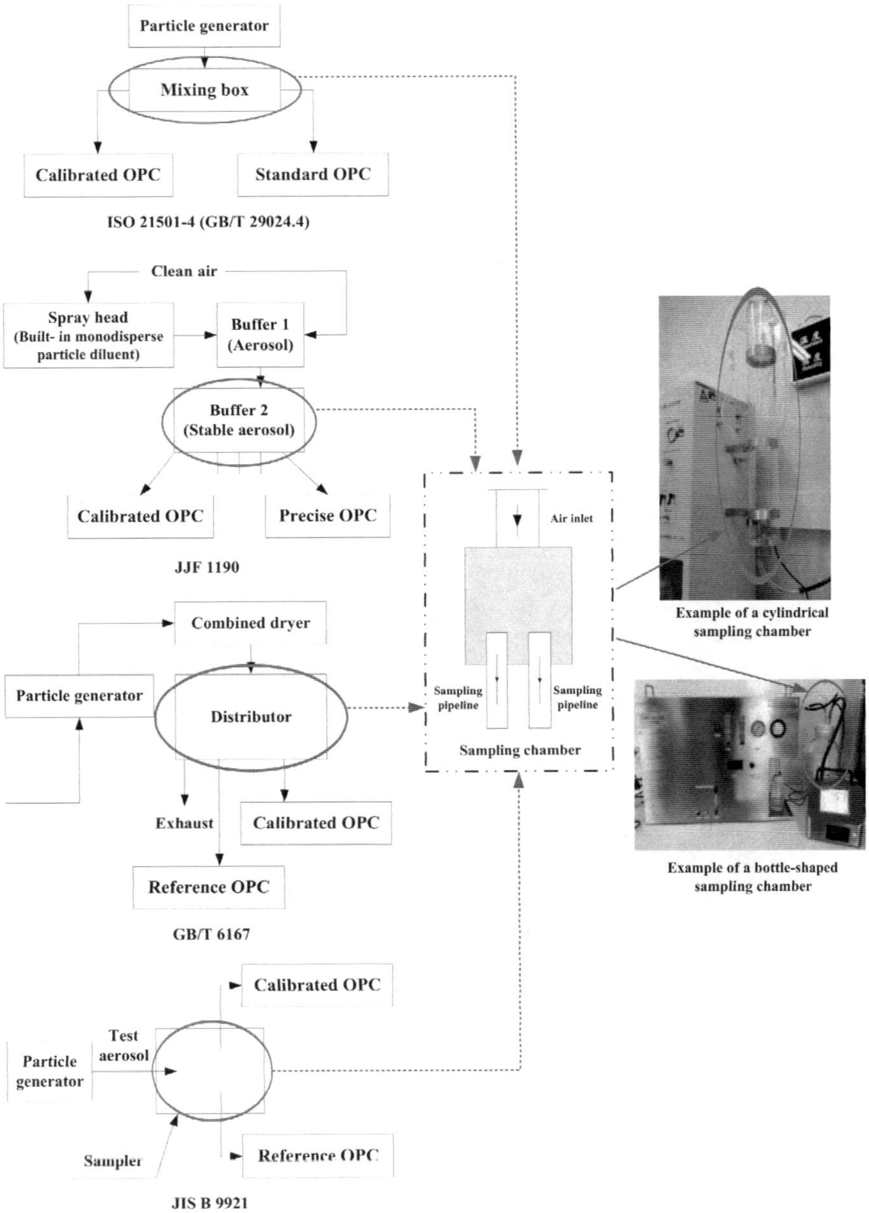

Figure 1. Sampling chamber for comparative method of airborne particle concentration.

So, the sampling chamber must ensure that the sampling flow of the two particle counters in the comparative test is consistent, also the relevant materials, sizes and positions of the sampling tubes, and the aerosol in the sampling chamber always maintains an ideal uniform distribution state, in order that we can try our best to ensure that the precise particle counter can provide a more accurate "true value" of the number of the standard particles for the calibrated OPC. However, in actual calibration process, the measured standard particles cannot be ensured completely similar as in theory between the precise particle counter and the calibrated OPC.

Therefore, whether the OPC or the CPC is used as the "precise particle counter", there is drawback of "uncertainty" in the "true value" of the number of the passing standard particles.

3. Non-traditional Calibration Method for Airborne Particle Concentration

In terms of the improvement of the calibration methods for airborne particle concentration, worldwide relevant metrology institutes put forward the method of measuring air cleanliness by filter (membrane) collection and electron microscope observation, and this method can still be used to count larger particles (particle size greater than 5μm) nowadays, such as GB 50591-2010, QJ 2214-1991 and IEST-RP-CC003.3-2003, but there is no general method to observe and statistical counting of submicron particles (0.5μm for example) under electron microscope in the world.

The research team of Beijing Orient Institute for Measurement and Test (BOIMT) tried to propose a new OPC calibration method for airborne particle concentration and its traceability scheme based on FESEM (Field Emission Scanning Electron Microscope) and AAO (Anodic Aluminum Oxide) template [25-27], building an inferred statistical calibration system of 0.5μm particle concentration (see figure 2), adding membranes at the rear end of the OPC light-scattering chamber's air outlet to collect the actual passing standard particles, realizing the same measurement gas path, equal gas flux and the same standard particles with the calibrated OPC, carrying out measurement of the settlement loss of standard particles in the light-scattering chamber, designing the inferred statistical counting method for 0.5μm particles based on the original method of microscopic observation and counting of larger particle size particles. According to the actual test results, the OPC calibration method based on AAO and FESEM makes sense in theory, which deviates from the inherent comparative calibration principle, and it realizes the collection, identification, inference statistical count of low flux and submicron particle (such as 0.5μm), and provides a theoretical and experimental basis for obtaining a more reliable "true value" of the number of the standard particles passing through the calibrated OPC.

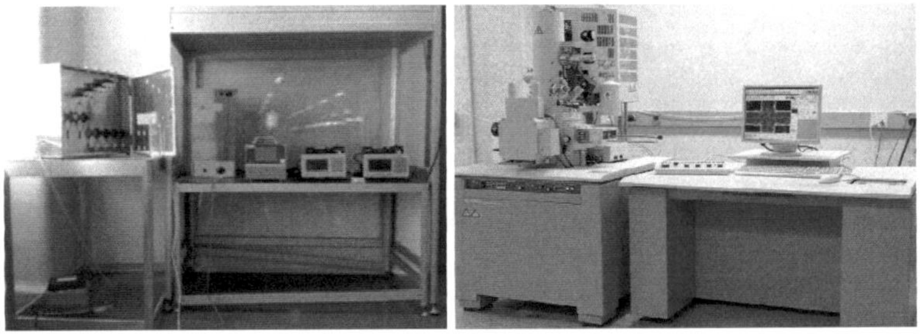

Figure 2. Composition of particle concentration measurement system.

4. Improvement Method for Airborne Particle Concentration Calibration

4.1. More Reliable "True Value" Measurement of the Number of the Passing Standard Particles

According to the theory of measurement error, the improvement of the comparative calibration accuracy of airborne particle concentration, also the further uncertainty reduction of measurement results, depends on the accuracy improvement of the "true value", so it must be ensured that the "precise particle counter" has the same measurement gas path, equal gas flux, the same or covering particle size range, the same measured standard particles with the calibrated OPC (see figure 3).

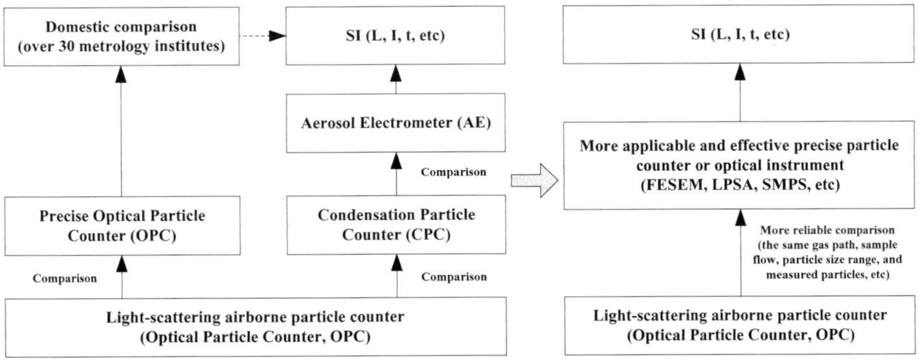

Figure 3. Framework of improved method for light-scattering particle concentration calibration.

4.2. *Inferential Statistics Measurement by the "Post, Collection and Check" Way*

On the basis of the previous research on airborne particle concentration calibration scheme based on FESEM and AAO template (see figure 4), we can build a more sophistic mathematical model of the calibrated OPC's standard particle flow impacting membrane according to the theory of discrete phase model and gas-solid fluid dynamics [28], obtain the probability distribution of the particles collected by the membrane and the pressure parameters of the membrane by simulation (see figure 5), increase the maximum working flow and other performance parameters of AAO template or other filter membranes, calculate the loss of standard particles in the calibrated OPC scientifically, optimize the calculation theory of the standard particle number based on statistical analysis under the scanning electron microscope (SEM), evaluate the uncertainty of measurement results, carry out iterative test, in order to improve and perfect the calibration theory and method based on inferential statistics for low flux and submicron size airborne particle concentration gradually .

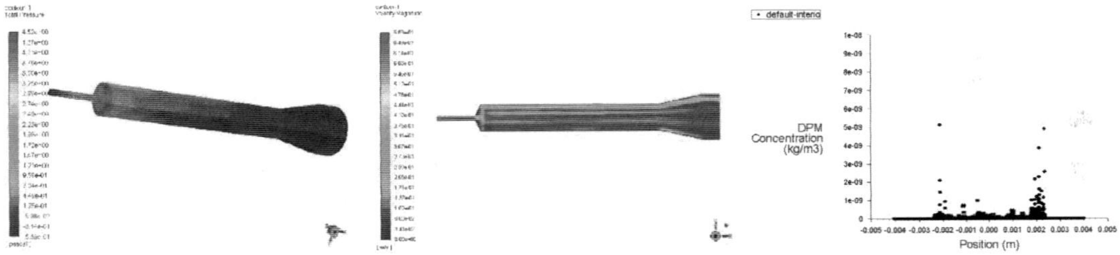

(a) Pressure distribution (b) Airflow velocity (c) Particle distribution

Figure 4. Aerodynamic simulation of light-scattering cavity's outlet structure.

4.3. *Direct Measurement by the "Front, Detect and Count" Way*

The "front, detect and count" way is more scientific and convenient than the "post, collection and check" one. The "front, detect and count" method mainly utilizes more suitable and effective particle counter or optical instrument, such as FESEM, laser particle size analyser (LPSA), scanning mobility particle sizer (SMPS) and so on [29]. This counter or instrument shall have an internal cylindrical particle signal detector, which can be installed at the front end of the calibrated OPC's air inlet, without affecting the normal air inlet operation under the normal working flow (see figure 6). What's more, the counter or instrument should have more accurate measurement accuracy of the particle size and concentration than the calibrated OPC, higher optical resolution or more accurate AD quantization error control, for example. It should be noted that the airborne particle concentration measurement result of the counter or instrument must be traceable to the SI.

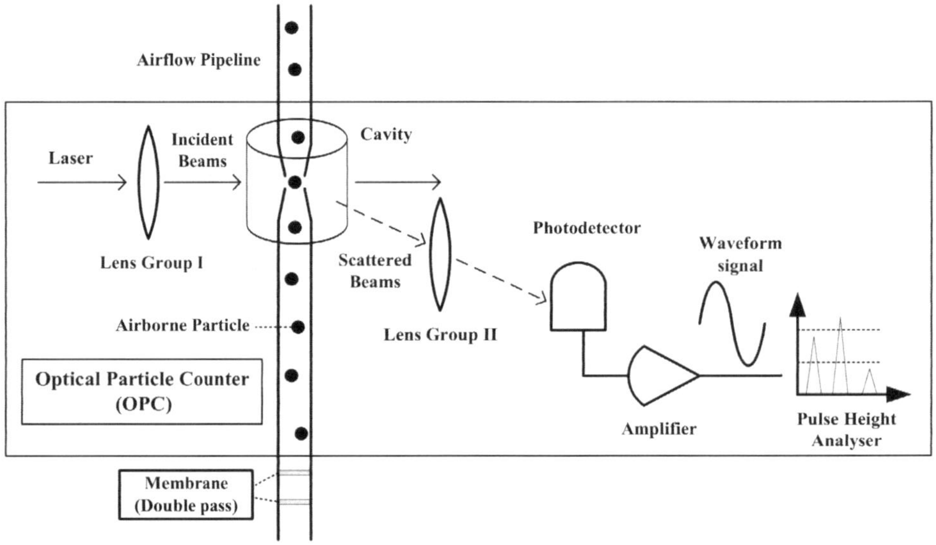

Figure 5. Scheme of standard particle number measurement by "post, collection and check".

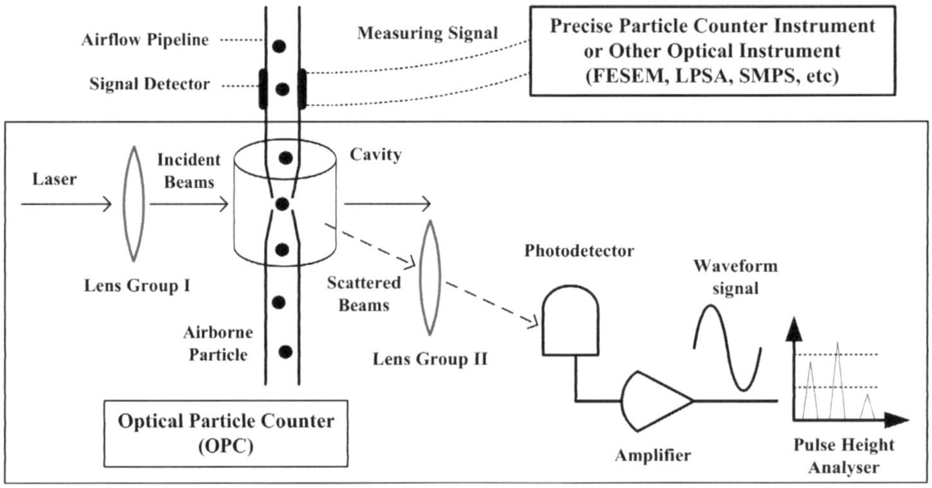

Figure 6. Scheme of standard particle number measurement by "front, detect and count".

5. Conclusion and Prospect

In recent years, more and more attention has been paid to the clean environment and its internal airborne particle quantity concentration in the fields of industrial production manufacturing and life, health and environmental safety, the same is true for the improvement of the measurement accuracy of airborne particle concentration, especially after the 2019-nCov virus outbreak, the demand for air aerosol measurement in confined spaces (such as elevators) and quality monitoring of microbial conditions in medical air environment. The improvement of OPC calibration method is one of the key factors to obtain the higher accuracy in measurement.

Analysing the standards and specifications of airborne particle concentration calibration methods worldwide, this paper points out that the traditional comparative calibration method, widely used at present, has drawback of "uncertainty" in principle, according to the theory of discrete phase model and gas-solid fluid dynamics, proposes two improved OPC calibration methods of "post, collection and check" and "front, detect and count" to obtain a more reliable "true value" of the number of the standard particles, in order to provide some theoretical and methodological innovation ideas for the research field of airborne particle concentration calibration and its traceability worldwide.

Acknowledgments
This research is supported by the Outstanding Young Talents Fund of China Academy of Space Technology (CAST).

References

[1] Pillai S G, Haldorai K, Seo W S and Kim W G 2021 COVID-19 and hospitality 5.0: Redefining hospitality operations *International Journal of Hospitality Management* **94** 102869

[2] Somsen G A, Rijn C M, Kooij S, Bem A and Bonn D 2020 Measurement of small droplet aerosol concentrations in public spaces using handheld particle counters *Physics of fluids* **32** 121707

[3] Laurentiu P and Daniel D 2021 Performance Evaluation of Particulate Matter and Indoor Microclimate Monitors in University Classrooms under COVID-19 Restrictions *International Journal of Environmental Research and Public Health* **18** 823-9

[4] Chau J P C, Liu X, Lo S H S, Chien W T and Wan X 2020 Effects of environmental cleaning bundles on reducing healthcare-associated Clostridioides difficile infection: a systematic review and meta-analysis *Journal of Hospital Infection* **106** 734-44

[5] Li Y, Wang B, Zhang P, Zhang Y and Zhu J 2020 A novel cleanliness control method for disk amplifiers *High Power Laser Science and Engineering* **8** 1-5

[6] Nowak N, Scheiber K, Pfeil J, Meyer J, Dittler A, Koch T and Kasper G 2020 Sampling and conditioning of engine blow-by aerosols for representative measurements by optical particle counters *Journal of Aerosol Science* **148** 105612

[7] Müller D, Glöckler F and Kienle A 2019 Application of Mie theory for enhanced size determination of microparticles using optical particle counters *Applied optics* **58** 4575-84

[8] Verdier N, Papy J M, Renard J B, Lefevre M and Agrapart C 2020 Enhanced detection and sizing algorithm to improve LOAC optical particle counter performances *Applied Optics* **59** 10892-901

[9] Li X H and Zhou Z M 2020 Discussion and application of calibration results of dust particle counter *Metrology & Measurement Technique* **47** 55-7

[10] Gao Z L, Ji Q Z, Yuan Y F and Wang H 2013 Analysis on measurement calibration specifications for light-scattering airborne particle counter *Standard Science* 42-5

[11] Wlasits P J, Stolzenburg D, Tauber C, Brilke S, Schmitt S H, Winkler P M and Wimmer D 2020 Counting on chemistry: laboratory evaluation of seed-material-dependent detection efficiencies of ultrafine condensation particle counters *Atmospheric Measurement Techniques* **13** 3787-98

[12] Steven T, Kenjiro I, Kumiko Y, Yoshiko M, Hiromu S and Jason S O 2020 Determining the cutoff diameter and counting efficiency of optical particle counters with an aerodynamic aerosol classifier and an inkjet aerosol generator *Aerosol Science and Technology* 2020 **54** 1335-44

[13] Wang W, Zhao X, Zhang J S, Yang Y X, Yu T Z, Bian J J, Gui H Q and Liu J G 2020 Design and evaluation of a condensation particle counter with high performance for single-particle counting *Instrumentation Science & Technology* **48** 212-29

[14] Peng J C, Zhou Z M and Li X H 2020 Analysis on the quantity value traceability of dust particle counter *Metrology & Measurement Technique* **47** 5-6, 11

[15] Liu J J, Qi T Y, Wang J W and Liu T 2018 Calibration technology and traceability system of particle counter (PNC) *Proceedings of the 22nd Symposium on Air Pollution Prevention and Control Technology* p 153-9

[16] Wang W, Zhao X, Zhang J, Yang Y, Yu T, Bian J, Gui H and Liu J 2020 Design and evaluation of a condensation particle counter with high performance for single-particle counting *Instrumentation Science & Technology* **48** 212-29
[17] Horender S, Auderset K, Vasilatou K 2019 Facility for calibration of optical and condensation particle counters based on a turbulent aerosol mixing tube and a reference optical particle counter *Review of Scientific Instruments* **90** 075111
[18] Geisler M and Dirscherl K 2020 Direct approach to determine the size setting error and size resolution of an optical particle counter *Review of Scientific Instruments* **91** 045105
[19] Huang Z H, Li J, Qi T Y and Yang G H 2019 Dilution control system in the traceability process of airborne particle counter *Metrology Science and Technology* 3-6
[20] Tran S, Iida K, Yashiro K, Murashima Y, Sakurai H and Olfert J S 2020 Determining the cut off diameter and counting efficiency of optical particle counters with an aerodynamic aerosol classifier and an inkjet aerosol generator *Aerosol Science and Technology* **54** 1335-44
[21] Horender S, Auderset K and Vasilatou K 2019 Facility for calibration of optical and condensation particle counters based on a turbulent aerosol mixing tube and a reference optical particle counter *The Review of Scientific Instruments* **90** 075111
[22] Sang-Nourpour N and Olfert J S 2019 Calibration of optical particle counters with an aerodynamic aerosol classifier *Journal of Aerosol Science* **138** 105452
[23] Sundararajan M, Devarajan M and Jaafar M 2020 Investigation of surface and mechanical properties of Anodic Aluminium Oxide (AAO) developed on Al substrate for an electronic package enclosure *Surface & Coatings Technology* **401** 126273
[24] Yatinkumar P, Janusas G, Palevicius A and Vilkauskas A 2020 Development of nanoporous AAO membrane for nano filtration using the acoustophoresis method *Sensors* **20** 3833
[25] Gao Z L, Ji Q Z, Chen J, Li G F, Wang R J, Wu J P, Zeng G L, Cheng Q D, Hu W Z and Tang X 2020 Metrological traceability methods for the light scattering airborne particle concentration *The International Society for Optical Engineering: 11th International Symposium on Precision Engineering Measurements and Instrumentation (ISPEMI 2020)* 37
[26] Gao Z L, Ji Q Z, Chen J, Zhang X B, Zhang W H, Tan J G and Wang C Y 2018 Statistical measurement method of the standard particles through airborne particle counter based on FESEM *Proceedings of SPIE (EI: 20191106638812) - The International Society for Optical Engineering: 10th International Symposium on Precision Engineering Measurements and Instrumentation (ISPEMI 2018)* **110530** 110530I-1-8
[27] Ji Q Z, Gao Z L, Zhang X B and Chen J 2014 Discuss on traceability method of light-scattering airborne particle counter's counting performance *Proceedings of SPIE (EI: 20151200656569) - The International Society for Optical Engineering: 9th International Symposium on Precision Engineering Measurements and Instrumentation (ISPEMI 2014)* **9446** 94464C-1-6
[28] Wang L B, Li D H, Chen Z G, Sun M Y and Zhu X S 2017 Numerical simulation of force and separation on particles in pulsing airflow *Journal of China University of Mining & Technology* **46** 162-168, 176
[29] Kong W M 2021 The nano-scanning electrical mobility spectrometer (nSEMS) and its application to size distribution measurements of 1.5-25nm particles *Atmospheric Measurement Techniques* **14** 5429-45

Study on Micro Recoil Mechanism of the Weapon with a Nozzle and Two Chambers Separated by a Partition

Longxu Ma[1], Ming Qiu[1*], Jie Song [1], Peng Si[1] and Zhenqiang Liao[1,2]

[1] School of Mechanical Engineering, Nanjing University of Science and Technology, Nanjing, 210094, Jiangsu, China
[2] School of Mechanical and Electrical Engineering, Global Institute of Softwaer Technology, Suzhou, 215163, Jiangsu, China
Email: qiuming@njust.edu.cn

Abstract. In order to improve the recoil reduction ability of the weapon without reducing the projectile velocity, the weapon with a nozzle and two chambers separated by a partition is proposed. Taking the 35 mm caliber grenade launcher as the research object, the physical model of its launching process is proposed, and the one-dimensional two-phase flow interior ballistic model is established. MacCormack difference scheme is used to calculate the coupling of multiple physical fields formed by front and rear chambers and nozzles. Compared with the calculation results obtained by using the classical interior ballistic model, the correctness of the two-phase flow interior ballistic model is verified. The effects of the charge amount in the rear chamber and the size of the air guide hole in the barrel on the internal ballistic performance of the weapon with a nozzle and two chambers separated by a partition. The calculation results show that compared with conventional weapons, the recoil impulse can be reduced with maintaining muzzle velocity, and the recoil reduction efficiency can reach 72.27%, which is of great significance for improving weapon performance.
Keywords. Nozzle, two chambers, recoil reduction, two phase flow interior ballistics.

1. Introduction

The weapon with a nozzle and two chambers separated by a partition is a new type of low recoil weapon, which can greatly reduce the recoil impulse with maintaining the muzzle velocity. Based on the expansion wave gun technology [1, 2], the weapon with a nozzle and two chambers separated by a partition is proposed. Although both of them reduce the recoil force of weapon by using the energy of propellant gas, compared with the expansion wave gun, the weapon with a nozzle and two chambers separated by a partition has the following advantages: 1) Because the propellant gas required for recoil reduction is provided by the second chamber rather than from the original chamber, more propellant gas is provided to reduce the recoil impulse of the weapon, and the recoil reduction performance is better. 2) The nozzle is connected with the rear chamber through the air guide hole in the barrel, so the expansion wave generated when the nozzle is opened will not be transmitted to the front chamber, thus affecting the motion of the projectile. 3) The front chamber and the rear chamber are connected by fire holes. The propellant gas in the rear chamber can not only ignite the front chamber, but also provide energy for the front chamber, so as to further improve the muzzle velocity without significantly increasing the maximum chamber pressure of the front chamber. 4) When the pressure of the rear chamber is lower than the front chamber pressure, the propellant gas in the front chamber will flow back to the rear chamber, slowing down the pressure gradient of the rear chamber, and maintaining the ability to reduce recoil. Interior ballistic phenomena are physical processes accompanied by high temperature and high pressure of chemical reaction. There are many chambers in the weapon with a nozzle and two chambers separated by a partition and the flow of propellant gas and powder particles between different chambers further increases the complexity of its interior ballistic process. The detailed interior ballistic data can not be obtained through the experiment, so the use of numerical simulation [3-5] to obtain accurate parameters is an effective method.

Jie, SONG et al. [6, 7] studied a dual chamber nozzle recoil reducing weapon, but there are limitations. On the one hand, it does not use two-phase flow numerical calculation, but uses lumped parameter method and Lagrange hypothesis to describe the interior ballistic cycle process with the average value of the variables in the space after the projectile, and does not consider the situation that the propellant gas and powder particles in the front chamber flow back into the rear chamber through the fire hole, on the other hand, because the gunpowder gas and solid particles in the rear chamber flow into the nozzle prematurely, the pressure in the rear chamber decreases too fast, which affects the recoil reduction ability of the weapon.

In this paper, a two-phase flow interior ballistics model of the weapon with a nozzle and two chambers separated by a partition is established. The MacCormack scheme [8-10] is used to solve the coupling field formed by the front and rear chambers, the air guide chamber and the nozzle. It is analyzed that the change of pressure difference between different chambers leads to critical and non critical flow, and even reverse flow. This change of flow will in turn affect the change of chamber pressure. This process of mutual coupling and interaction is difficult to be observed through experiments.

2. Physical Model

The launching process of the weapon with a nozzle and two chambers separated by a partition can be divided into five stages (figure 1): 1) In the first stage, after the primer is ignited, the propellant gas flows into the air guide chamber. When the propellant gas pressure in the rear chamber reaches the diaphragm breaking pressure, a part of the propellant gas flows into the front chamber; 2) In the second stage, When the pressure of the rear chamber is greater than the limit pressure of the rear spray diaphragm, the rear spray diaphragm is broken, the rear chamber is connected with the nozzle, and the gunpowder and gas in the rear chamber flows into the nozzle and flows out at a high speed through the nozzle to generate reverse thrust; 3)In the third stage, the gunpowder gas in the rear chamber ignites the gunpowder in the front chamber. After the gunpowder gas in the current chamber reaches the projectile starting pressure, the projectile is pushed into the rifling for accelerated movement; 4)In the fourth stage, the pressure of the front chamber is higher than that of the rear chamber, and the propellant gas in the front chamber flows back into the rear chamber; 5) In the fifth stage, the projectile goes out of the muzzle, that is, the aftereffect stage.

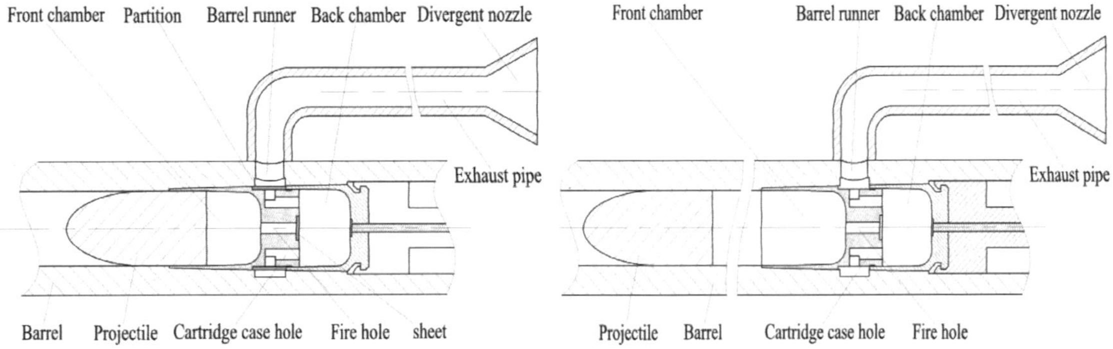

(a) Initial state (b) Post spray state

Figure 1. Schematic diagram of launch process of the weapon with a nozzle and two chambers.

3. Mathematical Model

3.1. Interior Ballistic Model of Two Phase Flow in Front Chamber

In the equations, ρ_g is the density of gas phase material; ϕ is the void ratio of gas phase; u_g is the velocity of gas phase movement; A is the cross-sectional area of gun bore; m_c, $\sum_k m_{ign,k}$ is the gas generation rate of powder in gas volume and the gas generation rate of various ignition elements; p is the gas pressure; $u_{ign,k}$ is the flow velocity of gas generated by various ignition elements in gun bore; f_s is the resistance between gas and solid phases in unit volume; e_g is the internal energy of gas ratio; e_p is the chemical energy of solid phase; f is the propellant force; k is the specific heat ratio of propellant gas; e_p is the stagnation enthalpy of igniter gas; q is the heat exchange between gas and solid phases in unit surface area.

$$\frac{\partial U}{\partial t}+\frac{\partial F}{\partial x}=\begin{cases} H+J' & \text{The charge from the rear chamber flows into the front chamber} \\ H-J & \text{The charge from the front chamber flows into the rear chamber} \end{cases} \tag{1}$$

$$U=\begin{bmatrix} \phi\rho_g A \\ (1-\phi)\rho_p A \\ \phi\rho_g u_g A \\ (1-\phi)\rho_p u_p A \\ \phi\rho_g A\left(e_g+u_g^2/2\right) \end{bmatrix} \quad F=\begin{bmatrix} \phi\rho_g u_g A \\ (1-\phi)\rho_p u_p A \\ \phi\rho_g A\left(u_g^2+p/\rho_g\right) \\ (1-\phi)A\left(\rho_p u_p^2+p+R\right) \\ \phi\rho_g u_g A\left(e_g+u_g^2/2+p/\rho_g\right) \end{bmatrix} \quad H=\begin{bmatrix} m_c A \\ -m_c A \\ m_c u_p A-f_s A+p\dfrac{\partial A\phi}{\partial x} \\ -m_c u_p A+f_s A+p\dfrac{\partial A(1-\phi)}{\partial x}+(1-\phi)R\dfrac{\partial A}{\partial x} \\ m_c A\left(e_p+u_p^2/2+p/\rho_p\right)- \\ f_s u_p A-AA_p q-p\dfrac{\partial A\phi}{\partial t} \end{bmatrix} \tag{2}$$

3.2. Auxiliary Equation

Other auxiliary equations, such as equation of state, burning rate equation, stress equation between particles, interphase heat exchange equation, gas-solid mass, momentum and energy exchange equation between two chambers, can be referred to [11, 12].

4. Analysis of Numerical Results

4.1. Model Validation

In order to verify the accuracy of the mathematical model and the reliability of the numerical calculation method, the chamber pressure curve and projectile velocity curve obtained by previous researchers according to the classical interior ballistic calculation are compared with the calculation results in this paper. As shown in figure 2 and figure 3, the curves are in good agreement, which indicates that the calculation results have a certain degree of credibility.

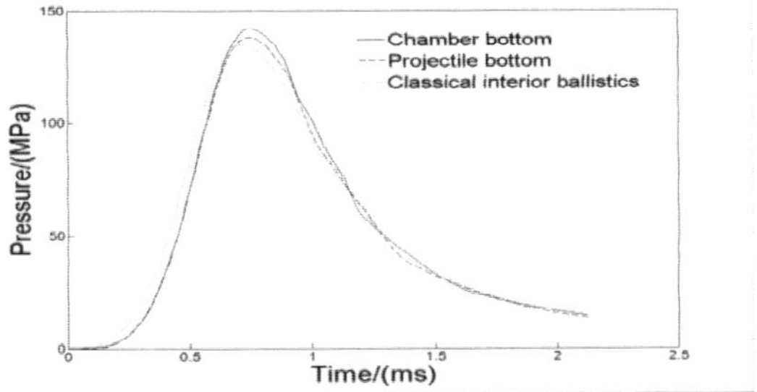

Figure 2. Pressure verification curve

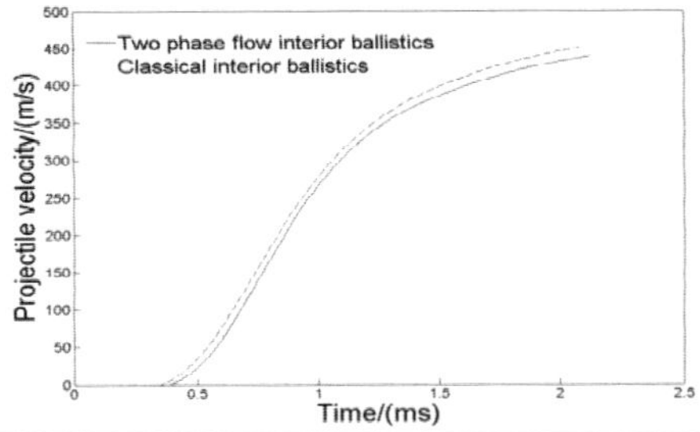

Figure 3. Velocity verification curve of projectile curve.

According to the data in table 1, the error of peak chamber pressure is 5.2% / 2.4%, and the error of muzzle velocity is 2.6%. Because the classical interior ballistics model assumes that all propellants are ignited instantaneously, and the ignition temperature of propellants is not considered, the calculated interior ballistics time is slightly less than that of two-phase flow interior ballistics.

Table 1. Comparison of numerical calculation results.

	Peak chamber pressure/MPa	Muzzle velocity/m.s^{-1}	Time of peak chamber pressure/ms	Interior ballistics time/ms
Two phase flow interior ballistics	141.88/138.00	438.88	0.76	2.14
Classical interior ballistics	134.76	450.62	0.72	2.07

4.2. Comparison of Interior Ballistic Performance of the Weapon with a Nozzle and Two chambers Separated by a Partition and Ordinary Weapon

It can be seen from figure 4 and figure 5 that the time for the pressure in the bore of the weapon with a nozzle and two chambers separated by a partition to reach the peak is 0.854 ms and the peak pressure is 152.171 mPa, while the time for the pressure in the bore of ordinary weapons to reach the peak is 0.764ms and the peak pressure is 141.884 mPa and that the starting time of the projectile of the weapon with a nozzle and two chambers separated by a partition is 0.502 ms and the muzzle velocity is 449.196 m/s, while the starting time of the projectile of the ordinary weapon is 0.401 ms and the muzzle velocity is 438.879 m/s. This is because the weapon with a nozzle and two chambers separated by a partition ignites the powder in the front chamber through the powder in the rear chamber and provides the powder gas energy for the front chamber, while the ordinary weapon ignites the powder by firing the primer. Because the powder gas energy provided by the ignition primer is less than that of the powder gas in the rear chamber, the chamber pressure of the weapon with a nozzle and two chambers separated by a partition is higher than that of the ordinary weapon and the muzzle velocity is greater than that of ordinary weapons.

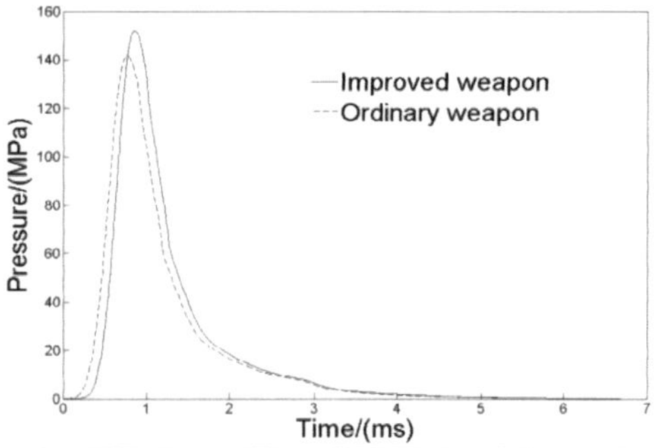

Figure 4. Pressure comparison curve.

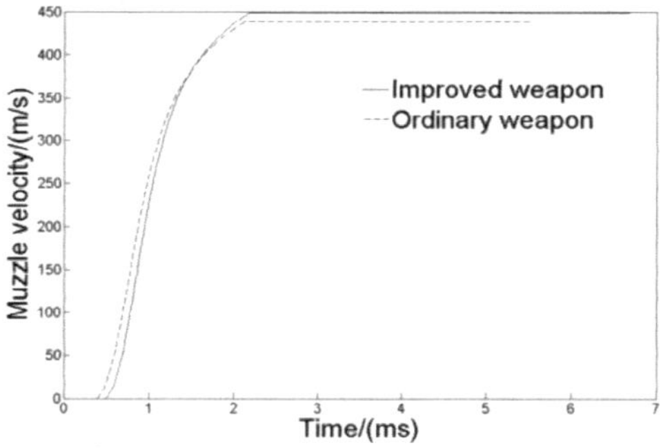

Figure 5. Velocity comparison curve of projectile.

It can be seen from figure 6 and figure 7 that the peak recoil force of ordinary weapons is -124.440 kN, and the recoil impulse is -106.893 N*s and the peak recoil force of the weapon with a nozzle and two chambers separated by a partition is -25.037k N, and the recoil impulse is -29.642 N*s. The recoil impulse curve of ordinary weapons shows a downward trend, and the degree of decline is very obvious. However, the recoil impulse curve of the weapon with a nozzle and two chambers separated by a partition shows a short rise at first and a slow decline trend. The recoil impulse is essentially the accumulation of recoil force in time. Because the recoil force of ordinary weapons is far greater than that of the weapon with a nozzle and two chambers separated by a partition, this leads to the decline of

recoil impulse curve of ordinary weapons is much greater than that of the weapon with a nozzle and two chambers separated by a partition.

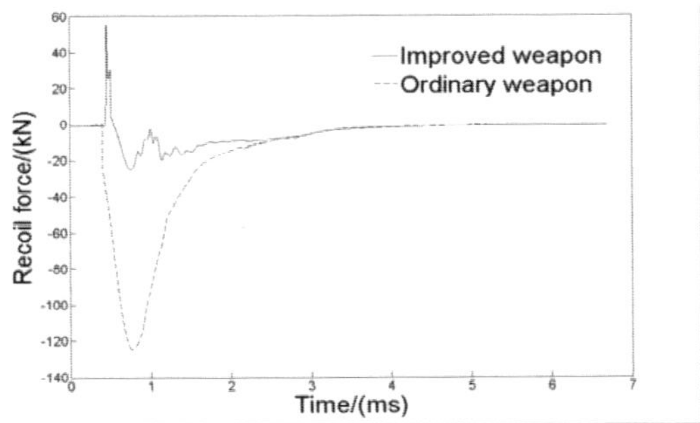

Figure 6. Recoil comparison curve.

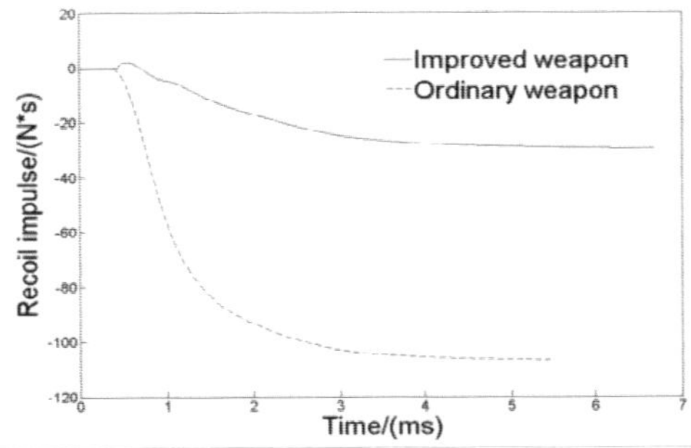

Figure 7. Recoil impulse comparison curve.

5. Factors Affecting the Interior Ballistic Performance of the Weapon with a Nozzle and Two Chambers Separated by a Partition

The structure of the weapon with a nozzle and two chambers separated by a partition is more complex than that of the conventional weapon, and the internal ballistic performance is also affected by many factors. Through numerical simulation, the effects of the charge amount in the rear chamber and the size of the air guide hole in the barrel on the interior ballistic performance of weapons are discussed.

5.1. Influence of Rear Chamber Charge

It can be seen from table 2 that with the increase of the charge in the rear chamber, the recoil reduction efficiency of the weapon increases significantly. However, the increase of the charge in the rear chamber also leads to the increase of the pressure difference between the front and rear chambers, which greatly affects the safety performance of the weapon.

Table 2. Comparison of effects of rear chamber charge on recoil reduction efficiency and pressure difference between front and rear chambers.

Charge amount /kg	Reduced recoil efficiency /%	Differential pressure between front and rear chambers /MPa
0.018	49.396	20.517
0.022	59.020	30.489
0.024	62.866	42.055
0.026	67.663	61.813
0.028	72.270	96.378

5.2. Influence of Barrel Air Guide Hole

It can be seen from table 3 that with the increase of the diameter of the air guide hole, the recoil reduction efficiency does not increase significantly, but the pressure difference between the front and rear chambers decreases greatly, This is because although increasing the diameter of the air guide hole of the barrel makes more gunpowder gas flowing into the nozzle, it will also make the pressure in the rear chamber drop faster, and then affect the combustion of gunpowder in the rear chamber, which in turn will affect the increase of the pressure in the rear chamber, resulting in the failure to improve the recoil reduction efficiency.

Table 3. Comparison of effects of barrel air guide hole diameter on recoil reduction efficiency and pressure difference between front and rear chambers.

Diameter of air guide hole/mm	Reduced recoil efficiency /%	Differential pressure between front and rear chambers /MPa
18	68.589	202.700
20	69.288	148.616
22	68.691	99.313
24	67.663	61.813
26	67.174	46.392

6. Conclusion

In order to improve the recoil reduction efficiency of a certain caliber grenade launcher, a one-dimensional two-phase flow interior ballistics model of the weapon with a nozzle and two chambers separated by a partition is established. Through the analysis of the calculation results, it can be proved that the established mathematical model can successfully simulate the interior ballistic process of the weapon with a nozzle and two chambers separated by a partition, which has practical significance for the theoretical analysis of the weapon with a nozzle and two chambers separated by a partition. It can be explained by the simulation of the interior ballistic process that the weapon with a nozzle and two chambers separated by a partition has the advantages of ensuring the initial velocity of projectile and reducing the recoil force significantly. Therefore, the interior ballistic performance of the weapon can be improved to a certain extent by adopting the technology of the weapon with a nozzle and two chambers separated by a partition.

Acknowledgments

This work was funded by the National Natural Science Foundation of China (Grant No. 12072161 and No. 51676099).

References

[1] Jianzhuang Z, Jian Z, Di C C and Lei Z 2010 Numerical simulation about rarefaction wave velocity and travel about rarefaction wave gun with two-door breech *Proceedings of the Third International Symposium on Test Automation & Instrumentation* vol 4 p 1374-7.

[2] Zhi J Z, Di C C, Zhang L and Jiang C C 2009 Modelling and simulation about rarefaction wave gun interior ballistics appling inertial breech *9th International Conference on Electronic Measurement & Instruments* p 1091-4

[3] Bougamra A and Lu H 2015 Interior ballistics two-phase reactive flow model applied to small caliber projectile-gun system *Propellants Explosives Pyrotechnics* **40**(5) 720-728.

[4] Cheng C and Zhang X 2019 Numerical investigation of two-phase reactive flow with two moving boundaries in a two-stage combustion system *Applied Thermal Engineering* **156** 422-431.

[5] Cheng C, Wang C and Zhang X 2019 A prediction method for the performance of a low-recoil gun with front nozzle *Defence Technology* **15**(5SI) 703-712.

[6] Song J, Liao Z Q and Li H Q 2018 Study on reducing recoil mechanism of jet gas with laval nozzles with double chamber *Journal of Harbin Institute of Technology* 501079-87.

[7] Song J, Qiu M and Liao Z Q 2018 Modeling of two jet gas devices and its firing efficiency comparison *Transactions of Beijing Institute of Technology* 386585-592.

[8] Jang J, Oh S and Roh T 2016 Development of three-dimensional numerical model for combustion-flow in interior ballistics *Journal of Mechanical Science and Technology* **30**(4) 1631-1637.

[9] Jang J, Oh S and Roh T 2014 Comparison of the characteristics of granular propellant movement in interior ballistics based on the interphase drag model *Journal of Mechanical Science and Technology* **28**(11) 4547-4553.

[10] Wang J, et al. 2018 Numerical simulation and optimized design of cased telescoped ammunition interior ballistic *Defence Technology* **14**(2) 119-125.
[11] Weng C S and Wang H 2006 *Computational Interior Ballistics* (Beijing: National Defense Industry Press).
[12] Jin Z M and Weng C S 2003 *Advanced Interior Ballistics* (Beijing: Higher Education Press).

On Simulation and Experiment of Flow Field of Combustion Coal Fallout Detection Instrument Based on CFD

Yaoshuo Sang [1,2], Qiunan Yan[1], Yong Liu[1], Long Zhang[1], Zhigang Li[1], Chao Ye and Chaohao Guo[1*]

[1] Hefei Institutes of Physical Science, Chinese Academy of Sciences, Hefei 230031, China.
[2] University of Science and Technology of China, Hefei 230026, China/P. R. China.
Email: 530108687@qq.com

Abstract. In order to further improve accuracy and stability of detection of combustion coal fallout propensity of cigarettes, author of the paper adopted computational fluid dynamics (CFD) technology for a three-dimensional numerical simulation of exhaust system of detection instrument, aiming to study characteristics of flow field near cigarettes. Moreover, a simulation model of eight-channel exhaust enclosure was established, obtaining vector diagram for flow velocity of flow field, velocity contour diagram, and pressure distribution cloud diagram. According to findings, flow field of eight channels is evenly distributed, with slow flow velocity around the instrument but furious inside channels. The wind velocity of cigarette monitoring channel is stable at about 200mm/s specified as per standard. However, there is significant change in pressure and flow velocity at the corners of channels, causing local turbulence. In experiments, average wind velocity of 8 monitoring channels was measured, and simulation results were compared with experiment data. Eventually, a conclusion is drawn that simulation result at cigarette monitoring channels changes consistently with the experimental data, with small errors as a whole. Therefore, the designed exhaust system complies with regulations on wind velocity stipulated by YC/T558-2018 Cigarettes—Determination of Combustion Coal Fallout Propensity of Burning Cigarettes. In a word, this paper is hoped to provide technical support for analogue simulation of exhaust system of cigarette detection instrument, and improve detection accuracy.
Keywords. Combustion Coal Fallout, CFD Analysis, Exhaust System, Experimental Verification

1. Introduction

Combustion coal fallout means that combustion coal falls off or is obviously skewed of cigarettes during smoking or pop-up soot, which will cause smoking interruption, with fire hazards. Moreover, it shall reduce consumer recognition of cigarette brands [1]. Existing researches on combustion coal have made achievements. Specifically, Mullin et al. [2] study the relationship between morphological parameters of combustion coal and combustion velocity, finding that free combustion velocity is positively correlated with cone length and area. LI Bin et al. [3] investigate and analyze behaviors of combustion coal, and determine a method for detecting fallout propensity. Additionally, mathematical regression equations are adopted by QIAO Yuemei et al. [4] to explore how combustion coal fallout is affected by parameters of cigarette paper. They point out in findings that faster combustion velocity of cigarettes requires higher for ash-wrapping effect of cigarette papers. Faster combustion velocity will result in "ashing burst" and "butt falling". According to standard requirements in YC/T 558-2018

Determination of Combustion Coal Fallout Propensity of Burning Cigarettes, the wind field of detection instrument should meet requirements of Appendix A of GB/T16450-2004 Routine Analytical Cigarette Smoking Machine- Definitions and Standard Conditions [5]: In the instrument, wind velocity of cigarette airflow was maintained at 200±30 mm/s. Wind velocity was the key factor affecting measurement results of combustion coal, so that when airflow moved too fast, free combustion of cigarettes would accelerate. There was a positive correlation between velocity and shape of combustion coal, which enlarged length and area of combustion coal, and improved probability of fallout, affecting detection accuracy of instrument. Therefore, in this paper, the developed CFP800A combustion coal propensity detection instrument was analysed in accordance with YC/T 558-2018 Determination of Combustion Coal Fallout Propensity of Burning Cigarettes. In addition, CFD technology was introduced to explore internal flow field of the instrument, so as to keep airflow velocity around the cigarette close to standard value. This is of great significance to improve detection accuracy and stability of the instrument.

In recent years, CFD (Computer Fluid Dynamics) technology is preferred by scholars to conduct a series of researches on flow field simulation of channels. Particularly, a 20-channel linear smoking machine model is designed by YAN Lihong et al. [6] to measured velocity of cigarettes at different positions; a rectifying net is installed to reduce wind velocity difference at starting position of the combustion. Constructing a negative pressure channel model of plug assembler, XU Jing et al. [7] analyze flow velocity, pressure and turbulence dissipation in negative pressure channel, and optimize negative pressure channel at entrance to improve the fluidity of airflow. Subsequently, scholars YANG Zhanping et al. [8] conduct a numerical simulation on airflow field distribution in wire nozzle, concluding that the larger the gap between inner and outer jackets, the more significant the asymmetrical distribution of inner airflow field. Internal flow field is dried by electrical heating in research of LIU Daoqi et al. [9], who aim to improve uniform flow field by adjusting inlet size, wind pressure and adding a wind baffle. At the same time, k-epsilon turbulence model is applied by Singh J et al. [10] to explore erosion and wear of pipeline by solid-liquid suspended solids, and analysed how erosion and wear are influenced by velocity, concentration, and size. Jeevahan Jeya et al. [11] propose three models - smooth channel, straight channel and oblique channel to cool turbine impeller, and after CFD analysis, determine optimal flow model under high turbulence. Furthermore, LI Ruping et al. [12] discuss influencing factor of fluid in jet-pipe servo valve, reveal characteristics of flow field under different jet-pipe's diameters, nozzle gaps and jet-pipe angles, and optimize parameters of jet-pipe and receiver. Most of above researches focus on flow field inside flow channel. However, there are few scholars who investigate overall flow field inside the instrument and the detection of wind velocity around cigarettes. In this paper, the author mainly studied wind field of the detection instrument, and introduced RNG k~ε turbulence model in light of influence of wind velocity on detection accuracy of combustion coal. In the meantime, the eight-channel exhaust flow field model was established to analyse vector diagram of flow velocity and pressure distribution cloud diagram, drawing the conclusion that model is uniformly distributed, with good gas fluidity, and stable wind velocity around the cigarettes. In the end, anemometer and special fixtures were used to carry out measurement experiments of wind velocity. According to results, simulation results are in good consistency with the experimental data.

2. Construction of Flow Field Analysis Model

2.1. Working Principle of Exhaust System

Exhaust system of detection instrument is composed of centrifugal fans, exhaust ducts, exhaust hoods, rotating disk, and cigarettes, as shown in figure 1(a). In working, gas is accelerated by centrifugal fan through high-speed rotating impeller, so that flue gas enters exhaust hood along eight channels, and exhausts by fan through the exhaust duct. According to standards, wind velocity of airflow around the cigarette should be between 170 mm/s~230 mm/s during each pumping. The velocity shall be greatly affected by structure of exhaust hood. For this reason, an eight-channel exhaust structure was designed in figure 1(b), and each channel corresponded to a cigarette, in order to ensure that wind velocity at each inlet was uniform and stable. The author verified rationality of design of the eight-channel structure and distribution of flow field inside the instrument through simulation analysis and experiments.

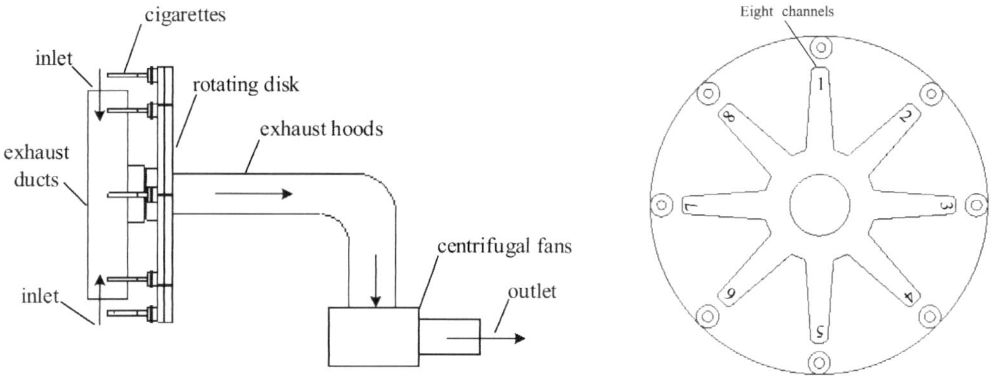

(a) Exhaust system (b) Eight channels

Figure 1. Schematic Diagram of Exhaust System.

2.2. CFD Calculation Theory

The analysis model for flow field of combustion coal fallout was established in accordance with the law of conservation of mass, and the law of conservation of momentum. Temperature change in working medium was ignored; for easy calculation, it was supposed that fluid was incompressible and constant Newtonian fluid, with mass conservation equation as follows:

$$\frac{\partial \rho}{\partial t} + div(\rho \vec{u}) = 0 \tag{1}$$

Wherein: $\vec{u}$ is velocity vector; ρ is density; t is time.

The momentum conservation equation is:

$$\begin{cases} \frac{\partial(\rho u)}{\partial t} + div(\rho u \vec{u}) = -\frac{\partial p}{\partial x} + \frac{\partial \tau_{xx}}{\partial x} + \frac{\partial \tau_{yx}}{\partial y} + \frac{\partial \tau_{zx}}{\partial z} + F_x \\ \frac{\partial(\rho v)}{\partial t} + div(\rho u \vec{u}) = -\frac{\partial p}{\partial y} + \frac{\partial \tau_{xy}}{\partial x} + \frac{\partial \tau_{yy}}{\partial y} + \frac{\partial \tau_{zy}}{\partial z} + F_y \\ \frac{\partial(\rho w)}{\partial t} + div(\rho u \vec{u}) = -\frac{\partial p}{\partial z} + \frac{\partial \tau_{xz}}{\partial x} + \frac{\partial \tau_{yz}}{\partial y} + \frac{\partial \tau_{zz}}{\partial z} + F_z \end{cases} \tag{2}$$

Wherein: ρ is density; τ_i is viscous stress; F_i is physical strength on micro-elements.

Common turbulence models in projects are Standard k-ε model and RNG k-ε model. Compared with standard k-ε model, the RNG model can better deal with flow field structure with great curvature, and three-dimensional flow field model cannot be calculated precisely by the standard k-ε model. Therefore, RNG k-ε turbulence model is adopted, with equation as below:

$$\frac{\partial(\rho\kappa)}{\partial t} + \frac{\partial(\rho\kappa u_i)}{\partial x_i} = \frac{\partial}{\partial x_j}[\alpha_k \mu_{eff} \frac{\partial k}{\partial x_j}] + G_k + \rho\varepsilon \tag{3}$$

$$\frac{\partial(\rho\varepsilon)}{\partial t}+\frac{\partial(\rho\varepsilon u_i)}{\partial x_i}=\frac{\partial}{\partial x_j}[\alpha_\varepsilon \mu_{eff}\frac{\partial\varepsilon}{\partial x_j}]+\frac{C^*_{1\varepsilon}\varepsilon}{k}G_k-C_{2\varepsilon}\rho\frac{\varepsilon^2}{k} \tag{4}$$

Wherein: k is turbulent kinetic energy, ε is turbulent dissipation rate, and G_k means production term of turbulent kinetic energy k caused by average velocity gradient. The model constants are obtained according to RNG theory: C_μ=0.0845, $C_{1\varepsilon}$=1.42, $C_{1\varepsilon}$=1.68, σ_k=1.0, σ_ε=1.3.

2.3. Construction of Numerical Model

Due to complicated internal structure of the instrument, it cannot be simplified into a two-dimensional mode through analysis and calculation. Therefore, three-dimensional calculations were used for simulation and solution. In order to reduce calculations and improve calculation efficiency, the author deleted smoking and pulling units inside the instrument that had a small impact on the flow field, and removed features such as rounded corners and chamfer. The simplified three-dimensional model is shown in figure 2(a), including three parts: outer cover, exhaust hood, and rotating disk. The box body is closed, and the gap between metal plate and profile refers to air inlet that communicates with atmosphere. It was simplified appropriately based on actual model, and 3 air inlets were established on front panel and both sides, as shown in figure 2(b). Moreover, the Boolean subtraction operation was used to cut the rotating disk and cigarettes, forming an enveloping solid that could realistically simulate airflow distribution around the cigarettes. The model was divided by unstructured tetrahedral grid. The author paid more attention to eight-channel exhaust hood, and grid in this area was encrypted to improve calculation accuracy. An expansion layer was added to the wall of model and flow field boundary was refined. At the same time, overall flow field inside the instrument was meshed, as shown in figure 2(b). A total of 2,648,499 nodes and 1,330,742 units were obtained.

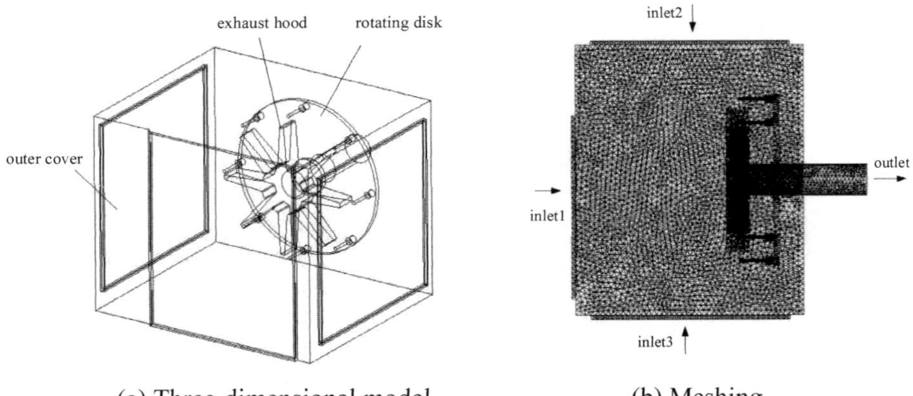

(a) Three-dimensional model (b) Meshing

Figure 2. Meshing of Flow Field.

Set boundary conditions of flow field: the air inlet was connected to atmosphere, and set as pressure inlet; reference value was standard atmospheric pressure. The outlet was set as velocity outlet, and wind velocity at outlet was calculated according to pipe diameter and fan parameters. The outer wall of flow field and other areas were all non-slip walls, with RNG k~ε turbulence model selected. The iteration steps of the solver were set to 500 steps, and convergence accuracy was 10-3. After iterative calculation based on 500 steps, the continuity convergence accuracy in residual curve was less than the set 10-3, indicating that the model convergence was reliable. *An Example.* In this example we can see that there are footnotes after each author name and only 5 addresses; the 6th footnote might say, for example, 'Author to whom any correspondence should be addressed.' In addition, acknowledgment of grants or funding, temporary addresses etc. might also be indicated by footnotes.

3. Results and Analysis

3.1. Analysis of Flow Field Velocity

Figure 3 (a) refers to flow velocity vector diagram of eight-channel flow field, and colors and arrows represent size and direction of velocity. The fluid enters the box through three air inlets and is discharged through eight channels. The flow field is evenly distributed as a whole, without obvious fluctuations. The outer cover is far away from the channel in four directions, and the gas flow rate is small. After it enters the channel, flow channel gets narrower and the flow velocity is fast. Figure 3 (b) is a vector diagram of flow velocity near one of the cigarettes. It shows that air velocity around the cigarette is stable, with good gas flow, and evenly distributed flow field. Flow velocity vector diagram in the central area is listed in figure 3 (c). The airflow of each channel converges in central area, and flow velocity increases dramatically. Due to eight-channel structure, flow velocity at arc area between two channels is slow, forming a slit. This signifies that airflow is drawn out immediately by the fan, and the flow field is reasonably distributed, no vortex formed here.

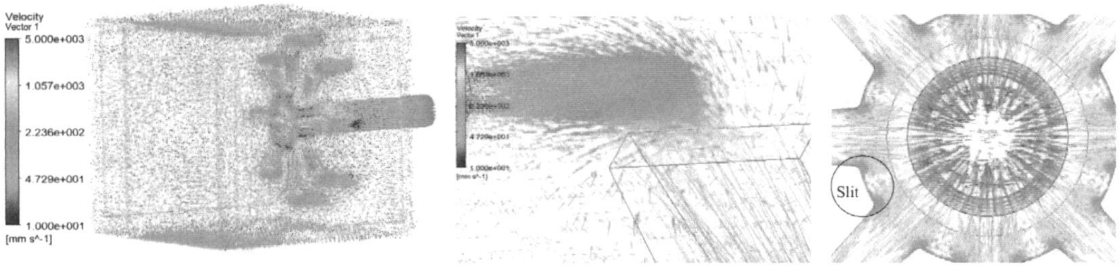

(a) Flow Velocity Vector Diagram (b) Part of the cigarette (c) Part of the slit

Figure 3. Flow Velocity Vector Diagram.

The x-y section is taken as reference plane to obtain the velocity contour cloud map, as shown in figure 4. Change of wind velocity around the cigarette is observed, and the contour line tells clearly that flow velocity in areas near the cigarette is about 200mm/s. The eight channels correspond to 8 cigarettes, so that flow velocity near the cigarettes is uniform and stable, meeting requirements of GB/T16450. In this section, wind velocity at channel entrance is probably 1800mm/s. As channel cavity becomes larger, flow velocity of gas gradually decreases, and changes uniformly, about 1500mm/s. After gas in eight channels flows into central area, flow velocity will experience collisions with significant fluctuations, causing tremendous energy loss, and decreasing velocity. The flow velocity of eight-channel model will be slightly turbulent inside the channel, but it is more uniform and stable around cigarettes, meeting standard requirements in the range of 200±30 mm/s.

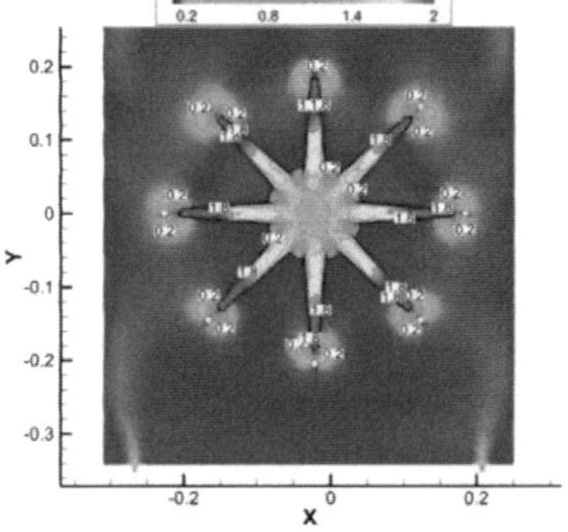

Figure 4. Analysis of X-Y Section Velocity at Z=300 mm.

3.2. Analysis of Flow Field Pressure

Figure 5 is cloud diagram of flow field pressure. The pressure is equally distributed as a whole, and judged according to colors. The inlet is atmospheric pressure, and outlet pressure is negative compared with inlet pressure; there is pressure difference at inlet and outlet, which discharges gas inside instrument along channels. The surrounding areas are farther away from suction opening, so the pressure is lower. After gas enters the channel, cavity becomes narrower, volume decreases, and pressure increases. Pressure will experience slight fluctuations at corners of the channels, and reaches the peak at air outlet. The pressure distribution of flow field is consistent with flow velocity that is relatively high in areas with high pressure. It is common to witness gradient change of pressure in eight-channel entrance and the corner inside the channel. The pressure of fluid inside channel tends to increase gradually. The reason is that inclination angle of channels gradually increases when fluid moves along eight channels, and converges to central area, leading to collision that builds up pressure. The pressure of fluid is uniform in outlet pipe area, with stable flow velocity. In summary, the pressure of flow field in eight-channel model is distributed uniformly. Pressure in the channels will increase slightly due to change of shape; pressure of flow field is better, and consistent with flow velocity, satisfying design requirements.

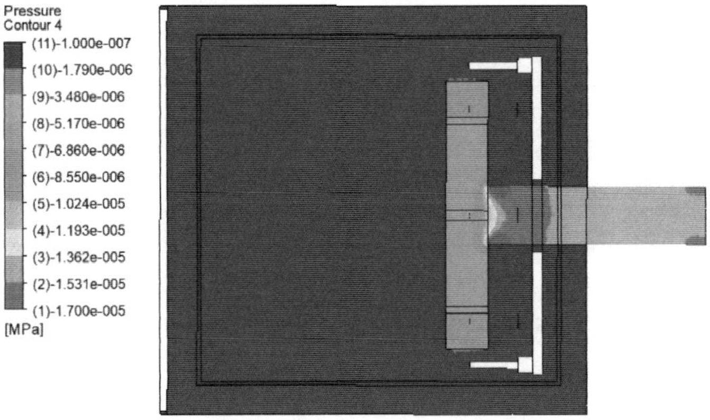

Figure 5. Cloud Diagram of Flow Field Pressure.

4. Experimental Verification

4.1. Experimental Design

Equipment: self-developed CFP800A combustion coal fallout detection instrument; Schiltknecht-ThermoAir6 Directional anemometer; AHLBORN-MA25902A data collector. Since it was necessary to measure wind velocity at outlet by inserting it into the pipe, the HT-9829 thermal anemometer of Laesent was used, with measurement range of 0.1~25 m/s, resolution of 0.01 m/s, and accuracy of ±5%; centrifugal fan was German ebmpapst-D2E146-HR93-01.

Methods: Please refer to figure 6 for experimental instrument and equipment. The wind field test was carried out on the detection instrument, in order to verify whether designed eight-channel exhaust hood met standard that wind velocity near the cigarette should reach 200±30 mm/s. Velocity of eight monitoring channels was measured by Schiltknecht-ThermoAir6 Directional anemometer, and AHLBORN-MA25902A data collector. For the purpose of improving measurement accuracy of wind velocity, a special fixture was designed to hold anemometer detector, so as to ensure accurate position of anemometer relative to each measurement point. Anemometer and fixture were introduced in figure 6(b). Eight red marks in figure 6(c) were selected as monitoring channels, and anemometer was adopted to measure average velocity per minute of each point. Each channel was measured eight times to reduce experimental errors, followed by a comparison between experimental data and simulation results.

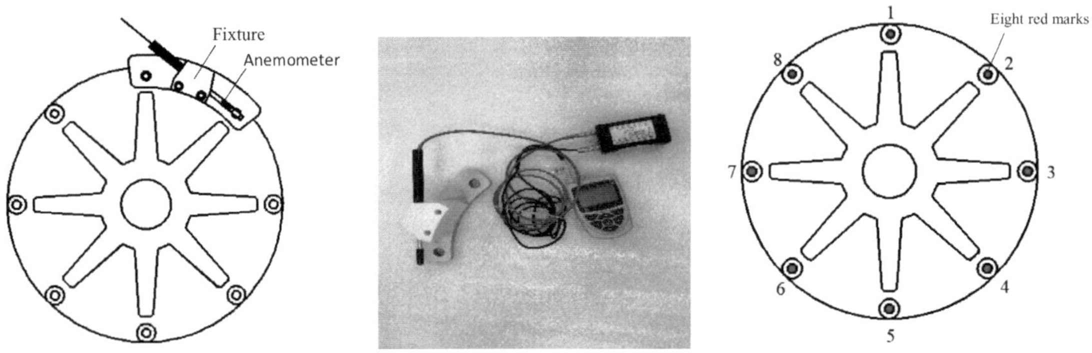

(a) Fallout detection instrument (b) Anemometer and fixture (c) Monitoring channel

Figure 6. Experimental Instruments and Equipment.

4.2. Data Analysis

Fixture and anemometer were adopted to measure wind velocity of 8 channels, and each channel was measured for 8 times. Measurement results were recorded in figure 7. According to measurement results, velocity of single channel was uniform and changed slightly. Wind velocity was different between channels, but it was within the deviation range required by the standard. The average wind velocity of 8 channels was 192.3 mm/s. In addition, designed eight-channel exhaust hood and exhaust system were effective to stabilize wind velocity around cigarettes, and reduce influence of wind velocity on fallout detection.

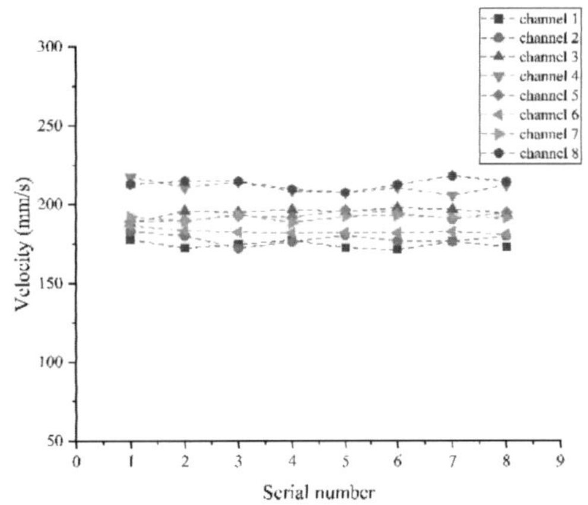

Figure 7 .Velocity measurement of channel.

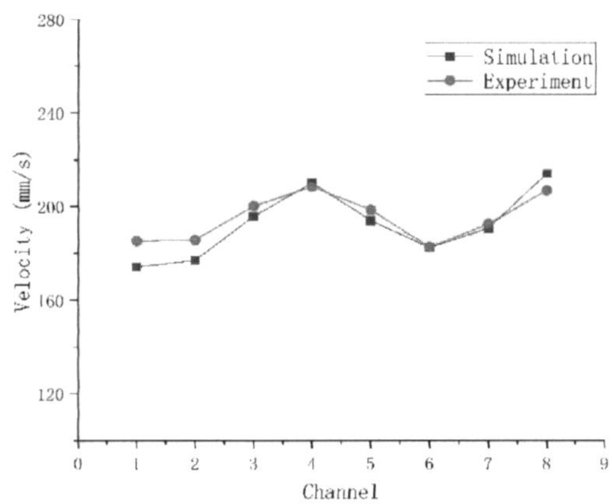

Figure 8. Comparison of simulation and experiment.

Average of 8 measurements for each channel was selected, and experimental data was compared with simulation results, obtaining a line chart as shown in figure 8. There were significant differences between simulation value and experimental value of monitoring channels 5 and 6. The main reason was that when the simulation model was established, smoking unit and lighting unit were deleted for easy calculation, and influence of falling through on wind velocity was ignored, resulting in a slight increase in wind velocity at points 5 and 6. However, experiment and simulation data in the figure changes in a consistent way. In general, simulation of flow field of the detection instrument can vividly reflect movement process of fluid inside the instrument and in channels, with consistent simulation results and experimental data.

5. Conclusions

In this paper, the author, based on CFD technology, introduces RNG k~ε turbulence model, and establishes simulation model of eight-channel wind field for instrument to detect fallout. At the same time, vector diagram of flow velocity, velocity contour diagram, and pressure distribution cloud diagram inside the instrument are obtained. According to findings, flow field is evenly distributed; gas flowability in eight-channel model is good. Surrounding area is farther from the channels, so, flow velocity is slow, but fast in channel. Wind velocity around the cigarettes fluctuates up and down around standard value.

In accordance with developed combustion coal fallout detection instrument, average velocity of 8 monitoring channels is measured through experiments, and the actual velocity on site meets standard of YC/T 558-2018 Cigarettes—Determination of Combustion Coal Fallout Propensity of Burning Cigarettes. Comparing experimental data with simulation results, the author concludes that change trend of both is consistent, and overall error is small. This verifies correct design of exhaust system model, reduces influence of wind velocity on fallout, and improves accuracy and stability of CFP800A fallout detection instrument.

References

[1] YC/T 558-2018 2018 Cigarettes-Detemination of Hot Coal Fallout Propensity of Burning Cigarettes.
[2] Mu L, Hui J Q, Tao S C, et al. 2019 Shape parameters of cigarette burning cone and their relationships with burning rate *Acta Tabacaria Sinica* **25**(01) 25-32.
[3] Li B, Gao L, Zhang Y, Hong L, et al. 2018 Observation and measurement of smokers' ash removal behavior in China *Beiträge zur Tabakforschung International/Contributions to Tobacco Research* **28**(1) 42-50.
[4] Qiao Y M, Zhao Q R, Li L Q, et al. 2017 Analysis of the influence of cigarette paper parameters on cigarette discharge rate *Farm Products Processing* **23** 44-8.
[5] GB/T 16450-2004 2004 Routine Analytical Cigarette-Smoking Machine-Definitions and Standard Conditons.
[6] Yan L H, Zhao Z H, et al. 2015 Improvement of air-flow control system in enclosure of linear smoking machine and its application *Acta Tabacaria Sinica* **21**(03) 25-33.
[7] Xu J, Zhou C J, et al. 2018 Flow field analysis of negative pressure passages for filter assembling in ZJ119-MAX *Tobacco Science & Technology* **51**(09) 79-84.
[8] Yang Z P, Hao X B, et al. 2019 Numerical simulation of air flow field inside tow transport jet in filter rod maker *Tobacco Science & Technology* **52**(07) 95-100.
[9] Liu D Q, Fan C H, et al. 2017 Analysis of internal flow field and optimization of mechanism parameters of negative pressure electric heating dryer based on fluent *Food & Machinery* **33**(09) 94-98+191.
[10] Singh J, Kumar S, Singh J P, et al. 2017 CFD modeling of erosion wear in pipe bend for the flow of bottom ash suspension *Particulate Science and Technology* 275-285.
[11] Jeevahan J, Durai Raj R B, Mageshwaran G, Sriram V, Joseph G B and Poovannan A 2019 Design and analysis of internal cooling passage of gas turbine using computational fluid dynamics *International Journal of Ambient Energy* **40**(1) 105-109.
[12] Li R P, Nie S L, et al. 2011 Simulation investigation on fluid characteristics of jet pipe water hydraulic servo valve based on CFD *Journal of Shanghai University (English Edition)* **15**(03) 201-206.

Numerical Investigation on Radiation Effect in Transpiration Cooling

Kang Qian[1], Taolue Liu[1], Fei He[1,*], Meng Wang[1], Longsheng Tang[2] and Jianxing Zhou[2]

[1] Department of Thermal Science and Energy Engineering, University of Science and Technology of China, Jinzhai Road 96, Hefei 230027, China
[2] Beijing Aerospace Technology Institute, Beijing 100074, China
Email: hefeihe@ustc.edu.cn

Abstract. This paper proposed a numerical strategy which could achieve the coupled modeling and solving of transpiration cooling with external high-temperature gas flow and especially take the radiation effect into account. Based on the numerical strategy, the heat and mass transfer characteristics of the transpiration cooling in a high-temperature gas channel were studied, and the radiation effect and corresponding influence factors were analyzed. The results indicated that the radiative heat flux takes an important role in the heat transfer between the transpiration cooling and external high-temperature gas flow which may reach 40% under the operating condition considered in this work, and the radiation absorption from the coolant is more obvious near the downstream wall. As the wall emissivity increases, the radiation heat transfer in the downstream area of the porous wall is enhanced significantly and thereby the wall temperature there increases, as the result, the uniformity of the temperature distribution on the whole porous wall is improved to some extent.

Keywords. Transpiration cooling, radiation, numerical.

Nomenclature			
T	temperature	α	specific area
$\boldsymbol{I}$	radiation intensity	ε	porosity
K	permeability	*Subscript*	
F	inertia coefficient	s	solid
$\boldsymbol{J}$	diffusive flux of species	f	fluid
k	thermal conductivity	eff	effective
		ms	mainstream

1. Introduction

Due to the high efficiency and uniform performance, transpiration cooling has been regarded as one of the most promising thermal protection techniques for the structures suffering extremely thermal environment, such as the leading edge and combustion chamber of high-speed vehicles. There have been many researches on the mechanism, influencing factors and configuration optimization of transpiration cooling. However, the coupled modeling and solving of the transpiration cooling with external thermal environment and gas flow are still insufficient, and especially the radiation effect from the external field is often excluded. As the result, the transpiration cooling performance is always overestimated under the fuel gas conditions where the radiation effect from the high-temperature gaseous mixture can't be ignored.

Recently, some researchers attempt to study the radiation effect in transpiration cooling by simplified method. Zhang [1] studied the effect of external radiation on a one-dimensional transpiration cooling problem by introducing a radiative thermal conductivity in the solid energy equation and calculating the radiative heat flux at the boundary through approximate radiation heat transfer equation, so there was no radiation field in his research. Jiang [2] studied the heat transfer in the rectangle channel with transpiration cooling and obtained the radiation heat transfer equations between the surfaces based on the view factor, but his work didn't include the radiation effect from the participating gaseous media. The above treatments of radiation effect are limited because the radiation from the surfaces and participating gases are mutually influenced and should be considered synchronously.

In this paper, a numerical strategy which could calculate the radiation heat transfer between transpiration cooling and external flow field has been proposed, and based on the proposed method, the heat and mass transfer characteristics of the transpiration cooling in a high-temperature gas channel were studied.

2. Modelling and Numerical Method

The physical model considered in this paper is sketched in figure 1(a). Fuel gas mixture after combustion flows across the upper surface of the porous channel wall, and heats the wall through convective and radiative heat transfer. Coolant is injected into pores from the bottom of the wall, flows and absorbs the heat stored in the porous matrix, and forms a protective film on the surface exposed to the hot gas after flowing out.

2.1. Mathematical Models

Different models are respectively developed to describe the fluid flow and heat transfer processes in the mainstream region and porous region.

2.1.1. Mainstream Region. For the steady state mainstream, the conservation equations concerned, i.e. the mass, momentum, species transport, energy, and radiative transfer equations are listed as follows:

$$\nabla \cdot (\rho \boldsymbol{u}) = 0 \tag{1}$$

$$\nabla \cdot (\rho \boldsymbol{u}\boldsymbol{u}) = -\nabla P + \nabla \cdot \bar{\bar{\tau}} + \frac{\partial}{\partial x_j}\left(-\overline{\rho u_i' u_j'}\right) \tag{2}$$

$$\nabla \cdot (\rho \boldsymbol{u} Y_i) = -\nabla \cdot \boldsymbol{J}_i \tag{3}$$

$$\nabla \cdot \left(\rho u \left(h + \frac{u^2}{2}\right)\right) = \nabla \cdot \left(k_{eff} T - \sum_i h_i \boldsymbol{J}_i + \bar{\bar{\tau}}_{eff} \cdot \boldsymbol{u}\right) + S_h \tag{4}$$

$$\nabla \cdot (I(\boldsymbol{r}, \boldsymbol{s})\boldsymbol{s}) + (a + \sigma_s) I(\boldsymbol{r}, \boldsymbol{s}) = a n^2 \frac{\sigma T^4}{\pi} + \frac{\sigma_s}{4\pi} \int_0^{4\pi} I(\boldsymbol{r}, \boldsymbol{s}') \Phi(\boldsymbol{s} \cdot \boldsymbol{s}') d\Omega' \tag{5}$$

where the Reynold stress is calculated based on the SST $k - \omega$ turbulent model, and the radiation calculation is based on the DO radiation model.

2.1.2. Porous Region. The mass and momentum equations for the fluid in the porous region are based on the volume-averaged method and Brinkman-Forchheimer extended Darcy's law, and expressed as:

$$\nabla \cdot (\rho_f \boldsymbol{u}) = 0 \tag{6}$$

$$\nabla P = -\frac{\mu}{K} \boldsymbol{u} - \frac{F}{\sqrt{K}} \rho_f |\boldsymbol{u}| \boldsymbol{u} + \nabla (\mu_{eff} \nabla \boldsymbol{u}) \tag{7}$$

The local thermal non-equilibrium characteristic, i.e., the convective heat transfer between the solid and the fluid in the pore is taken into account, and thus the energy equations for the fluid and the solid matrix are respectively expressed as:

$$\nabla \cdot (\rho_f \boldsymbol{u} h_f) = \nabla \cdot (k_{f,eff} \nabla T_f) + Q_{sf} \tag{8}$$

$$\nabla \cdot \left(k_{s,eff} \nabla T_s\right) = Q_{sf} \tag{9}$$

The heat transfer between the fluid and the solid is calculated by [3]:

$$Q_{sf} = \alpha(0.933\varepsilon^2 - 0.245\varepsilon + 0.0165)Re_d^{0.8} Pr^{1/3} k_f \left(T_s - T_f\right)/d_p \tag{10}$$

2.2. Numerical Strategy

As shown in figure 1(a), the coolant outflow mixes with the mainstream, so besides heat transfer, there is also mass transfer between the mainstream region and the porous region. A numerical strategy that could achieve the coupled solution of the two regions is proposed here by defining the physical parameter transfer at the interface.

The pressure and velocity of fluid in the two regions are consecutive at the interface as suggested by Betchen [4] and Ochoa [5]:

$$P_p = P_{ms}, \boldsymbol{u}_p = \boldsymbol{u}_{ms} \tag{11}$$

An equivalent wall temperature [6] is introduced at the interface due to the temperature deviation between the solid and fluid:

$$T_{ms} = \varepsilon T_f + (1 - \varepsilon)T_s \tag{12}$$

The total heat flux entering the porous region consists of radiation and convection /conduction. The former can be obtained from the calculation of radiation field in the mainstream region, and the latter can be calculated by [7]:

$$q_c'' = -c_p\left(\frac{\mu}{Pr_t} + \frac{\mu_t}{Pr_t}\right)\nabla T_{ms} \tag{13}$$

According to Wang [8], the heat flux distribution between the solid matrix and fluid at the interface can be written as:

$$q'' = q_s'' + q_f'' \tag{14}$$

$$\frac{q_s''}{q_f''} = \frac{(1-\varepsilon)k_s}{\varepsilon k_f} \tag{15}$$

Numerical simulations were conducted by Ansys Fluent, and UDF was embedded to achieve the coupled calculations between porous and mainstream regions. As shown in figure 1(b), the external flow and radiation fields were calculated in case-1 by receiving the interfacial temperature and velocity distributions from the case-2, the flow and heat transfer inside the porous medium were calculated in case-2 by applying the hot side heat flux and pressure distributions extracted from the case-1. Iterations were carried out through the interfacial data exchange from the results of the two cases and continued until the convergence criteria have been achieved.

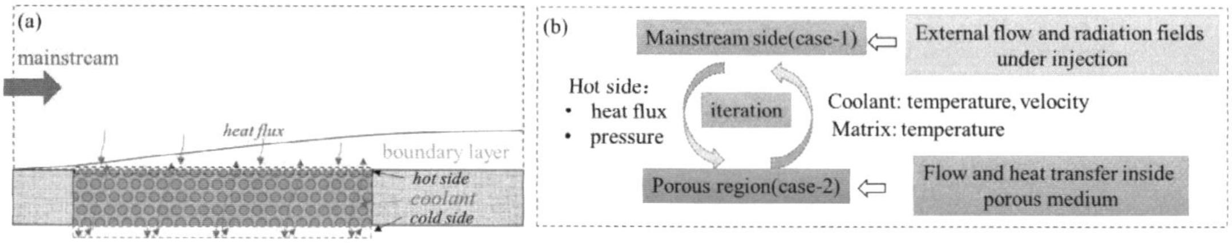

Figure 1. The schematics of (a) transpiration cooling and (b) coupled solution process.

2.3. Geometry and Test Conditions

The geometrical model is shown in figure 2 with the coordinate system and scales marked on.

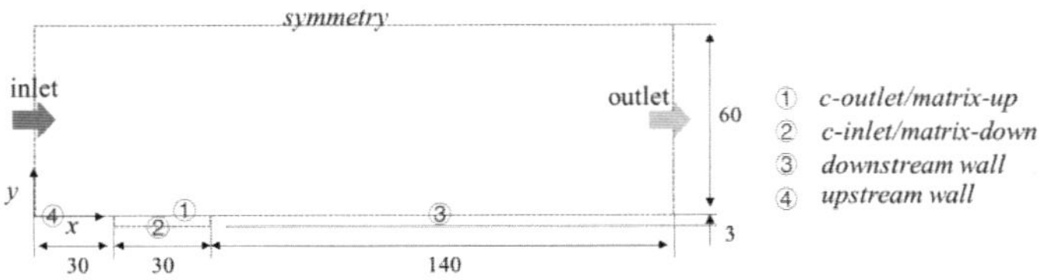

Figure 2. Geometrical model.

The mainstream was a high-temperature gaseous mixture, the coolant was superheated steam, and their components are listed in table 1. In the simulations, the total temperature of the mainstream and the coolant was 2000 K and 500 K respectively, the mass flux of the mainstream was 200 kg/m^{2}s, the mass flow rate of the coolant was 0.03 kg/s, the operating pressure was 0.5 MPa, and the turbulent intensity in the mainstream was 10%.

Table 1. Components and their mass fraction in the mainstream and coolant.

	Mainstream			Coolant
Component	H_2O	CO_2	N_2	H_2O
Mass fraction	0.08	0.21	0.71	1

Several cases were designed and in the basic case, the upstream wall and downstream wall were set as adiabatic, all the wall emissivity was set as 0.8. The parameters of the porous medium are listed in table 2. At the cold side, a convective boundary condition was used to account for the heating process of coolant before entering the matrix and a non-dimensional coefficient, Rc [9], was introduced as follows:

$$R_c = h_c/\rho u c_p = 0.25 \quad (16)$$

Table 2. Parameters of the porous medium.

ε	K	F	α	k_s
0.33	8.3E-13m^2	0.75	33500m^{-1}	20W/mK

The computational domains were separated into two parts, whose meshes were both generated with ICEM. The grid independence of numerical results was checked by three mesh strategies respectively with the grid number 144000, 210000 and 285000. Under the same boundary conditions, the temperatures on the surface of porous wall were calculated with the three mesh strategies, and the results indicated that the mesh with grid number 210000 was enough for the calculation.

3. Analyses and Discussions of Numerical Results

The pressure, steam mass fraction, and temperature distributions in the mainstream and porous regions are presented in figure 3, where the temperature in the porous region on the figure 3(a) is the averaged value ($\varepsilon T_s + (1-\varepsilon)T_f$). A low-temperature film adherent to the wall could be observed in the mainstream region and it is more apparent near the end of the porous wall(x$\approx$ $60mm$). In the porous region, the fluid temperature is very low at the cold side and increases rapidly as the fluid flowing towards the hot side. The temperature of solid matrix is higher than that of fluid, and the thermal non-equilibrium characteristic is remarkable at the cold side and the hot side. Besides, both the temperature distributions of the fluid and solid at the hot side are not uniform, the temperature difference over the surface exceeds 400K.

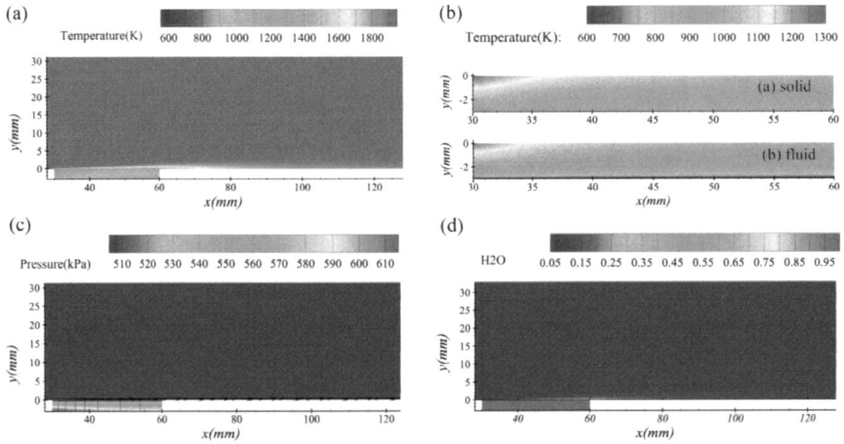

Figure 3. Contours: (a) full temperature (b) porous temperature (c) pressure (d) H_2O mass fraction.

To account for the radiation effect of the coolant outflow, the net spherical mean radiation is introduced which is defined as the deviation of the spherical mean absorbed radiation and emitted radiation. A positive value means a radiative sink related to the radiation absorption. The distribution of the net spherical mean radiation in the mainstream region is presented in figure 4. In most area, the value of the net spherical mean radiation is negative, and is particularly lower above the porous wall, indicating that a large amount of radiant heat is dissipated from the gaseous mixture in the mainstream and enters the porous wall. In the film layer formed from the coolant outflow, the value of the net spherical mean radiation is positive, which implies that the coolant outflow has a positive effect on weakening the mainstream-side radiative heat transfer. In addition, as the more coolant flows out and accumulated, the radiation absorption by the coolant outflow is more significant in the downstream region.

The temperature and heat flux distributions on the interface between the two regions are shown in figure 5. At the interface, the solid temperature is a bit higher than that of the fluid and their deviation gets smaller along with the mainstream direction, which more specifically, is 56K at the front and 24K at the end. As shown in figure 5(b), the radiative heat flux takes a large proportion in the total heat flux especially when x>45mm where the ratio of radiative heat flux to the total heat flux reaches 38%~ 40%.

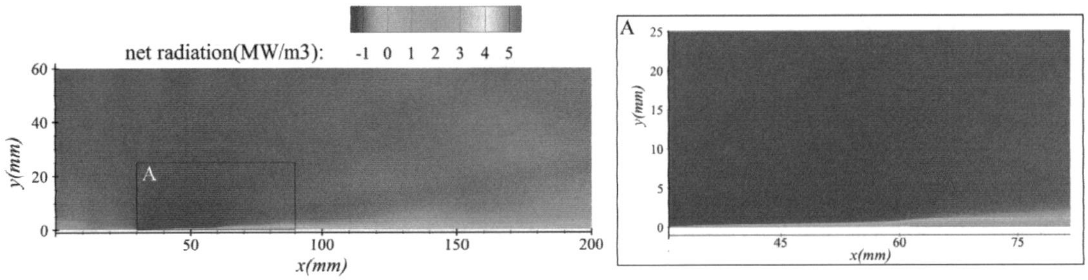

Figure 4. Net spherical mean radiation.

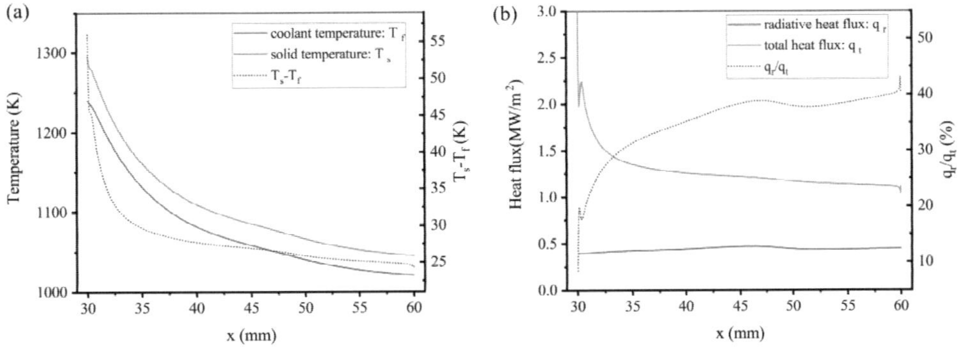

Figure 5. Temperature and heat flux distributions at the interface.

To analyze the effect of radiant heat transfer on the transpiration cooling quantitatively, two additional cases were calculated where the wall emissivity of 0.4 and no radiation was considered respectively, and the results were compared with the basic case where the wall emissivity was 0.8 and shown in figure 6. Under the same operating condition, the surface temperature of the porous wall increases with the wall emissivity, and the maximum increment reaches 105K at the end of the porous wall. 1) the temperature difference over the porous wall decreases as the wall emissivity increases, which indicates temperature distribution on the surface of channel wall is more uniform at high wall emissivity. All those phenomena can be explained with the heat flux distribution presented in figure 6(b). With the increase of wall emissivity, the heat flux enters the porous wall from the high-temperature mainstream increases and the heat flux distribution is more uniform.

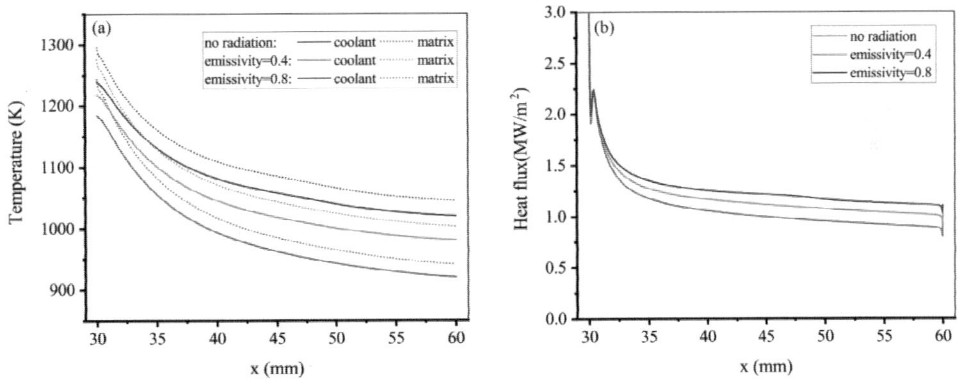

Figure 6. Temperature and heat flux distributions under different emissivity.

4. Conclusions

The radiation effect from the external flow field on the transpiration cooling has been studied in this paper, and some results could be concluded as follows:

(1) The radiative heat flux takes a large proportion in the total heat flux entering the transpiration cooling structure, which can reach 40% under the operating conditions considered in this work.

(2) The radiation absorption effect from the coolant outflow is more obvious near the downstream wall.

(3) The higher the wall emissivity, the higher the hot-side temperature and the smaller the amplitude of temperature variation on the porous wall.

Acknowledgement

This work is supported by the National Natural Science Foundation of China (contract no. 51806206).

References

[1] Zhang W H, Yi Y, Bai X H and Nakayama A 2020 A local thermal non-equilibrium analysis for convective and radiative heat transfer in gaseous transpiration cooling through a porous wall. *International Journal of Heat and Mass Transfer* **162** 120389.

[2] Jiang P X, Yu L and Ren Z P 2004 Influence of variable thermophysical properties and thermal radiation on convection heat transfer in transpiration cooling *Journal of Aerospace Power* 1000-8055 02-0184-07.

[3] Liu Y Q, Xiong Y B, Jiang P X, Wang Y P and Sun J G 2013 Effects of local geometry and boundary condition variations on transpiration cooling *International Journal of Heat and Mass Transfer* **62** 362-372.

[4] Betchen L, Straatman A G and Thompson B E 2006 A nonequilibrium finite-volume model for conjugate fluid/porous/solid domains *Numerical Heat Transfer, Part A: Applications* **49**(6) 543-565.

[5] Ochoa-Tapia J A and Whitaker S 1998 Momentum jump condition at the boundary between a porous medium and a homogeneous fluid: Inertial effects *Journal of Porous Media* **1**(4) 201-217.

[6] Phanikumar M S and Mahajan R L 2002 Non-Darcy natural convection in high porosity metal foams *International Journal of Heat & Mass Transfer* **45**(18) 3781-3793.

[7] Dahmen W, Müller S, Rom M, Schweikert S, Selzer M and von Wolfersdorf J 2015 Numerical boundary layer investigations of transpiration-cooled turbulent channel flow *International Journal of Heat and Mass Transfer* **86** 90–100.

[8] Wang J H and Shi J X 2008 Discussion of boundary conditions of transpiration cooling problems using analytical solution of LTNE model *Journal of Heat Transfer* **130** 014504-1.

[9] Xiong Y B 2013 *Study of the Mechanism of Flow and Heat Transfer on Supersonic Transpiration Cooling* (Beijing: Tinghua University).

Study on the Expansion of Primary Flow for the Mixing Uniformity in the Two-Phase Ejector

Lixing Zheng*, Hongwei Hu and Changning Mi

School of Electric Power, Civil Engineering and Architecture, Shanxi University, Taiyuan, Shanxi, China.
Email: lxzheng@sxu.edu.cn

Abstract. The expansion of primary flow in the suction chamber of the CO_2 two-phase ejector is investigated and its influences on the mixing characteristics are analyzed. An ejector model is developed, by constructing differential equations for mass, momentum and energy then get the governing equation. In the suction chamber, the expansion of primary flow and the compression of secondary flow are modeled along the flow path. Based on the constant-pressure mixing theory, the pressure equilibrium positions of two stream (namely at the inlet and inside of mixing chamber, respectively) are considered. The mass and energy transfer in the mixing chamber were analyzed by using the double-flow model formulation. The ejector performance parameters are obtained for the different operation conditions, and the distributions of temperature and velocity of two streams in the mixing chamber are presented. The simulation results showed the influence of primary flow expansion on the pressure lift ratio was relatively obvious, and the larger expansion distance was helpful to improve the mixing efficiency and decrease the thermodynamic entropy change during the mixing. Moreover, the temperature of secondary flow for lower primary flow pressure presented larger descent rates at the initial of mixing. This work is helpful for the improvement of ejector theoretical model and the optimization design.
Keywords. Suction chamber, CO_2 ejector, expansion, mixing uniformity.

1. Introduction

Nowadays, energy and environmental problems are becoming increasingly prominent. In the field of energy, research is developing towards energy conservation and efficiency improvement. Using CO_2 fluid as working medium can improve system performance and is widely used [1]. CO_2 is considered as a very promising working fluid substitute because of its low price, zero ozone depletion potential, almost zero GWP effect, non-toxic and nonflammability advantages [2]. At the same time, due to the special thermal and physical properties of CO_2, CO_2 gradually becomes the working fluid that people are keen to study.

In most applications of transcritical carbon dioxide refrigeration systems, the CO_2 two-phase ejector has significant advantages. It can recover the expansion work lost during the throttling process, so it has great potential as an alternative method to improve cycle efficiency [3]. Although the structure of the ejector is simple, its internal flow characteristics are very complex, including supersonic flow, phase change, and mass transfer, momentum transfer and energy transfer between primary and secondary flows. Therefore, it is necessary to study its flow characteristics for the ejector design and optimization.

Many studies were carried out by establishing theoretical models and experiments tests to several related performance parameters of the ejector were studied. Haida et al. [4-5] reported a series of researches on the CO_2 two-phase ejector of the homogeneous equilibrium model and establishment the homogeneous relaxation model. After that, the model was continuously improved to better describe the internal flow field of the CO_2 ejector, thereby improving of ejector efficiency and other related performance. Zhu et al. [6] studied the specific flow state mechanism of supersonic two-phase flow of CO_2 ejector and the influence law of ejector performance through visual experiments. Through photography could intuitively see the characteristics of primary flow. At the same time, the influence of primary flow expansion angle on ejector performance was also studied. Li et al. [7] also used the visualization experiment in studying the two-phase flow characteristics of the trans-critical CO_2 ejector. For the isentropic expansion process of primary flow, the change of phase transition position was obtained on the basis of experiment and theoretical analysis. In these literatures, there are few studies on the fluid mixing between the suction chamber and the mixing chamber in the CO_2 ejector. Zheng and Deng [8] used the lumped parameter method and took the isentropic flow as the energy conservation equation to analyze the suction chamber. Wang and Yu [9] used the lumped parameter method as the control equation of the suction chamber and developed a thermodynamic model of the ejector.

In terms of ejector design theory, Keenan et al. proposed mixing theory with the same pressure and mixing theory with the same area. For the same pressure mixing theory, the equal pressure was kept and mixed the primary flow and secondary flow; the constant-area mixing theory is to mix the primary flow and the secondary flow with a constant cross area in the ejector mixing chamber. In specific applications, the constant-pressure mixing theory is better than the constant-area mixing theory. Taleghani et al. [10] came up with a newly ejector model where it was assumed that the two flows reached the same pressure before the inlet of the mixing chamber. Zhu [11] assumed that the pressure of the primary flow at the inlet of the mixing chamber was the same as the inlet pressure of the suction, and then the two flows maintained a constant pressure. However, for some working conditions, the primary flow may continue to expand in the mixing chamber after passing through the suction chamber. As a consequence, it is indispensable to study the primary flows expansion inside the suction chamber and mixing section to analyze the mixing characteristics.

This paper mainly studies the pre-mixing and mixing characteristics of CO_2 two-phase ejector. The governing equations are expressed using differential equations for mass, energy, and momentum. The mixing chamber is modeled by using a double-flow model, and the mass and energy transfer during mixing are considered. Two mixing conditions (I two streams reach the constant pressure at the mixing inlet, П two streams reach the constant pressure inside the mixing chamber) are discussed, and the effect for the mixing performance of ejector efficiency is analyzed. The entrainment ratio, pressure lift ratio, mixing efficiency and thermodynamic entropy change under different working conditions are obtained. The temperature and velocity distribution of the primary flow and the secondary flow in the mixing chamber are described in order to study the mixing characteristics of the ejector.

2. Mathematical Model Formulation

An ejector model was presented in which the circulation path is divided into many elements, and the governing equation were expressed through the differential equations of mass transfer equation, momentum equation and energy exchange equation. In the suction chamber, the differential equations of the ejector fluid primary flow and the secondary flow were formulated separately. In the model, there are different assumptions, the mixing principle and the equation solving method are different, so the assumptions used to facilitate the problem in the presented CO_2 two-phase ejector fluid mixing model are as follows:

- Flow of the entire ejector was one dimensional and steady.
- The two-phase flow was assumed to be a homogeneous equilibrium model.
- The influence of thermal diffusion and the turbulent viscous heating was not considered for two flows of fluid in the ejector.

•The motive nozzle was the type of converging and the fluid was chock at the nozzle throat position.

•The non-isentropic oblique wave and shock wave were neglected and the two streams were assumed no interaction along the flow path.

•Coefficient ψ was assumed to represent the primary flow expansion and the secondary flow acceleration.

•The primary flow and secondary flow reached the uniform pressure and then mixed with the same pressure.

•The mass transfer between the two streams was calculated by the condensation and entrainment mechanisms [12].

•The momentum transfer was determined by the resistance of interface between the primary flow and the secondary flow these two fluids and the mass transfer is the main cause of momentum gain or loss.

•The energy transfer was denoted by the enthalpy and the mass transfer leaded to the kinetic energy gain or loss.

•The friction between the two fluids and the wall causes a pressure drop inside the ejector,.

•The heat transfer process between the fluid medium and the wall was ignored.

2.1. Single-Flow (Motive Nozzle, Suction Chamber)

Based on the previous set of assumptions, the transfer equation between masses, momentum change equation and governing conservation equation of energy, along with the state equation of primary flow and secondary flow these two fluids in the motive nozzle and suction chamber represented a system of ordinary differential equations, the specific equation is as follows:

$$\frac{1}{V}\frac{dV}{dl}+\frac{1}{A}\frac{dA}{dl}+\frac{1}{\rho}\frac{d\rho}{dl}=0 \tag{1}$$

$$A\frac{dp}{dl}+\rho VA\frac{dV}{dl}+f\rho\frac{V^2}{2}\frac{dF}{dl}=0 \tag{2}$$

$$\frac{dh}{dl}+V\frac{dV}{dl}=0 \tag{3}$$

$$\frac{d\rho}{dl}-\left.\frac{\partial\rho}{\partial s}\right|_p\frac{ds}{dl}-\left.\frac{\partial\rho}{\partial p}\right|_s\frac{dp}{dl}=0 \tag{4}$$

The corresponding integral of the established differential equation is obtained in matrix form was shown in equation 5.

$$\begin{bmatrix} 0 & \frac{1}{V} & 0 & \frac{1}{\rho} \\ A & A\rho V & 0 & 0 \\ 0 & V & 1 & 0 \\ \left.\frac{\partial\rho}{\partial P}\right|_{is}-\frac{1}{T\rho}\left.\frac{\partial\rho}{\partial s}\right|_p & 0 & \frac{1}{T}\left.\frac{\partial\rho}{\partial s}\right|_p & -1 \end{bmatrix}\begin{bmatrix}\frac{dp}{dl}\\ \frac{dV}{dl}\\ \frac{dh}{dl}\\ \frac{d\rho}{dl}\end{bmatrix}=\begin{bmatrix}-\frac{1}{A}\frac{dA}{dl}\\ -f\rho\frac{V^2}{2}\frac{dF}{dl}\\ 0\\ 0\end{bmatrix} \tag{5}$$

Among them, the derivative of pressure, velocity, enthalpy and density are the matrix of the unknowns, $X=[dp/dl\ dV/dl\ dh/dl\ d\rho/dl]^T$. During the calculation of primary flow expansion, we firstly assumed the velocity of primary flow inlet $V_{p,in}$, and the inlet parameter including the primary flow pressure p_p, temperature T_p, inlet area $A_{n,in}$, throat area A_t and axis length L_n of the motive nozzle were the known parameters. The partial derivatives of $\partial p/\partial s|_p$ and $\partial p/\partial p|_s$, According to the state parameters and geometric structure of each element, the changeable cross-sectional area along the flow path dA/dl and the derivative of lateral surface area dF/dl were obtained. Owing to the expansion of the flow in the nozzle, two-phase flow will occur and develop, so the sound velocity was dealt with each case on its merits referring to the literature [13]. In addition, the friction factor f was predicted by literature [14]. The system differential equations were resolved through the fourth order Runge-Kutta method, and the fluid pressure p_n, specific enthalpy h_n, velocity V_n and density ρ_n were obtained. When an element was completed, the next continued to compute until the nozzle throat position, and the Mach number was equal to 1.0 as the iteration condition.

For the calculation in the suction chamber of CO_2 ejector, which was similar that in the motive nozzle. Herein, the expansion angle of primary flow at the motive nozzle outlet γ_n was assumed. The inlet parameter of the secondary flow including pressure p_s, temperature T_s, inlet area $A_{s,in}$, velocity $V_{s,in}$ as well as the outlet parameters of motive nozzle were the known parameters. The exit position (NXP) of the motive nozzle with respect to the mixing chamber was utilized to calculated the cross area of two streams along the passage. For this paper, we discussed the performance parameters and flow distribution when the position of two streams reaching the constant pressure was at the inlet and inside of mixing chamber, respectively. To be more specific, the separation from the motive nozzle outlet position to the position of two streams reaching constant pressure L_s was 8.4mm and 9.4 mm, respectively. The former was equal to the value of NXP which meant the pressure equivalent position was at the inlet of mixing chamber while the later was larger than NXP and the position inside the mixing chamber. To combine the known structure parameters of suction chamber, the flow parameters are determined and then iterated the expansion angle γ_n. Then the entrainment ratio μ ($\mu=\dot{m}_s/\dot{m}_p$) of ejector performance parameter and the expansion coefficient of primary flow ($\psi=A_{p,mix}/A_n$) was obtained, where $A_{p,mix}$ was the cross area of primary flow occupied at the pressure equivalent position for the two streams. Figure 1 shows the iteration of the primary flow and secondary flow in the suction chamber.

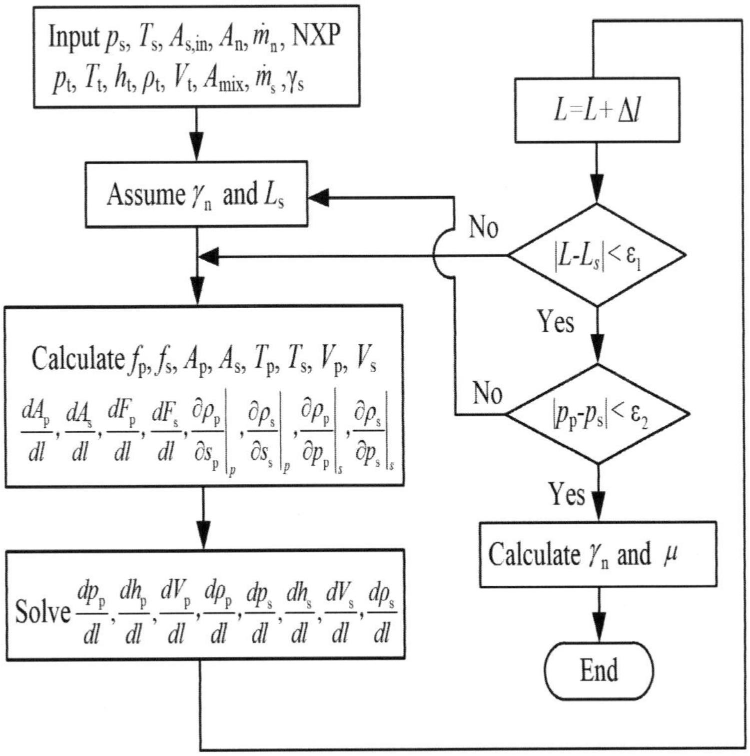

Figure 1. Flowchart of the calculation in the suction chamber.

2.2. Double-Flow (Mixing chamber, Diffuser)

The molding of mixing chamber and diffuser are created by using a double-flow model formulation. The coaxial double-fluid flow was formed through the primary flow (α) and secondary flow (β) , where the primary flowed in the middle and secondary flow in a loop. The equations of mass, momentum as good as energy for mixing chamber and diffuser were similar, while the geometric relationship was different. The detailed governing equations and the variable parameter could refer to the previous literature [8].

The nine-order system variation was obtained: $X=\left[\frac{dp}{dl}\ \frac{dV_\alpha}{dl}\ \frac{dV_\beta}{dl}\ \frac{dh_\alpha}{dl}\ \frac{dh_\beta}{dl}\ \frac{d\rho_\alpha}{dl}\ \frac{d\rho_\beta}{dl}\ \frac{dA_\alpha}{dl}\ \frac{dA_\beta}{dl}\right]^T$.

The inlet conditions were the results of previous calculations, which included pressure p, velocities V_α, V_β, specific enthalpies h_α, h_β, densities ρ_α, ρ_β, and the cross-sectional area A_α, A_β of the flow passage. Figure 2 shows the calculation procedure of the mixing chamber.

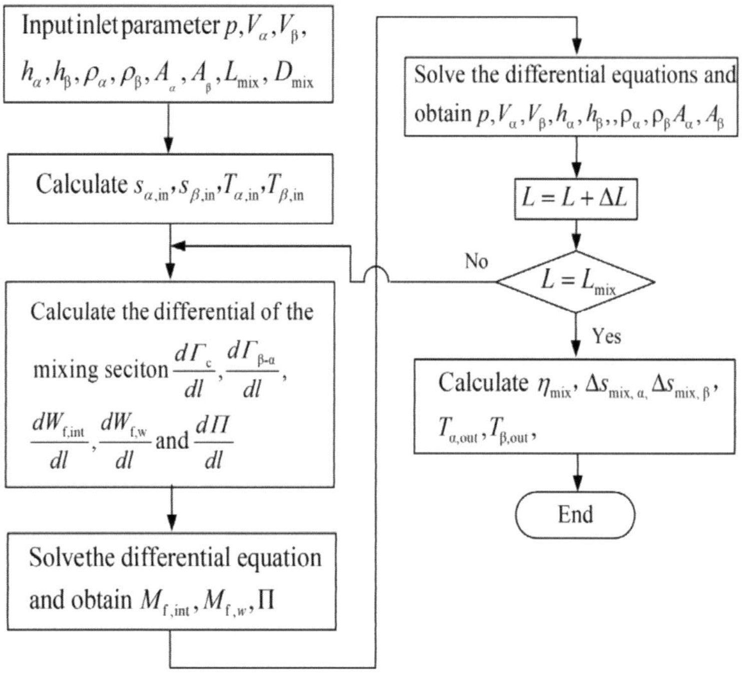

Figure 2. Flowchart of the calculation of the mixing chamber.

The main losses in the mixing chamber were from the influence of friction factors and the change in the transfer of momentum. Therefore, the mixing chamber efficiency η_{mix} was proposed for the loss, which involves the monument transfer inefficiency of primary flow and secondary flow, the friction impact at the duct wall and the secondary flow as well as the imaginary mixing layer between two streams of fluid. Therein, we referred to the expression proposed by the literature [8]. In addition, due to the above mix efficiency describing the monument conservation, lacking the representation of energy loss, while the thermodynamic entropy change more effective to represent the irreversibility generated in the fluid core. Thus the entropy change in the mixing chamber was also calculated:

$$\Delta S_{mix} = \dot{m}_p \left(s_{p,ou} - s_{p,in}\right) + \dot{m}_s \left(s_{s,ou} - s_{s,in}\right) \tag{6}$$

After calculation of the mixing chamber, the similar procedure was applied to the diffuser and the pressure lift ratio ($\lambda = p_{dif}/p_s$) could be obtained.

3. Simulation Conditions

3.1. CO_2 Ejector Structure

In this paper, the ejector structure referred to the previous published article [8], which proposed an adjustable ejector and carried out the experiment test to analyze the ejector performance. Herein on this basis the simulation study was conducted. The basic structure and dimensions of the ejector used in this model are shown in figure 3, and the corresponding size values of the ejector are listed in table 1.

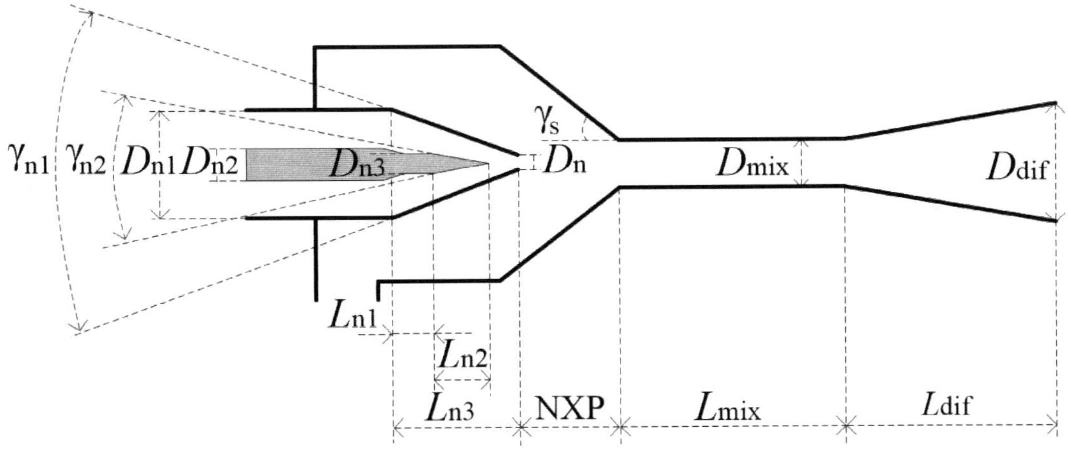

Figure 3. Basic dimensions for the ejector geometry.

Table 1. Ejector geometry configurations (the dimension notation referenced to figure 3).

Signature	Value	Signature	Value
D_n, mm	1.8	γ_{n1}, °	50.0
D_{n1}, mm	12.0	γ_{n2}, °	20.0
D_{n2}, mm	5.0	D_{mix}, mm	2.5
D_{n3}, mm	2.1	L_{mix}, mm	25.0
L_{n1}, mm	3.8	D_{dif}, mm	10.0
L_{n2}, mm	6.4	NXP, mm	8.4
L_{n3}, mm	10.5	γ_s, °	40.0

3.2. Working Conditions

This paper studies the ejector performance under various working conditions. The primary flow parameters: pressure p_p are 9.0 MPa, 10.0 MPa, and 11.0 MPa respectively; temperature T_p is equal to 313.5 K. The secondary flow parameters: pressure p_s increases from 3.2 MPa to 3.9 MPa with an interval of 0.1 MPa; temperature $T_s=T_{sat}(p_s)+5.0$ K. The diameter of the nozzle throat is 0.546 mm. In order to compare the difference of mixing performance between the expansion distance L_s equaling to NXP(8.4mm) and L_s larger than NXP, the simulation calculates were performed. When the two streams reached the constant pressure inside the mixing chamber, the position was located at the one fifth of mixing chamber length, namely L_s was equal to 9.4mm. In addition, during the formulation mentioned in section 1.2, the friction factor ε referenced for the object of two-phase flow was 0.2 for mixing section and diffuser, respectively. The scale coefficient a in the interface drag coefficient presented in the differential of the interface momentum transfer rate, which was 0.5. The flow process of primary flow and secondary flow is always accompanied by the mutual transfer of heat, and the heat transfer coefficient between the two flow was referred to literature by Hwang [15].

4. Results and Discussions

Figure 4 indicates how the entrainment ratio μ changes as the secondary flow inlet available pressure p_s under different primary pressures p_p corresponding to the two expansion distances. It can be seen from the curve variation trend in this figure, as the secondary flow pressure of the ejector increase, the entrainment ratio μ also gradually increases; the results show that, under the same working conditions, the entrainment ratio μ corresponding to the larger expansion distance is also larger, which means that the fluid is the expansion inside the mixing chamber is larger than the entrainment ratio μ produced by the expansion at the inlet; the magnitude of the primary flow pressure will also produce the same result. This is because an increase in the expansion distance will cause the expansion coefficient of the primary flow to increase, which in turn increases the secondary flow velocity. Even if the cross-sectional area occupied by the secondary flow in the ejector is gradually reduced, the velocity increases faster under this working condition, which ultimately leads to an increase in the entrainment ratio μ.

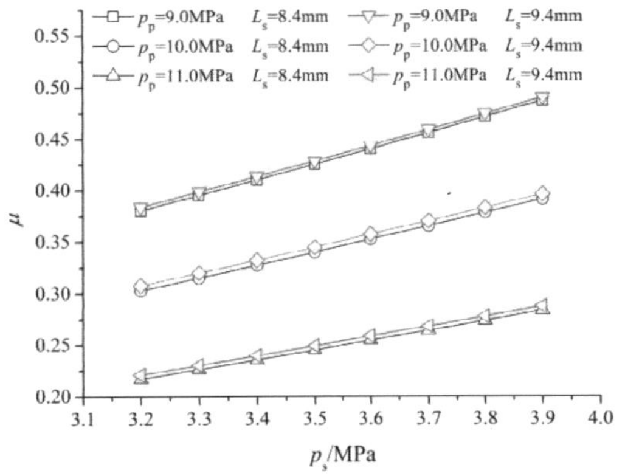

Figure 4. μ vs p_s for different p_p and L_s.

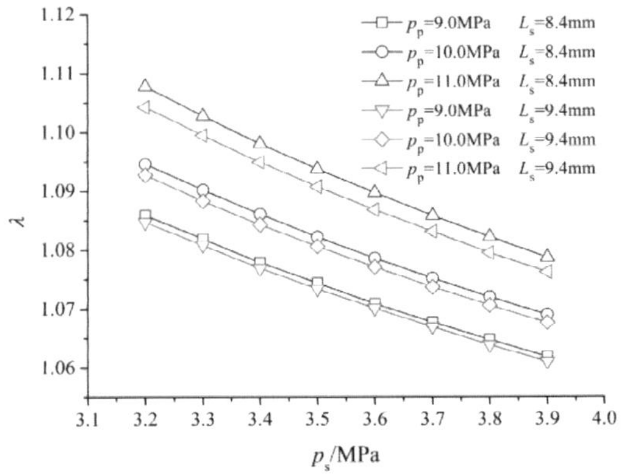

Figure 5. λ vs p_s for different p_p and L_s.

From figure 5, we can see the relationship between the pressure lift ratio λ and the secondary flow available pressure, and at the same time, it can also reflect the corresponding pressure lift ratio λ relationship between the expansion distance and the primary flow change of pressure level. The results show that the pressure lift ratio λ decreases as the fluid flows increase of secondary flow pressure. It could be found that when the expansion distance of primary flow L_s was larger than NXP (8.4 mm), λ decreased and it was evident when the p_p was higher value. This is mainly because higher secondary flow pressure is more conducive to carrying more mass secondary flow, and for the primary flow, the higher the pressure, the more energy the primary flow has, and the more mass secondary flow can be sucked up. Expansion in the mixing chamber is beneficial to improve the ejector coefficient.

Figure 6 and figure 7 show the simulation results of variation of efficiency and entropy change respectively in the mixing chamber. As we can see, with the build up secondary flow inlet pressure p_s, the mixing efficiency increases gradually, while the entropy change shows the opposite change. The mixing efficiency of L_s equal to 9.4 mm is obviously higher than that of L_s equal to 8.4 mm, this is because the friction and momentum transfer between primary flow and secondary flow is less than L_s equal to 8.4 mm, which leads to the increase of mixing efficiency, but for entropy change, the simulation result of expansion distance of 8.4 mm is higher than that of 9.4 mm. When the expansion distance is 8.4 mm, under the same secondary flow inlet pressure, the mixing efficiency with low primary flow pressure is relatively higher, but the entropy change is opposite. On the whole, the expansion distance can affect the efficiency of the mixing chamber in the ejector, and the efficiency is also closely related to the pressure of the two fluids.

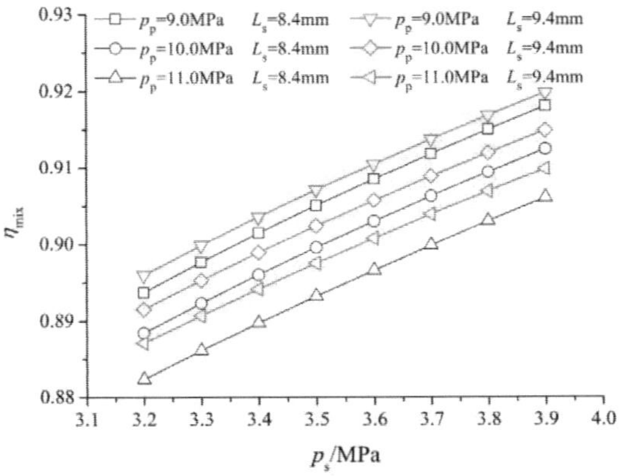

Figure 6. η_{mix} vs p_s for different p_p and L_s.

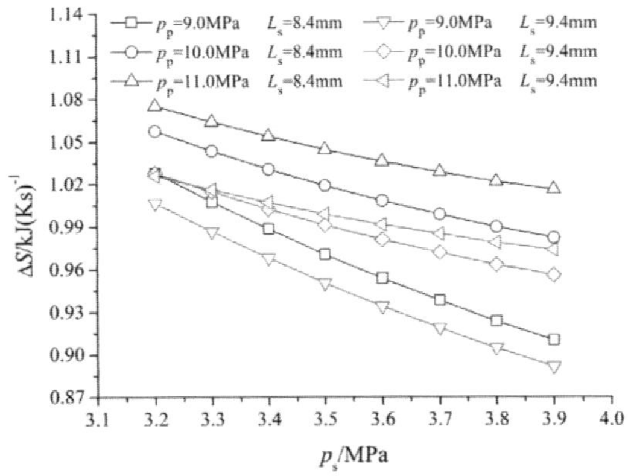

Figure 7. $\triangle S$ vs p_s for different p_p and L_s.

Figure 8 and figure 9 show the temperature and velocity distributions of the two streams of fluid in the mixing chamber when the evaporation pressure is 3.9 MPa. Among them, the parameters of the primary flow were represented by solid lines, and the parameters of the secondary flow were represented by dashed lines. It can be seen from figure 8 that with the extension of the flow path, the temperature of the primary flow increases and the temperature of the secondary flow decreases, but the temperatures of the two flows are finally close. When the primary flow pressure was low, the temperature of the leading secondary flow decreases rapidly at the beginning of mixing. As can be seen in figure 9, along the flow path, the primary flow velocity gradually decreases while the secondary flow velocity increases, which is mainly due to the mass and energy transfer between the two streams.

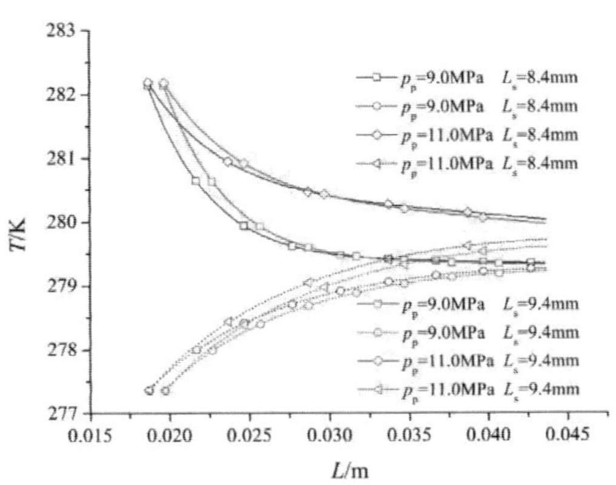

Figure 8. Temperature distribution along the axis of mixing chamber (p_e=3.9MPa).

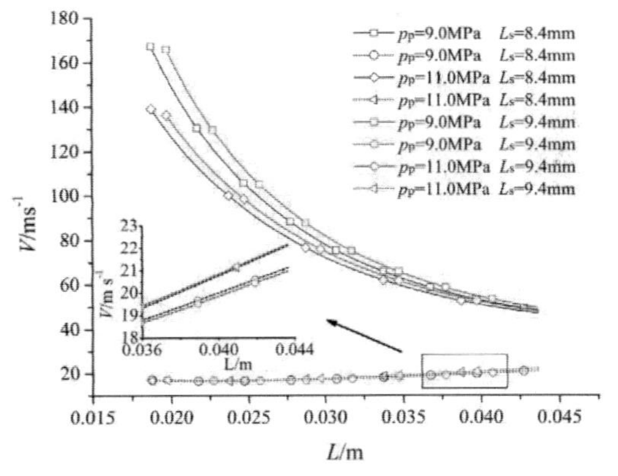

Figure 9. Velocity distribution along the axis of mixing chamber (p_e=3.9MPa).

5. Conclusions

The effect of primary flow expansion on mixing performance in the suction chamber of CO_2 two-phase ejector was studied in this study. The ejector model was established by using the distributed parameter method, especially for the suction chamber, the expansion of the primary flow and the compression of the secondary flow are established by using the differential equations of mass, momentum and energy conversion. The conditions under which the two kinds of flow reach equal pressure at the inlet of the mixing chamber and inside the mixing chamber are analyzed, and their effects on the performance differences of the ejector under different working conditions are compared.

When the primary flow and secondary flow reached the same pressure inside the mixing chamber, the entrainment ratio μ was larger than that when the pressure equivalent position of two streams was at the mixing chamber inlet, while the pressure lift ratio λ presented inverse trend. Moreover, the efficiency of mixing chamber η_{mix} for expansion distance L_s larger than NXP was larger than that when L_s was equal to NXP. Due to the transfer of mass, momentum and energy, these irreversible losses lead to entropy change in the mixing chamber, and for L_s larger than NXP, it was less than that for L_s equaling to NXP. This study helps to understand the flow distribution inside the ejector and to know the efficient design of the ejector When modeling the suction chamber, the distributed parameter method was used, during which the interaction between the two fluids in the flow process, the non-isentropic oblique wave and the shock wave were ignored, so the model should continue to be improved in the future research.

Acknowledgments
This work was supported by the National Natural Science Foundation of China (Grant No. 51806132), Application Basis Research Project Foundation of Shanxi Province (Grant No. 201801D221353) and Scientific and Technological Innovation of colleges and universities of Shanxi Province (Grant No. 201802011).

Reference

[1] Liu Y X, Zhao Y Y, Yang Q C, Liu G B and Li L S 2020 *Case Studies in Thermal Engineering* **21**
[2] Dai B M, Zhao P, Liu S C, Su M Q, Zhong D, Qian J B, Hu X W and Hao Y 2020 *Energy Conversion and Management* **222**.
[3] Taleghani S T, Sorin M, Poncet S and Nesreddine H 2019 *Energy Conversion and Management* **185** 442–54
[4] Haida M, Smolka J, Hafner A, Ostrowski Z, Palacz M, Madsen K B, F€orsterling S, Nowak A J and Banasiak K 2018 *Energy* **153** 933-948
[5] Haida M, Smolka J, Hafner A, Palacz M, Banasiak K and Nowak A J 2018 *International Journal of Refrigeration* **85** 314-33
[6] Zhu Y H,Wang Z C,Yang Y P and Jiang P X 2017 *International Journal of Refrigeration* **74** 354-61
[7] Li Y F, Deng J Q, Ma L and Zhang Y Z 2018 *Energy Conversion and Management* **71** 729-41
[8] Zheng L X and Deng J Q 2017 *Energy Conversion and Management* **142** 244-56
[9] Wang X and Yu J L 2016 *International Journal of Refrigeration* **71** 26-38
[10] Taleghani S T, Sorin M and Poncet S 2018 *International Journal of Refrigeration* **87** 91-105
[11] Zhu Y H and Jiang P X 2018 *International Journal of Refrigeration* **86** 218-27
[12] Banasiak K and Hafner A 2011 *International Journal of Thermal Sciences* **50** 2235-47
[13] Cardemil M J and Colle S 2012 *Energy Conversion and Management* **64** 79-86
[14] Revellin R and Thome J R 2007 *Experimental Thermal and Fluid Science* **31** 673-85
[15] Hwang Y and Rademacher R 1998 *HVAC&R Research* **4** 245-63

Journal of Physics: Conference Series **2097** (2021) 012013 doi:10.1088/1742-6596/2097/1/012013

Analysis of Characteristics of Damping Valve Based on High-Order Discrete Scheme

Shichao Pan* **and Weimin Li**

Liaoning University of Technology School of Mechanical Engineering and Automation, Jinzhou, 121001, Liaoning, China.
Email: panshichao1025@163.com

Abstract. In this paper, based on the high-order discrete scheme, a two-way fluid-solid coupling numerical simulation is for the damping valve plate. According to the discrete method, the governing equations of fluid structure coupling of damping valve plate are studied, including the basic conservation laws; Meanwhile, it analyzes the discretization of the control equation, including the discretization method and the high-order discretization format when the finite volume method is adopted. And based on this discrete format, a numerical simulation was performed on the damping valve, the oil flow condition is analyzed, and the velocity of the throttling hole at different time points and the throttling pressure are analyzed.
Keywords. High-order discrete scheme, fluid-structure interaction, throttle slice.

1. Introduction

Damper damper plate is the most important throttling element in throttle valve system. It plays a decisive role in regulating the flow rate and direction of oil, then regulating the throttling pressure and controlling the damping force. Chen y et al. [1] Explored the equivalent thickness splitting theory of superimposed throttle disc, and carried out numerical simulation analysis based on flow theory; Based on the fluid resistance theory; Lian Z et al. [2] Analyzed the throttle disc under different disc opening, and obtained the optimized wedge throttle disc; Tezduyar T.E et al. [3] carried out fluid structure coupling analysis based on stable space-time fluid structure coupling (SSTFSI) technology, which improved the convergence of fluid structure coupling numerical simulation. In the process of two-way fluid structure coupling, the discretization of the calculation region is its core content. The discretization results have a decisive impact on the generation of dynamic structural grids, and the grid quality has an important impact on the convergence of numerical simulation results. Therefore, this paper analyzes the discretization method and discretization scheme of the calculation region, and uses the high-order discretization scheme to analyze the characteristics of the damping valve plate.

2. Two-Way Fluid-Structure Coupling Numerical Simulation Control Equation

The flow of fluid follows the laws of conservation of physics [4, 5]. The basic laws of conservation include the law of conservation of mass, momentum and energy. The governing equation is a mathematical description of the law of conservation.

2.1. Mass Conservation Equation

The outflow in the mass of the fluid micro-element in a unit time is equal to the mass of the flowing into the same time. The mass conservation equation can be derived from the law of mass conservation:

$$\frac{\partial \rho}{\partial t}+\frac{\partial \rho u}{\partial x}+\frac{\partial \rho v}{\partial y}+\frac{\partial \rho w}{\partial z}=0 \tag{1}$$

where, ρ is the density of hydraulic oil; t is the unit time; u is the vector of speed.

2.2. Momentum Conservation Equation

The rate of change of momentum of a micro element with unit time is equal to the sum of all external forces acting on the micro element, and the momentum conservation equation about *x,y,z* three directions can be written:

$$\begin{aligned}
\frac{\partial \rho u}{\partial t}+div(\rho u\boldsymbol{u}) &= -\frac{\partial p}{\partial x}+\frac{\partial \tau_{xx}}{\partial x}+\frac{\partial \tau_{yx}}{\partial y}\frac{\partial \tau_{zx}}{\partial z}+F_x \\
\frac{\partial \rho v}{\partial t}+div(\rho v\boldsymbol{u}) &= -\frac{\partial p}{\partial y}+\frac{\partial \tau_{xy}}{\partial x}+\frac{\partial \tau_{yy}}{\partial y}\frac{\partial \tau_{zy}}{\partial z}+F_y \\
\frac{\partial \rho w}{\partial t}+div(\rho w\boldsymbol{u}) &= -\frac{\partial p}{\partial z}+\frac{\partial \tau_{xz}}{\partial x}+\frac{\partial \tau_{yz}}{\partial y}\frac{\partial \tau_{zz}}{\partial z}+F_z
\end{aligned} \tag{2}$$

where p is the pressure, $\tau_{xx},\tau_{xy},\tau_{xz}$ is the viscous stress on the surface of the infinitesimal body. F_x,F_y,F_z is the volume force. When the volume force has only gravity and only acts in the z-axis direction then $F_x=0,F_y=0,F_z=-\rho g$.

2.3. Energy Conservation Equation

The law of energy conservation is the basic law of energy exchange in flow system. The theory is that the energy growth rate in a micro element is equal to its net thermal efficiency plus the work done by its volume force and area force, which can be written as:

$$\frac{\partial(\rho T)}{\partial t}+div(\rho uT)=div(\frac{k}{c_p}\cdot gradT)+S_T \tag{3}$$

where, c_p is the specific heat capacity, T is the temperature, k is the thermal conductivity, S_T It is the heat generated inside the fluid and the heat energy converted by the mechanical energy of the fluid due to the viscous effect.

3. Discretization of Governing Equations

3.1. Discretization Method

There are three commonly used discretization methods in bidirectional fluid structure coupling analysis. The finite difference method divides the solution area into grid cells, uses a finite number of grid nodes instead of the solution domain, and then derives a system of difference equations containing discrete points. In the finite element method, the solution domain is arbitrarily divided into several small elements, and the interpolation function is constructed by using the extreme value principle to transform it into a finite element equation. The finite volume method divides the calculation area into structural grids, and there is a control volume integral near each structural grid node to obtain the control equation. Using the discretized control equation obtained by the finite volume method, the characteristic variables in the finite control volume have accurate integral conservation.

3.2. High-Order Discrete Scheme

When the finite volume method is used to establish the discrete equation, the field variable and its derivative value at the control volume interface can be obtained by interpolation through the corresponding value on the node. The purpose of introducing the interpolation operation is to establish a discrete equation. The interpolation method is different, and the obtained results are also different, so the interpolation method is also called discrete format. In order to improve the calculation accuracy, more relevant nodes are considered when calculating the parameter value of the control volume interface, and a higher-order interpolation formula is used for calculation, which is a higher-order discrete format.

3.2.1. Second-Order Welcome Style. In the second-order upwind style, the field variables on the control volume interface are determined by the upstream node value and the upstream far neighbor node value. The schematic diagram of the second-order upwind style is shown in figure 1.

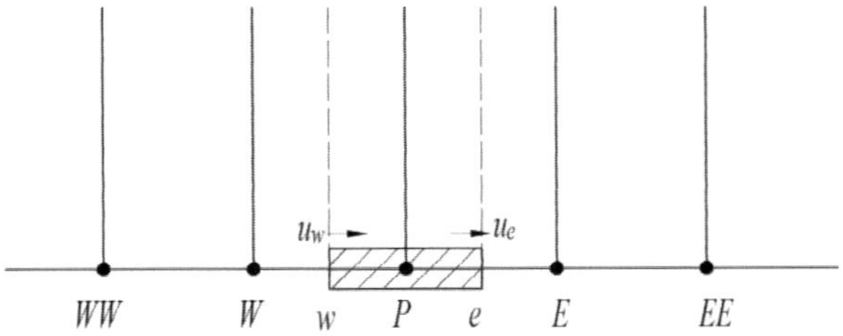

Figure 1. Schematic diagram of the second-order welcome style.

The second-order upwind formula has second-order calculation accuracy. The discrete equation includes not only the field variables at adjacent nodes, but also the field variables at other nodes at adjacent nodes. The discrete equation is normalized by the field variable coefficients as follows:

$$a_P\varphi_P = a_W\varphi_W + a_{WW}\varphi_{WW} + a_E\varphi_E + a_{EE}\varphi_{EE} \tag{4}$$

where, $a_W = D_w + \frac{3}{2}\alpha F_w + \frac{1}{2}\alpha F_e, a_{WW} = -\frac{1}{2}\alpha F_w, a_E = D_e - \frac{3}{2}(1-\alpha)F_e - \frac{1}{2}(1-\alpha)F_w$

$a_{EE} = \frac{1}{2}(1-\alpha)F_e, a_P = a_W + a_E + a_{WW} + a_{EE} + (F_e - F_w)$.

3.2.2. QUICK Format. The QUICK format means the second-order upwind interpolation of the convection term, and it is used to calculate the second-order interpolation calculation format of the control volume interface value. The QUICK format introduces correction coefficients on the basis of linear interpolation, and uses the three node values on both sides of the control volume boundary to perform interpolation calculation. And two nodes are located on both sides of the interface, and the other node is the distant neighbor node on the windward side. After normalizing the field variable coefficients, the discrete equation is:

$$a_P\varphi_P = a_W\varphi_W + a_{WW}\varphi_{WW} + a_E\varphi_E + a_{EE}\varphi_{EE} \tag{5}$$

where, $a_W = D_w + \frac{6}{8}\alpha F_w + \frac{1}{8}\alpha_e F_e + \frac{3}{8}(1-\alpha_w)F_w$, $a_{WW} = -\frac{1}{8}\alpha F_w$

$a_E = D_e - \frac{3}{8}\alpha_e F_e - \frac{6}{8}(1-\alpha_e)F_e - \frac{1}{8}(1-\alpha_w)F_w$, $a_{EE} = \frac{1}{8}(1-\alpha_e)F_e$

$a_P = a_W + a_E + a_{WW} + a_{EE} + (F_e - F_w)$.

4. Fluid-Solid Coupling Analysis of Damping Valve

4.1. Modeling Assumptions and Simplification
The amount of calculation for the fluid-structure coupling numerical analysis of the damping valve is huge. For this reason, assumptions and simplifications are made on the basis of ensuring the reliable analysis results.

(1) The influence of temperature on the density of the oil should be ignored.

(2) When the oil flows through the orifice at high speed, the heat generated by friction should be ignored.

(3) The oil is assumed to be a completely incompressible liquid.

4.2. Build Fluid Model
According to the recovery valve, the numerical model of fluid field is established, as shown in figure 2. The fluid model includes an inlet, an outlet, a fluid-solid coupling surface for bidirectional data exchange, and a smooth wall without slippage. The fluid-solid coupling surface is defined as the contact point between the liquid and the damping valve plate.

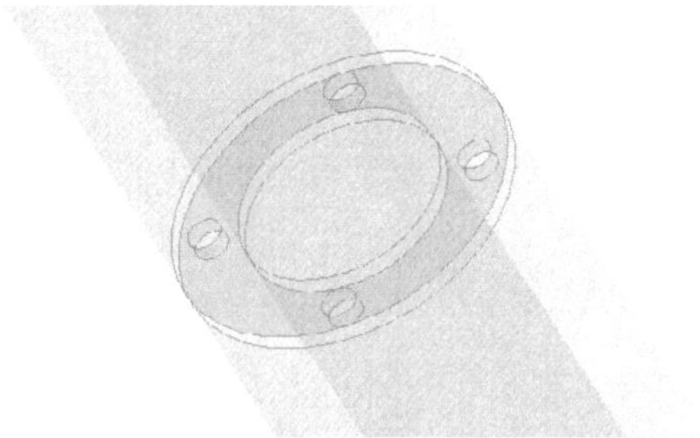

Figure 2. Fluid model.

4.3. Discretization of Fluid Model of Damping Valve System
The fluid model of the damping valve system needs to be discretized. In the process of two-way fluid structure coupling numerical simulation, the damping valve plate produces large deformation. In order to improve the accuracy of numerical simulation and further enhance the convergence of the calculation process, the damping valve plate is meshed and encrypted in the Mesh module, the encryption coefficient is set to 3 times encryption, and the fluid model grid is established, as shown in figure 3.

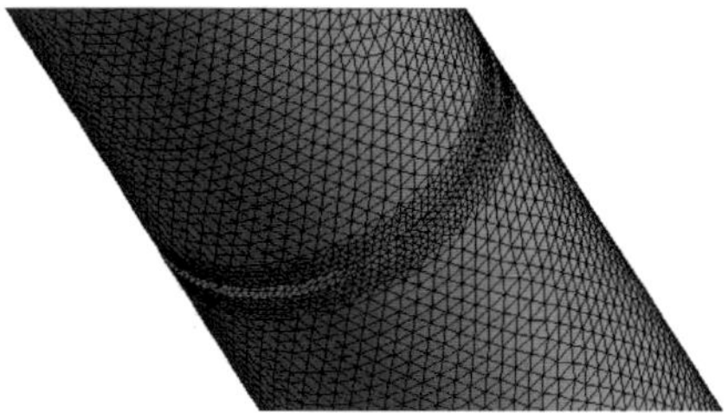

Figure 3. Discretization model of fluid.

Check the quality of the established mesh. In the fluid structure coupling analysis, the element quality of the grid has a decisive impact on the numerical simulation results, which is related to the convergence of the numerical simulation results and the reliability of the convergence results [6, 7]. The warpage of single mesh and the orthogonality of discrete mesh are the most important. The mesh distortion mass of the fluid model is shown in figure 4, and the mesh orthogonal mass is shown in figure 5. Warpage is one of the basic component quality inspection standards. The optimum value of warpage is zero. As can be seen from the grid warpage quality diagram in figure 4, most of the grid distortion quality of the fluid model is less than 0.3, and the grid quality is very good. For the fluid structure coupling numerical simulation analysis, the orthogonal quality of the grid should be greater than 0.1. It can be seen from figure 5 that more than 98% of the grid elements have orthogonal mass values greater than 0.1, which meets the requirements of grid warpage and grid orthogonality in fluid structure coupling numerical simulation.

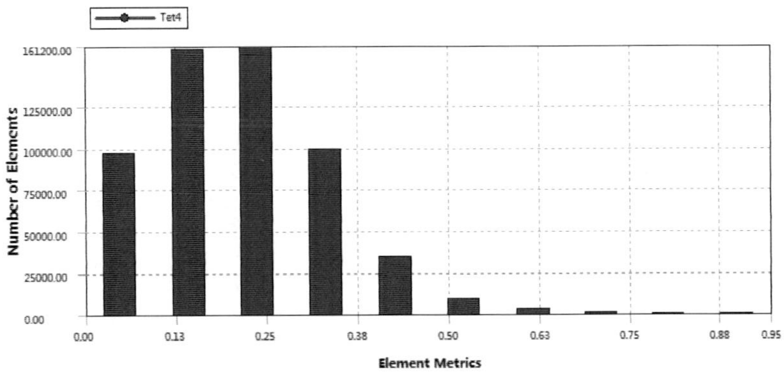

Figure 4. Mesh warpage quality.

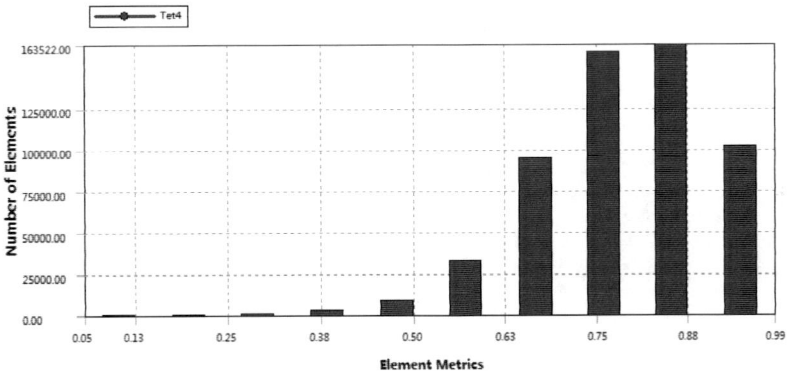

Figure 5. Orthogonal quality of the grid.

4.4. Materials and Boundary Conditions

The damping characteristics of the damping valve are analyzed in the FLUENT module of ANSYS Workbench. In this paper, the density of the oil used is 890kg/m^3, the kinematic viscosity is 12.2mm^2/s, and the viscosity index is 202 mm^2/s. The boundary condition at the inlet is 0.5m/s*t, which is set as the inlet flow velocity of oil, and the inlet velocity increases with time. The oil flow state at the outlet is set to outflow, and the liquid at the outlet is free-flowing.

4.5. Solution Strategy

In the FLUENT module, a density-based solver is used to define the solution strategy, and the solution state is defined as transient [8]. In order to accelerate the convergence of fluid structure coupling numerical simulation analysis, the relaxation factor in the iterative process is set to 0.35, the total solution time is defined as 2s, the time step is 0.05, and the total solution steps are 40.

4.6. Analysis of Numerical Simulation Results

After solving the two-way fluid-solid coupling numerical simulation, it converges after 100 iterations. The iteration convergence curve is shown in figure 6, and the force convergence curve area is shown in figure 7.

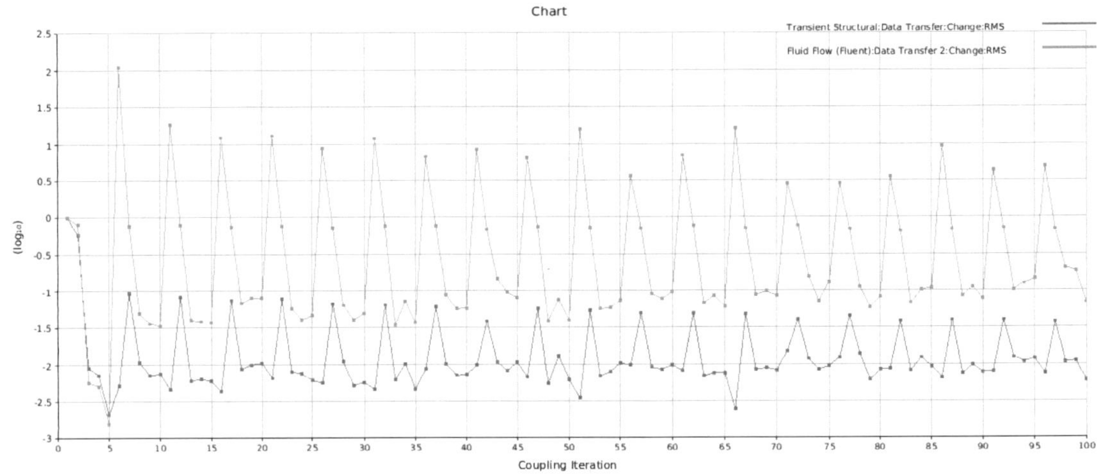

Figure 6. Iterative convergence curve.

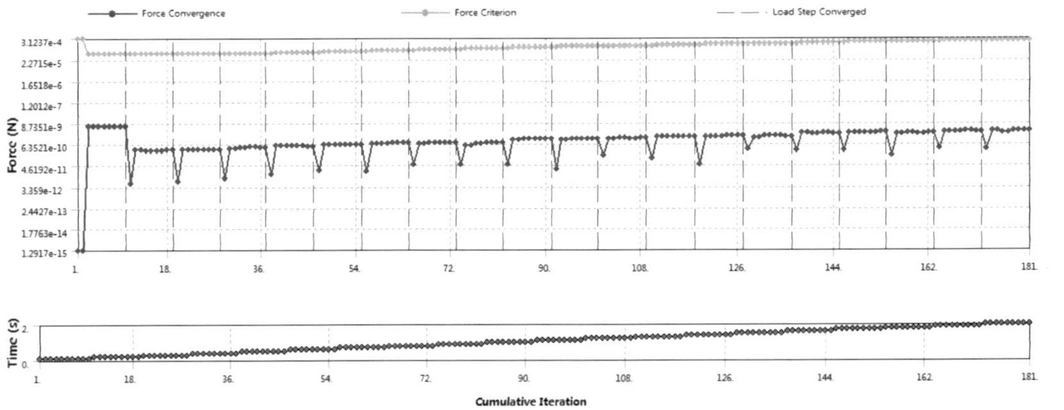

Figure 7. Force convergence curve.

After the two-way fluid-structure coupling numerical simulation converges, the streamline diagram is shown in figure 8. From the streamline diagram, it can be seen that the speed of the oil changes suddenly when it flows through the orifice of the damping valve plate.

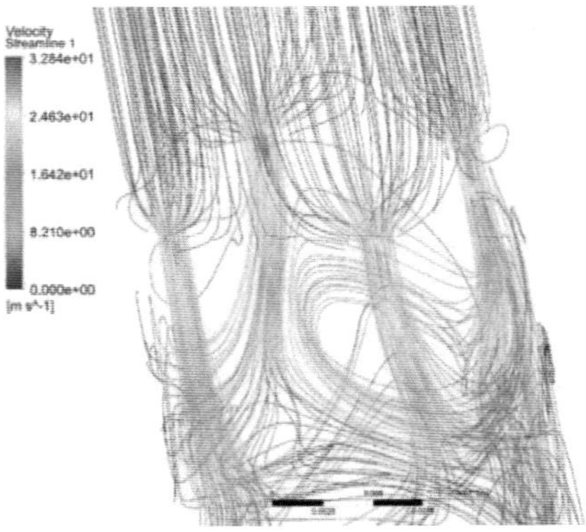

Figure 8 Velocity streamline diagram

The flow velocity of the oil flowing through the damping valve is analyzed. The flow velocity of the oil at 0.3s, 1.0s and 2.0s is shown in figure 9.

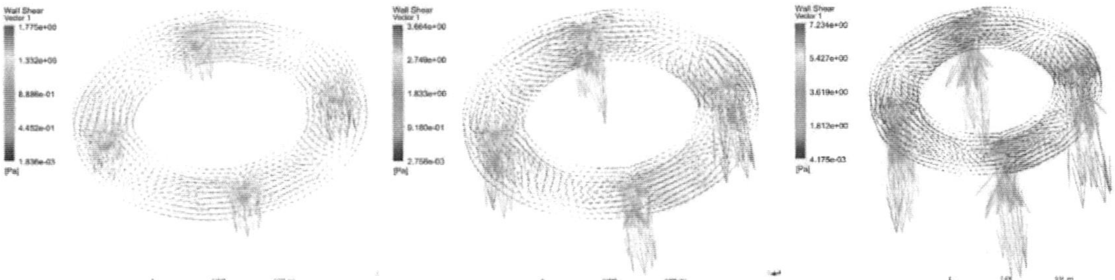

(a) (b) (c)

Figure 9. The velocity at (a) 0.3 s, (b) 1.0 s, (c) 2.0 s.

It can be seen from the figure that when the oil flows through the orifice of the damping valve plate, the flow rate changes suddenly, and the flow area at the orifice has a sudden change, resulting in the throttle pressure, and with the increase of time, the oil velocity at the inlet is also keep increasing. And the maximum speed point is constantly changing, the maximum speed is 7.775m/s in 0.3s, the maximum speed is 3.664m/s in 1.0s, and the maximum speed is 7.234m/s in 2.0s. Different throttling pressures are produced due to changes in speed points. The throttling pressures at different time points are shown in figure 10.

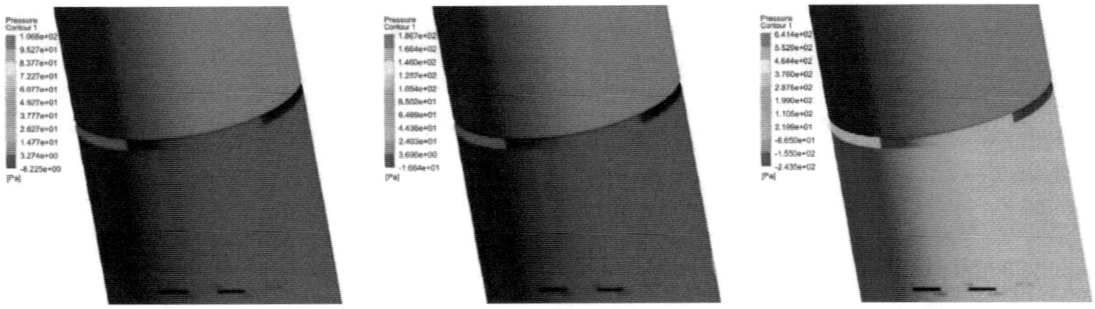

(a) (b) (c)

Figure 10. The pressure at (a) 0.3 s, (b) 1.0 s, (c) 2.0 s.

5. Conclusion

Based on the numerical simulation of two-way fluid-structure coupling, the paper conducts an in-depth study on the discretization of its core content calculation area. On the basis of discretization method, the finite difference method, finite element method and finite volume method are studied, and the most widely used finite volume method is fully analyzed, and the method of discretization scheme of finite volume method is proposed. Based on the method, in order to explore the flow state of oil and the throttling pressure, the two-way fluid structure coupling numerical simulation is performed on the damping valve, and it analyzes the oil flow path, the velocity characteristics of the orifice at different time points and the throttling pressure produced.

Acknowledgments

Scientific research fund of education department of LiaoNing province: Research on Lightweight of Automotive Components Based on Multidisciplinary Optimization Design (JZL202015401).

References

[1] Chen Y, Han X, Xiang L, et al. 2020 Research on equivalent thickness of oil and gas spring damping valve *IOP Conference Series: Materials Science and Engineering* **926** 012007.
[2] Lian Z, Meng Y, Gan L, et al. 2004 Fluid field analysis of high pressure throttle valve and it's structure improvement *China Petroleum Machinery*.
[3] Tezduyar T E, Sathe S, Schwaab M, et al. 2010 Arterial fluid mechanics modeling with the stabilized space–time fluid–structure interaction technique *International Journal for Numerical Methods in Fluids* **57**(5) 601-629.
[4] Bazilevs Y, Takizawa K and Tezduyar T E 2013 *Computational Fluid-Structure Interaction: Methods and Applications* (John Wiley & Sons Inc.).
[5] Antoci C, Gallati M and Sibilla S 2007 Numerical simulation of fluid-structure interaction by SPH *Computers & Structures* **85**(11-14) 879-890.
[6] Gupta M and Manohar R P 2015 A critique of a second-order upwind scheme for viscous flow problems *AIAA Journal* **16**(7) 759-761.
[7] Kurganov A, Prugger M and Wu T 2017 Second-order fully discrete central-upwind scheme for two-dimensional hyperbolic systems of conservation laws *SIAM Journal on Scientific Computing* **39**(3) A947-A965.
[8] Cheng H M 2021 A fully local hybridised second-order accurate scheme for advection-diffusion equations (in press) https://arxiv.org/abs/2103.08551.

Simulation Study on the Influence of Injector Coupling Leakage on Fuel Injection

Ziwei Zhang

China North Engine Research Institute, Tianjin 300400, China
Email: 751334315@qq.com

Abstract. In order to study the effect of fuel leakage of an ultra-high pressure common rail injector control valve coupling on fuel injection performance, a simulation model was established by AMESim and the accuracy was verified by fuel injection test data. The leakage law of couples with different clearances was analyzed by using numerical simulation method and then the influence of control valve coupling on fuel injection performance was analyzed. The results demonstrate that the increase of the matching clearance of the slide valve coupling makes the start time of needle valve advanced and delay its end time. The injection rate and injection duration increase with the increase of the matching clearance of slide valve coupling. The increase of the matching clearance of the control plunger coupling keeps the start time of the needle valve unchanged at first, and then delay slightly, while the end time remains unchanged at first, and then show the trend of advance. The injection rate and injection duration decrease with the increase of the matching clearance of plunger coupling.

Keywords. Fuel leakage, Common rail injector, Injection performance, Slide valve coupling, Plunger coupling.

1. Introduction

In order to reduce the fuel consumption and emission of diesel engine to a greater extend and improve its power performance, the high pressure common rail system should improve the performance of ultra-high pressure, injection precision and multi waveform injection controllability [1]. The fuel injector works in the harsh environment of high pressure, high temperature and vibration in multi working cycles. Unreasonable design of precision coupling can accelerate friction and wear, and even affect the sealing performance. The injector coupling has high matching accuracy with high motion speed. With the increase of the working cycle of the injector, the wear of coupling intensifies. Under the combined action of wear on the mating surface and fuel pressure in the matching clearance, the fuel leakage increases excessively [2]. Excessive leakage rate can affect the fuel injection performance, reduce fuel injection energy and increase the loss of driving power.

The research on fuel leakage of injector coupling can provide theoretical guidance for the design and performance optimization of an ultra-high pressure injector. This paper takes an ultra-high pressure injector as the research object, in which the control valve parts include control plunger coupling and slide valve coupling. The injection process was tested and simulated. The leakage characteristics of injector control plunger and slide valve coupling and the influence of leakage under different coupling clearance on injection performance were studied by numerical simulation.

2. Structure and Working Process of Fuel Injector

Figure 1(a) is a schematic diagram of the structure of the common rail injector, which mainly includes a solenoid valve part, a hydraulic servo part and a nozzle assembly. Figure 1(b) is a schematic view of the structure of the injector control valve assembly comprising the slide valve coupling and the control plunger coupling.

The working process of the fuel injector is described as follows: as the ECU sends an injection command, the solenoid valve coil is energized to make the electromagnetic valve lift, and the fuel in the control chamber leaks out through the small hole at the lower end of the slide valve and flows into the low-pressure cavity. The needle valve is raised by upward force, and high pressure fuel is ejected from the jet hole [3].

When the ECU sends fuel injection stop signal, the solenoid is powered off and the armature returns to its original position. The high pressure fuel flows into the pressure storage chamber through the small hole on the slide valve [4]. When the fuel in the pressure storage chamber exceeds a certain pressure, the slide valve falls under the action of hydraulic pressure and then the fuel injection cycle ends.

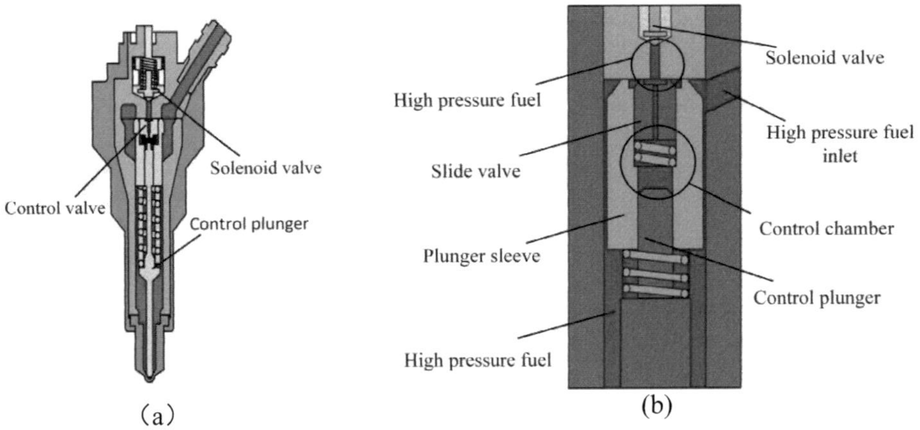

Figure 1. Structural diagram of common rail injector.

3. Simulation Model and Experimental Verification

According to the structure and working principle of high pressure injector, the simulation model of the injector is established by using software AMESim. Figure 2 shows the one dimensional simulation model, including electromagnetic force control module, fuel supply module, solenoid valve control module and fuel injection control module [5]. The parameters of simulation model are illustrated in table 1.

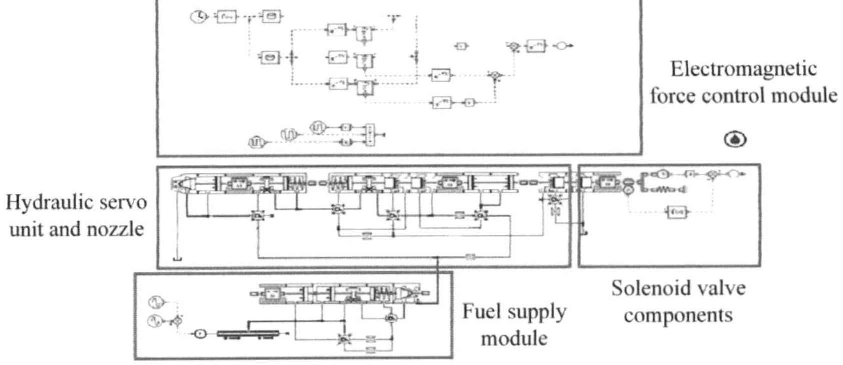

Figure 2. Simulation model of injector.

Table 1. Parameter setting of simulation model.

Parameter	Value
Common rail pressure	200 MPa
Cylinder pressure	4 MPa
Injection frequency	33 Hz

In order to confirm the validity of the model, the injection test of the injector was carried out. The single cycle fuel injection quantity was measured under the conditions of 140 MPa and 200 MPa rail pressure and injection pulse width of 0.6 ms, 0.7 ms, 0.8 ms, 0.9 ms and 1 ms respectively. Figure 3 shows the comparison between simulation results and test results. It can be seen from figure 3 that the error between the test results and the simulation results is less than 10%, and the error is within the engineering error range [6]. The simulation model can effectively simulate the injection process.

In order to confirm the validity of the model, the injection test of the injector was carried out. The single cycle fuel injection quantity was measured under the conditions of 140 MPa and 200 MPa rail pressure and injection pulse width of 0.6 ms, 0.7 ms, 0.8 ms, 0.9 ms and 1 ms respectively. Figure 3 shows the comparison between simulation results and test results. It can be seen from figure 3 that the error between the test results and the simulation results is less than 10%, and the error is within the engineering error range [6]. The simulation model can effectively simulate the injection process.

After the reliability of the model is verified, the fuel leakage of the control plunger and the slide valve with different coupling clearance, and the influence of fuel leakage on the hydraulic response and fuel injection performance of the injector are simulated and analyzed.

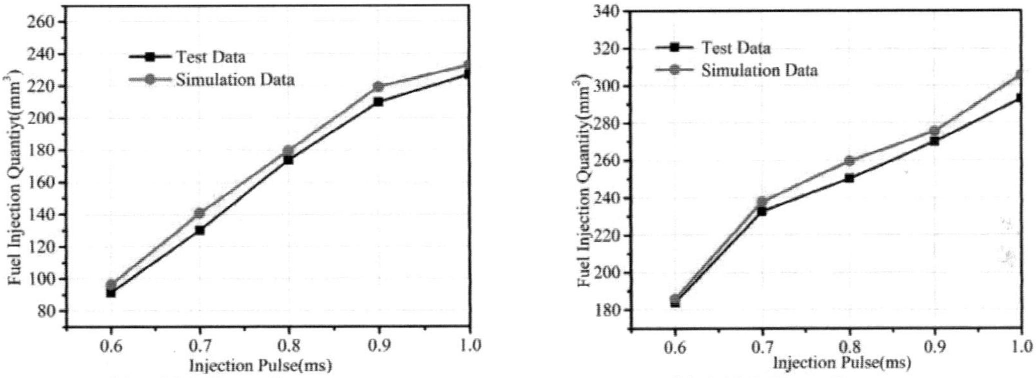

(a) 140 MPa rail pressure (b) 200 MPa rail pressure

Figure 3. Comparison of experimental and simulation results.

4. Results and Discussion

4.1. The Influence of Clearance on the Fuel Leakage

Figures 4 and 5 show the variation of flow rate of the plunger couples and slide valve couples with different coupling clearance. The range of coupling clearance is between 0.002 mm and 0.006 mm. It can be seen from the figure that the leakage law of coupling parts under different coupling clearances is similar, but the flow rate increases greatly with the increase of clearance.

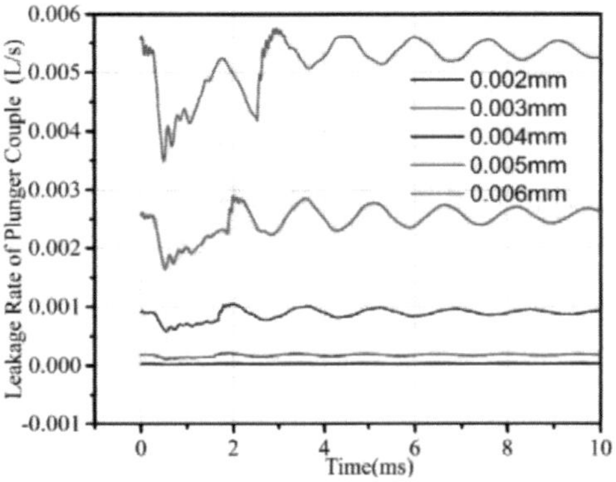

Figure 4. Leakage rate of plunger Coupling depending on clearances.

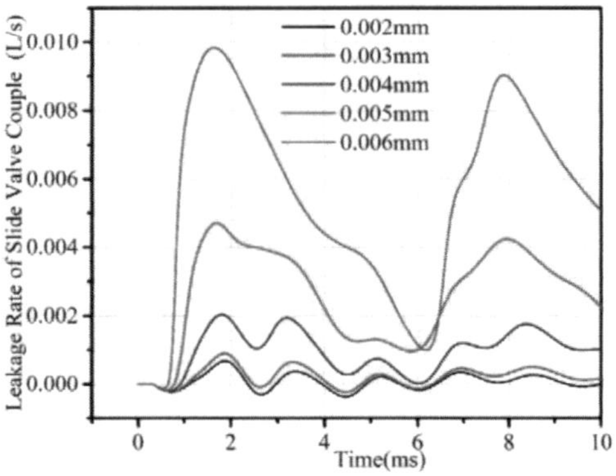

Figure 5. Leakage rate of slide valve coupling depending on clearances.

4.2. Effect of Slide Valve Coupling Clearance on Injection Performance

The response of components in the injector to the electromagnetic signal is realized through the transmission of hydraulic pressure. The movement of the needle valve is driven by the pressure difference between the needle valve chamber and the control chamber, and the opening and closing of the control slide valve is driven by the pressure difference between the control chamber and the annular chamber at the top of the slide valve. As mentioned above, the size of the injector precision coupling clearance can produce different leakage rate, and which affects the establishment of fuel pressure in control valve chamber. Thus, the response time of the injector components to the electromagnetic signal is changed [7].

As shown in figure 6, when the clearance of slide valve coupling changes from 0.002 mm to 0.022 mm, the start time of needle valve is decreased and the end time is longer, resulting in the increase of injection duration. As shown in figure 7, when the clearance of slide valve coupling changes from 0.002 mm to 0.022 mm, the fuel injection duration increases from 1.02 ms to 1.18 ms, with an increase of 15.68%.

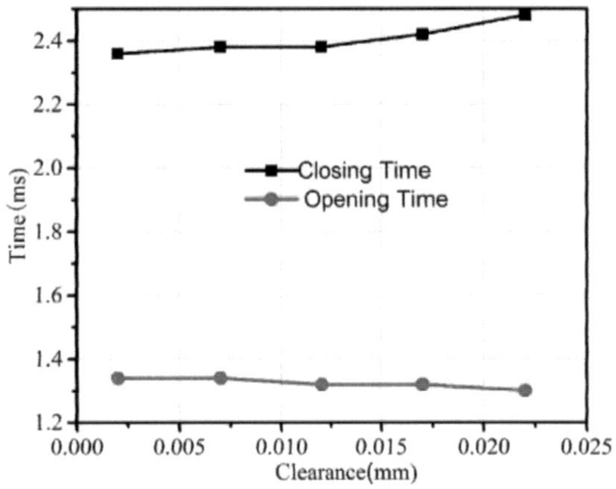

Figure 6. Start and end time of needle valve depending on slide valve coupling clearance.

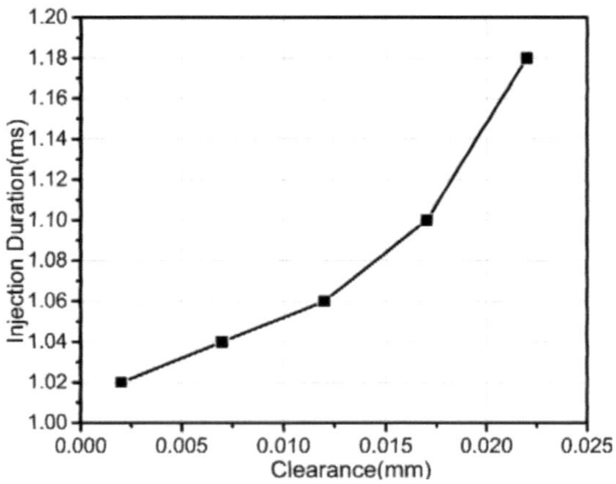

Figure 7. Fuel injection duration depending on slide valve coupling clearance.

The reason for this phenomenon is that when the solenoid valve is energized, the fuel in the control chamber flows out through the small hole at the lower end of the control slide valve and into the low-pressure chamber, and the needle valve is lifted under the action of upward force. The fuel leakage of the slide valve coupling accelerates the pressure drop in the control chamber, while the pressure fluctuation in the needle valve chamber is almost unchanged, resulting in the early opening of the needle valve. When the solenoid valve is powered off, high pressure is established in the control chamber, and the needle valve is closed and then fuel injection ends [8].

As shown in figure 8, when the clearance of the control slide valve coupling is 0.022 mm, the lift of the needle valve is increased by about 0.2 mm compared with that when the clearance is 0.002 mm. The clearance not only makes the needle valve open in advance and close in delay, but also makes the needle valve lift increase significantly. As can be seen from figure 9, when the clearance of slide valve coupling changes from 0.002 mm to 0.022 mm, the maximum injection rate increases from 0.17 L/s to 0.23 L/s, with an increase of 35.29%; The injection duration increased from 1.2 ms to 2.22 ms, with an increase of 85%.

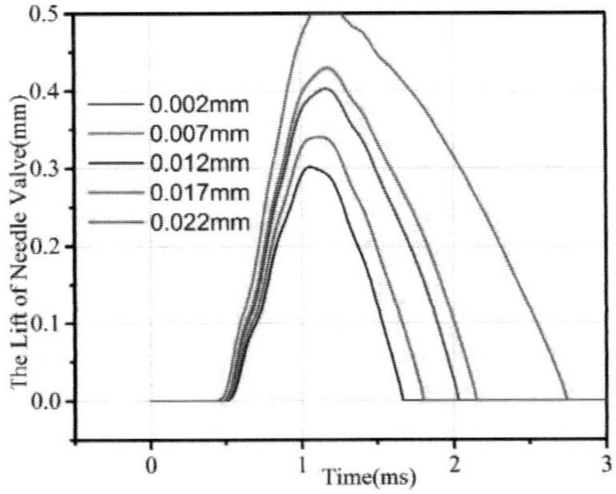

Figure 8. Needle valve lift depending on slide valve coupling clearance.

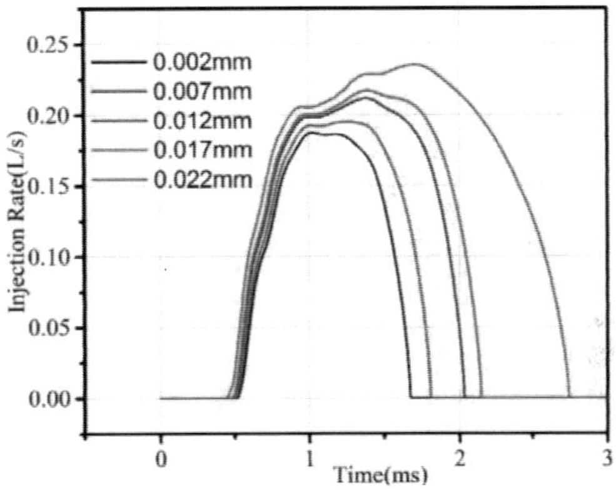

Figure 9. Injection rate depending on slide valve coupling clearance.

As shown in figure 10, when the clearance of slide valve coupling changes from 0.002 mm to 0.022 mm, the cycle fuel injection quantity increases from 230 mm^3 to 280 mm^3, with an increase of 21.7%. Figure 11 shows the change curve of the ratio of the single cycle leakage of the slide valve coupling with different clearances to the single cycle return fuel when the slide valve coupling clearance is 0.002mm. It can be seen from the figure that the slide valve coupling clearance has little impact on the performance of the injector.

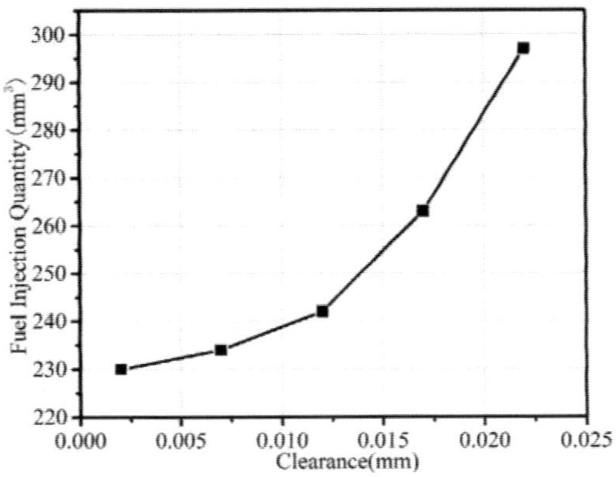

Figure 10. Cycle fuel injection quantity depending on slide valve coupling clearance.

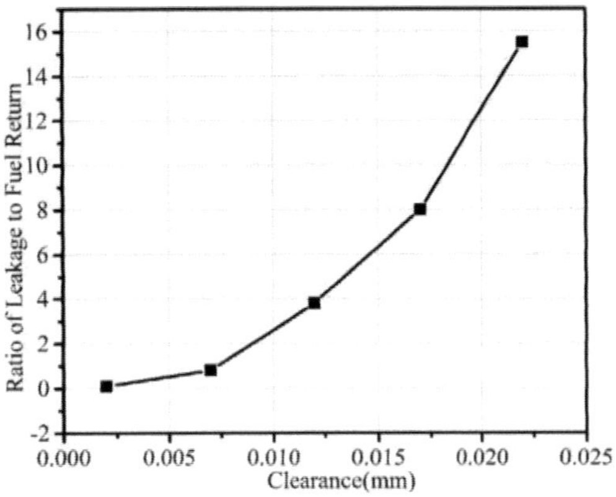

Figure 11. The ratio of leakage to return fuel depending on slide valve coupling clearance.

4.3. Effect of Plunger Coupling Clearance on Injection Performance

As shown in figure 12, when the clearance of the control plunger coupling changes from 0.002 mm to 0.012 mm, the start time of the needle valve remains unchanged at 1.32 ms and the end time remains unchanged at 2.38 ms. When the clearance of the control plunger coupling is greater than 0.012 ms, the start time of the needle valve is prolonged and the end time is shortened. As shown in figure 13, when the coupling clearance of the control plunger changes from 0.002 mm to 0.012 mm, the injection duration remains unchanged for 1.02 ms. When the coupling clearance changes from 0.017 mm to 0.022 mm, the injection duration decreases from 1 ms to 0.96 ms, and the injection duration of the injector shows a trend of first unchanged and then shortened.

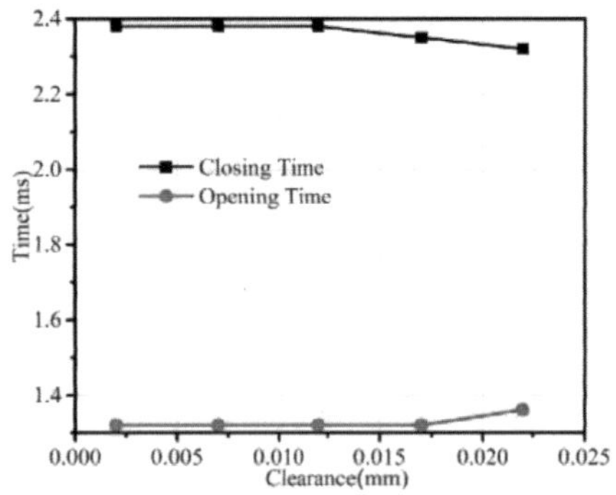

Figure 12. Start and end time of needle valve depending on plunger coupling clearance.

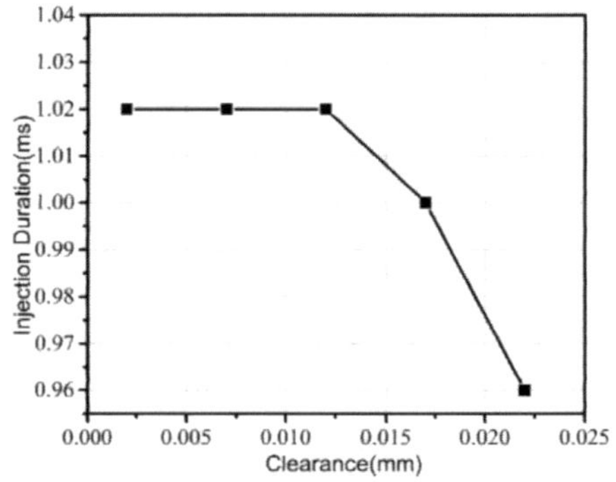

Figure 13. Fuel injection duration depending on plunger coupling clearance.

As shown in figure 14, when the control plunger coupling clearance is 0.022 mm, the needle valve lift is about 0.1mm lower than that when the clearance is 0.002 mm. The increase of the control plunger coupling clearance not only delays the opening and early closing of the needle valve, but also significantly reduces the needle valve lift. As shown in figure 15, when the clearance of the control plunger coupling changes from 0.002 mm to 0.022 mm, the maximum injection rate decreases from 0.185 L/s to 0.175 L/s, with a decrease range of 5.4%. The injection duration decreased from 1.7 ms to 1ms, with a decrease of 41.18%. From the above analysis, as the matching clearance of the control plunger coupling increases to a certain extent, the fuel injection duration can be shortened and the fuel injection quantity is reduced. By comparing figure 8 and figure 14, figure 9 and figure 15, we know that the change of the control plunger coupling clearance has a greater impact on the fuel injection performance compared with the change of the slide valve coupling clearance.

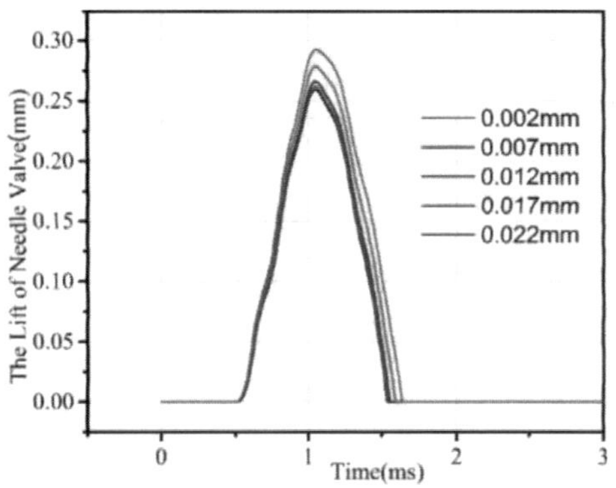

Figure 14. Needle valve lift depending on plunger coupling clearance.

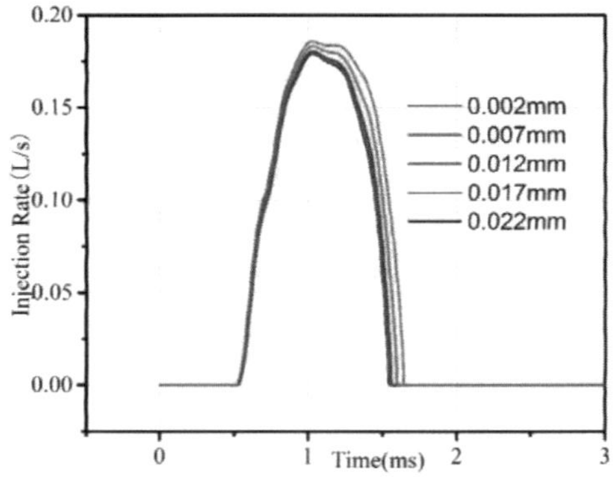

Figure 15. Injection rate depending on plunger coupling clearance.

As shown in figure 16, when the clearance of the control plunger coupling changes from 0.002 mm to 0.022 mm, the cycle fuel injection quantity decreases from 230 mm^3 to 202 mm^3, with a decrease range of 12.17%. Figure 17 shows the curve of the ratio of the single cycle leakage of the control plunger coupling with different control clearance to the single cycle return fuel when the plunger coupling clearance is 0.002 mm. It can be seen from the figure that when the single cycle leakage of the couple is small compared with the reference oil return, the coupling clearance has little impact on the injection performance.

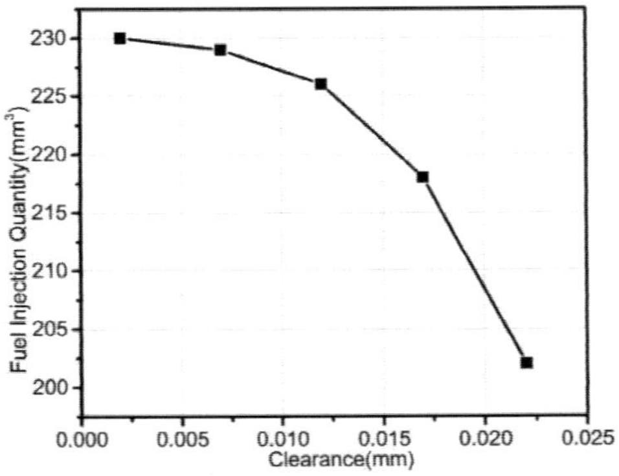

Figure 16. Cycle fuel injection quantity depending on plunger coupling clearance.

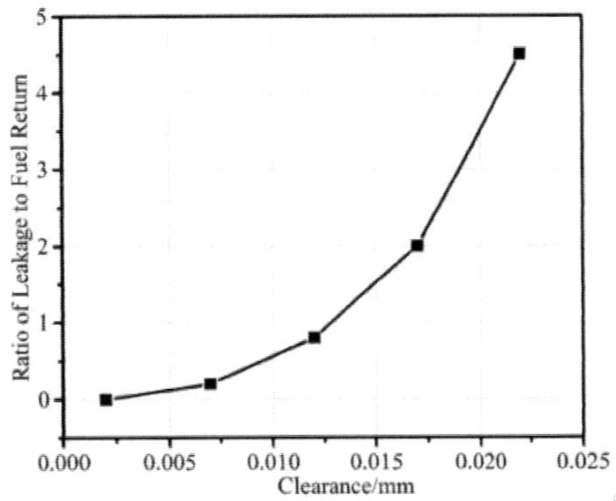

Figure 17. The ratio of leakage to return fuel depending on plunger coupling clearance.

5. Conclusion

A one-dimensional simulation model of ultra-high pressure common rail injector was established, which was used to simulate injection process. The influence of fuel leakage of precision coupling on hydraulic response and injection performance is analyzed. The main conclusions are as follows.

With the increase of the matching clearance of the slide valve coupling, the start time of the needle valve is advanced and the end time is delayed, and the fuel injection duration of the injector increases rapidly. The change of the hydraulic response of the injector caused by the increase of the clearance of the slide valve coupling leads to the increase of the fuel injection rate and fuel injection quantity. Besides, when the single cycle leakage of the slide valve is too small compared with the reference return fuel, the impact of the coupling clearance on the fuel injection performance of the injector is also small.

With the increase of the matching clearance of the control plunger coupling, the start time of the needle valve remains unchanged first and then slightly delayed, while the end time remains unchanged first and then shows an early trend. In addition, the injection duration of the injector shows a trend of unchanged first and then shortened. Moreover, the increase of the matching clearance of the control plunger coupling leads to the decrease of the injection rate and fuel injection quantity. The amount of cycle fuel injection quantity decreases with the increase of the matching clearance of the control plunger coupling.

References

[1] Belchev S, Ivanov Z and Uzuntonev T 2020 Determination of the stages of the injection process for common rail injectors using vibration pulses *IOP Conf. Ser.: Mater. Sci. Eng.* **977(1)** 1-8.

[2] Chao Q U, Fei H Z, Wei Y P et al. 2020 Fluctuation law of two injection fuel injection quantity of high-pressure common rail system and its combined prediction model *Chin. Intern. Combust. Engine Eng.* **41(6)** 46-54.

[3] Du Y, Qian C, Zhao Y et al. 2019 Numerical simulation of pressure fluctuation in high-pressure common-rail fuel injection system *American Journal of Mechanical and Industrial*

Engineering **4(3)** 45-51.

[4] Dong Q, Yang X Y, Ni H et al. 2020 An on-line measurement method of injection rate of high pressure common rail system *Measurement* **170(3)** 440-451.

[5] Lee C H 2019 A computer simulation of injection rate characteristics of solenoid type common rail injector according to injector driving current patterns *Journal of ILASS-Korea* **24(3)** 114-121.

[6] Zhang Z H, Yang Q and Sun B G 2019 Experimental study on pressure fluctuation characteristics and fuel property parameters of 250 MPa common rail system *Trans. Beijing Inst. Technol.* **39(11)** 1113-1116.

[7] Krivtsov S N, Yakimov I V and Ozornin S P 2018 Numerical analysis and experimental studies on solenoid common rail diesel injector with worn control valve *IOP Conf. Ser.: Mater. Sci. Eng.* **327(4)** 1-9.

[8] Zhang J H and Chen M 2018 After-sale data based common rail injector remanufacturability analysis *Journal of Shanghai Jiaotong University* **23(3)** 337-344.

Investigation of Huge Wave in Gas-Liquid Two-Phase Churn Flow

S Zhang[1], Q Q Shao[1], B Hu[2] and K Wang[2*]

[1] China National Petroleum Pipeline Network Group Co., Ltd South China Branch, Guangzhou 510620, Guangdong, China;
[2] School of Mechanical and Storage Engineering, China University of Petroleum-Beijing, Beijing 102249, China.
E-mail: wang_ke@cup.edu.cn

Abstract. Churn flow frequently occurs in power plants, chemical engineering, petroleum, and other industrial applications. Due to its chaotic nature, churn flow has a significant influence on safety and management control. As one of the essential characteristics of churn flow, depth knowledge of the huge wave is crucial for a better understanding churn flow. However, relevant studies on these issues are still in shortage because it is difficult to capture its behaviours experimentally. In this study, we employed the high-speed camera to capture the evolution and properties of huge waves under churn flow conditions in a vertical pipe. The inner diameter of the pipe is 19 mm. Based on the observation, the flooding of the falling film in churn flow is demonstrated to be the slug/churn transition mechanism. Additionally, the liquid distribution in the cross-section of the tube is provided and discussed in detail. Compared with the existing experiment data, we carefully analyze the properties of huge waves, such as frequency and amplitude.
Keywords. Churn flow, two-phase flow, entrainment, pressure drop.

1. Introduction

The critical heat flux (CHF) is significant for system security in chemical and engineering applications, such as nuclear reactors, power plants, boilers, etc. Generally, when excessive liquid entrainment occurs in the annular flow, the so-called "dry-out" phenomenon occurs with the liquid film being utterly exhausted from the pipe wall. The deterioration of heat transfer caused by the dry-out would cause the pipe to overheat and even explode. Therefore, it is essential to predict the occurrence of dry-out to avoid such severe accidents.

Whalley et al. [1] provided a classic physical model for predicting the occurrence of dry-out under the steady-state circumstance. Note that the integration process is dependent on the initial entrained fraction Eo at the start of the annular flow, which varies between zero and unity. Generally, Eo is empirically assumed due to its difficulty in measurement. Table 1 illustrates the assumed value of entrained fraction Eo at the onset of annular flow from the literature [1-9]. Although these assumed values provide good predictions on the CHF, the assumptions do not correspond with reality. It is worth mentioning that

Ahmad et al. [10] put forward that the integration of the entrainment and deposition process from the beginning of churn flow instead of the annular flow is more logical. Thus, the characteristic of churn flow is vital for CHF prediction.

Table 1. Assumed value of *Eo* by different authors.

Reference	Assumed value
Whalley et al. [1] Saito et al. [2]	0.9
Hewitt and Govan [3]	0.99
Lee et al. [4] Okawa et al. [5-7]	Entrained fraction at equilibrium state
Govan et al. [8]	0.99 0.91
Sugawara [9]	0.001 0.05

Churn flow is a typical intermittent upward flow, which is normally observed in the vertical and near-vertical tubes. Due to its chaotic nature, the definition of the churn flow still has controversy. Zuber and Findley [11] considered the churn flow as a category of bubbly flow. Taitel et al. [12] argued that the churn flow is just a developing slug flow. Based on the experimental observation, Hewitt and Hall Taylor [13] believed that churn flow occurs after the crush of slug flow. They defined it as an intermediate flow regime between the slug and annular flow. Generally, there are four points of view on the formation of the churn flow, i.e., bubble coalescence mechanism, entrance effect mechanism, wake effect mechanism and flooding mechanism. Jayanti and Hewitt [14] reviewed these four models and proposed that the flooding mechanism is more proper to interpret the slug-churn transition. Additionally, Jayanti et al. [15] observed and confirmed that huge waves, also called large waves, appear on the falling film when the transition is approaching.

A typical feature of the churn flow distinguishes from the annular flow is the existence of huge waves travelling up and down periodically over a falling film throughout the regime. In the photographic study of Hewitt et al. [16] with chromium dye, it was found that there is a periodic reversal of bulk liquid. Note that the huge wave is the primary phenomenon. Ultimately, by increasing the gas velocity, the liquid downward film ceases, resulting in an upward annular flow. According to the work of Hewitt and Wallis [17] and Wallis [18], the critical condition for the flow reversal can be described by a dimensionless superficial gas velocity U_{sg}^{*} as:

$$U_{sg}^{*} = u_{sg}\sqrt{\frac{\rho_G}{gd(\rho_L - \rho_G)}} \approx 1 \qquad (1)$$

where u_{sg}, d, ρ_L, ρ_G are the superficial gas velocity, pipe diameter, densities of liquid and gas, respectively.

Owen [19] investigated the variation of the pressure gradient, including bubbly flow, slug flow, churn flow and annular flow. Based on his research, there was a sudden increase in the pressure gradient when the slug flow breaks down. After that, the pressure gradient gradually decreases during the churn flow and eventually increases when the annular flow occurs. To explore the relationship between the flooding of the falling film and the formation of churn flow, Govan et al. [20] employed the method for the flooding measurement. They compared the pressure gradient and liquid holdup in the cases of with or without the falling liquid film below the liquid injector. According to their observation, there was no discernible difference between these two cases, demonstrating that the flooding phenomenon is closely related to the slug/churn transition. Moreover, Barbosa et al. [21] experimentally and numerically investigated the characteristics of churn flow. They employed a high-speed camera to record the process of wave formation and motion in a 31.8 mm i.d. tube under churn flow conditions. By analyzing the forces acting on the control volume, a mathematical model for wave levitation was also built. Enrico and Davide [22] used the VOF method to numerically investigate the properties of churn flow with the fluids of air-water and R134a vapour-liquid mixtures. They also proposed a simplified model to investigate the effects of liquid and gas superficial velocities and pipe diameters on the suspension process of the typical ring-type waves in churn flow.

The generation of the entrained droplets in churn flow results from undercut and sheared-off of the interfacial waves [23]. The droplet deposition in churn flow is often dominated by the radial velocities imparted at the generation of droplets. Unlike annular flow, the variation of the entrained fraction is inversely related to the superficial gas velocity [24-26]. Wallis et al. [27] used a uniaxial positioning sampling probe to measure the entrainment rate of churn and churn-annular transition. They indicated that a considerable amount of liquid is entrapped as droplets in the churn flow, and the entrained fraction reaches the lowest value at the churn-annular transition. Barbosa et al. [26] used an isokinetic probe to measure the entrained fraction both under churn and annular flow conditions. They proved that the entrained fraction has an inverse ratio to the gas velocity in the churn flow region, passing through a minimum value at the churn/annular transition, and increasing in annular flow again. Accordingly, an empirical equation for Eo under adiabatic and equilibrium conditions was proposed as:

$$Eo(\%) = 0.95 + 342.55\sqrt{\frac{\rho_L \dot{m}_L}{\rho_G \dot{m}_G} d_T^2} \tag{2}$$

where $\dot{m}_G$, $\dot{m}_L$, d_T, ρ_L, ρ_G are the local gas mass flux, the local liquid mass flux, pipe diameter, liquid density, and gas density, respectively. The minimum of *Eo* occurs at the transition to annular flow, when dimensionless superficial gas velocity $U_{sg}^* = 1$.

Moreover, Azzopardi and Wren [28] investigated the entrained fraction with a T-junction. The results indicate that entrained fraction has a weak dependency with pipe diameter and the inlet gas flowrate but a strong relationship with liquid flowrate. Ahmad et al. [10] proposed an empirical equation for entrainment rate in churn flow. Accordingly, the CHF in both uniform and non-uniform heating tubes was predicted, and the predicted results are all less than ± 20%.

$$\frac{E_{churn}}{E_{annular,local}} = -8.73U_G^* + 9.73 \tag{3}$$

In the present paper, the phase distributions in the axial direction and cross-section of the tube under churn flow conditions are experimentally investigated. By analyzing high-speed photography, the evolution of the huge waves and mechanism of entrainment are qualitatively studied. Since the film oscillates in the flow direction, the conventional method for measurement of entrained fraction (extracting the liquid film) is not feasible in this pattern. Therefore, the liquid distribution in the cross-section of the tube is still measured by the optical method.

2. Experiment System and Method

2.1. Experimental System

Figure 1 depicts the experimental system consisting of the test section, gas and liquid loops. The air is separated into two ways after the regulator valve, i.e., one way of developed gas is fed into the test section at the end of the tube, and the other flows into the branch gas channel. The water flows into the test section by a 0.55 kW pump through a porous wall. It can be observed that the water in the test section can also be removed in two ways: one is to discharge upward water from the upper outlet, and the other is to extract falling film from the extraction sinter.

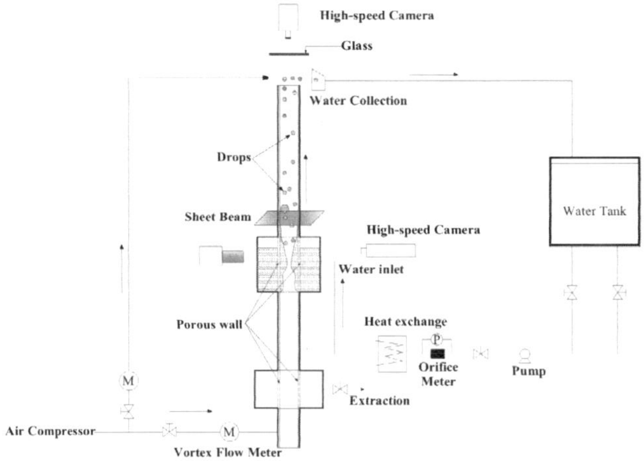

Figure 1. Diagram of the test facility.

The test section is made of transparent polymethylmethacrylate with 2 m in length and 19 mm in diameter. For the convenience of extracting the liquid from the test section, there are two porous parts in the test section. The water inlet section, which is designed with 300 holes of 1 mm diameter staggered in 15 rows, is equipped with a water chamber, while the water extraction section is designed with 250 holes of 1 mm diameter. Note that the spacing of the holes is about 2 mm. The air and water are metered by a rotor flow meter and an orifice flow meter, respectively. In addition, the liquid mass flux carried upward is measured by the weighing method. For the flow visualization, a Memrecam fx K3 high-speed camera is used to obtain the movement and growth of the huge waves and the liquid distribution in the tube's cross-section. In this study, the sampling frequency is set to be 100 frame/s. A sheet beam and backlight illumination are employed to light up the test section both in the axial and the cross-section parts. To prevent the pollution of the CCD lens by entrained droplets in the gas flow, a strong lateral direction wind is introduced to avoid these injected liquids.

2.2. Experimental Procedure

The experimental method proposed by Barbosa et al. [21] is adopted in the present study. The gas first flows into the test section at a relatively smaller value, while the liquid enters the liquid chamber at an expected value. Under these circumstances, the liquid is extracted into the test section to form a falling film. Finally, the falling film would be completed extracted out of the test section from the outlet sinter. By increasing the gas flow rate to a certain value, the falling film ceases near the outlet sinter.

Accordingly, the so-called huge waves repeatedly form and eventually move upwards to the out of the pipe. During wave evolution, the huge waves break up into droplets of different sizes, causing the upper flow to oscillate wildly. Therefore, the region located above the water inlet part is the expected churn flow.

In the present study, all the experiments are carried out under atmospheric pressure and room temperature. The gas flow rates are 5.2 ~ 8.5 m^3/h, and the liquid flow rates are 14.9×10^{-3} ~ 43.5×10^{-3} kg/s. As illustrated in figure 2, all the experiments are performed over a churn flow condition.

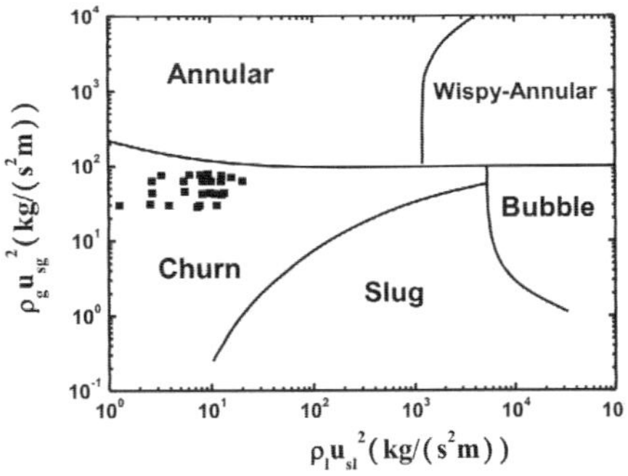

Figure 2. Distribution of the test points on the flow pattern map [29].

3. Results and Discussions

3.1. Evolution of Huge Waves

Figure 3 illustrates the evolution of a typical huge wave. It is assumed that an individual wave has already been generated in the water inlet section at t=0 ms, as seen in figure 3(a). Figure 3(b) shows that the wave grows in radial and axial directions and moves slightly down the porous section. At t=24 ms, the wave crest closely reaches its maximum amplitude or even forms a liquid bridge at this circumstance. In the following evolving process, the wave that picks up liquid from the film falls above and starts to move upwards, as seen in figure 3(c). Eventually, the wave is torn by the coming gas flow to form a number of entrained droplets. It is worth mentioning that the upper liquid film oscillates along the flow direction. As shown in figure 3(d), the wave almost moves out of the frame, and a newly born wave appears around the inlet section. Thus, it is demonstrated that the flooding of the falling film leads to the onset of churn flow.

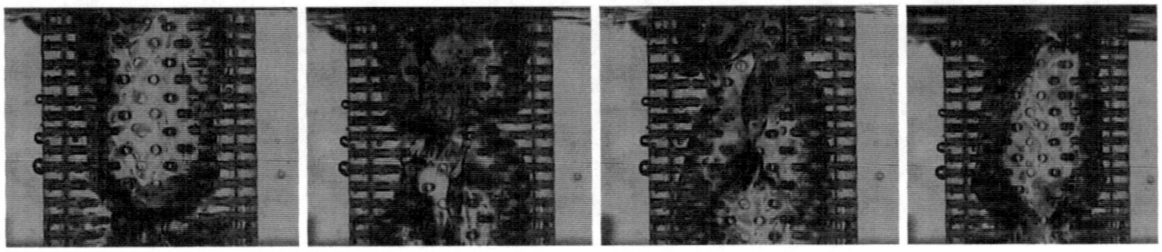

(a) t=0 ms (b) t=12 ms (c) t=24 ms (d) t=30 ms

Figure 3. Evolution of huge waves in churn flow.
(u_{sg}=6.17 m/s, u_{sl}=74.85×10^{-2} m/s, U^*=0.49)

Due to gravity, the huge wave would first move downwards for a distance to a "balanced" position and finally reverse upward. Figure 4 shows the positions that huge waves change to travel upward under different flow conditions. The results indicate that the lower the superficial gas velocity, the wave travels farther downward. Additionally, the critical amplitude of huge waves decreases with the increasing gas velocities, which can be attributed to the promotion of the entrainment caused by the increasing interfacial shear stress. Note that the liquid film performs quite differently after the collapse of the huge wave. As seen in figure 5, the wave becomes more distorted at the lower superficial gas velocity because of the lower shear stress.

(a) u_{sg}=4.80m/s (b) u_{sg}=5.98m/s (c) u_{sg}=7.25m/s (d) u_{sg}=7.84m/s

Figure 4. The positions that huge waves reverse to travel upward at different superficial gas velocity (G_l=16.25×10^{-2} kg/s).

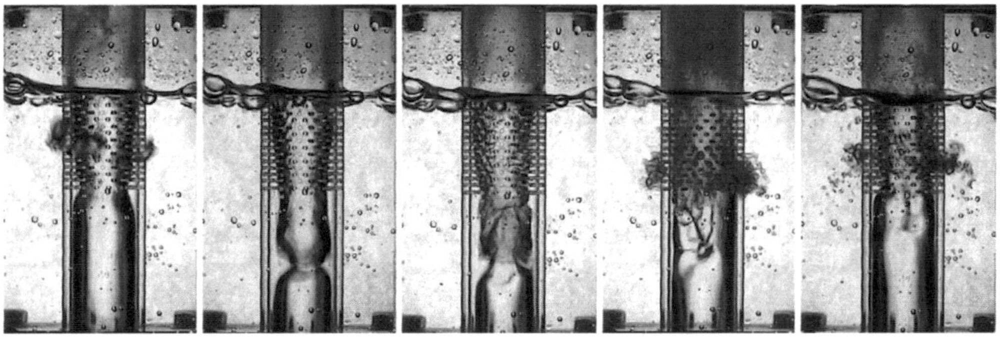

t=0 ms t=120 ms t=130 ms t=150 ms t=190 ms

Figure 5. Comparison of oscillation at different flow conditions (u_{sg} =4.99 m/s, u_{sl} =8.84×10^{-2} m/s, U_g^* =0.40).

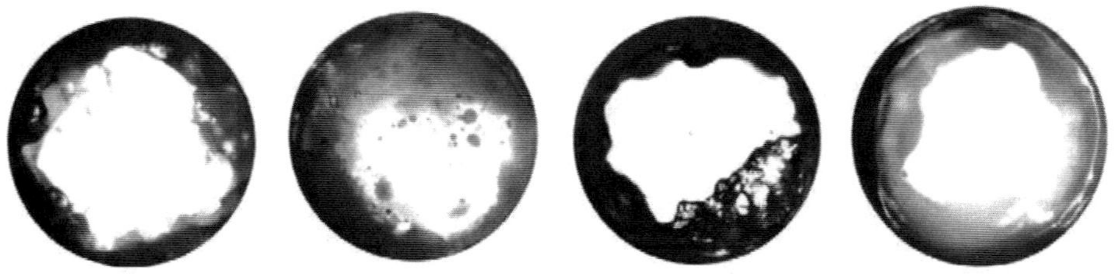

(a) $M_l = 2.03\times10^{-2}$ kg·s^{-1}

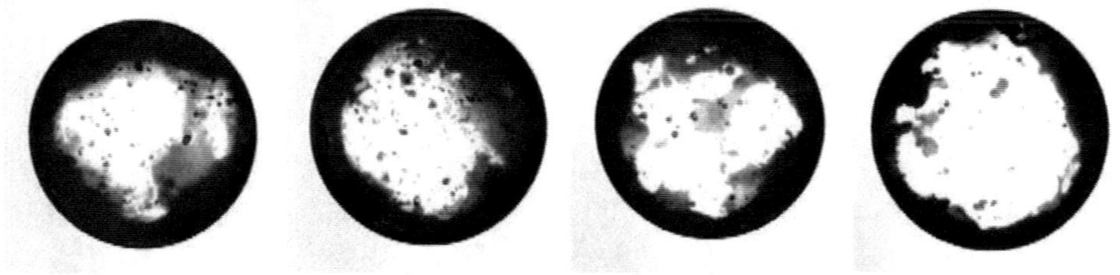

(b) $M_l = 2.39\times10^{-2}$ kg·s^{-1}

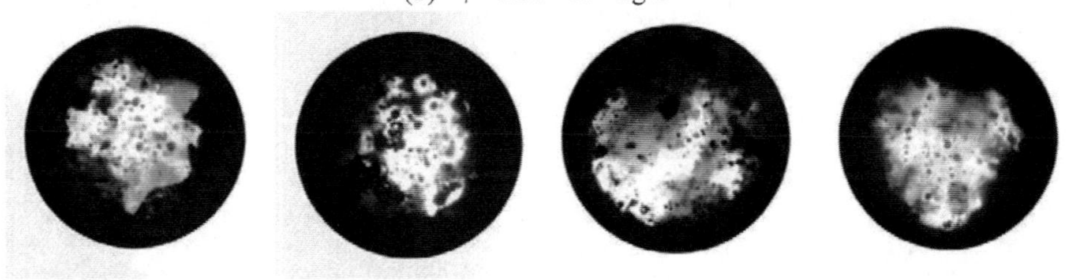

(c) $M_l = 5.59\times10^{-2}$ kg·s^{-1}

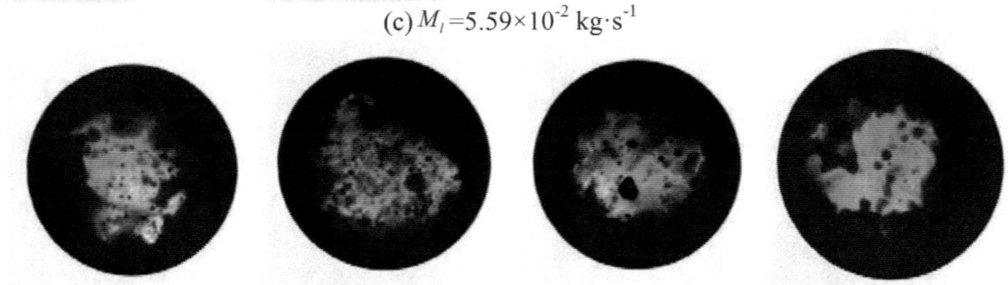

(d) $M_l = 7.64\times10^{-2}$ kg·s^{-1}

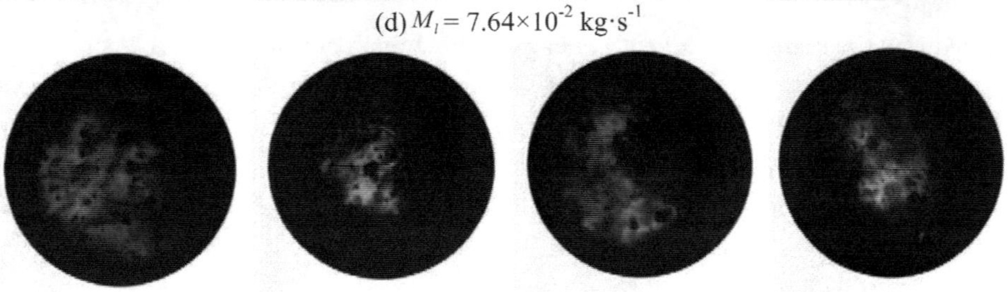

(e) $M_l = 9.40\times10^{-2}$ kg·s^{-1}

Figure 6. Axial view of churn flow. ($U_g^* = 1.02$)

Figure 6 illustrates the liquid distribution in the cross-section of the tube from churn flow to annular flow. Unlike the uniform distribution of the liquid film along the tube's circumference in the annular flow (figure 6(a)), which is consistent with the work of [13, 30], the circumferential distribution of the liquid film in churn flow is non-uniform. Remarkably, the oscillation of the liquid film is attributed to this irregular distribution. Therefore, 2D simplification may not work correctly in the simulation under churn flow conditions, which cannot reflect the natural characteristics of this particular flow pattern. Another essential characteristic of churn flow is the drop size, which varies in time and space. Due to the mechanism of entrainment in churn flow (undercut or bag break-up) differing from that in annular flow (ligament break-up), the drop sizes are quite different. The qualitative result inferred from figure 6 is that drop sizes are larger than annular flow, and there seem to be some liquid blocks in the gas core. Relevant research on drop size will be studied in the future.

3.2. Frequency of Huge Waves

Based on the experimental observation, part of the liquid is transported upwards during the movement of interfacial waves, while the liquid film between the waves travels downwards. Thus, the net fluid mass flow rate G_{nl} (the liquid that waves carried upward) may vary from zero to total liquid mass flow rate. Parts of experimental data are listed in table 2.

Table2. Data bank of various experiments in different conditions.

V_g (m^3/h)	$G_l \times 10^{-3}$ (kg/s)	$G_{nl} \times 10^{-3}$ (kg/s)	u_{sg} (m/s)	$u_{sl} \times 10^{-3}$ (m/s)	U^*	Re_g	f (s^{-1})
5.1	20.51	10.15	5.00	35.90	0.40	6079.04	8.36
5.1	28.52	17.74	5.00	62.75	0.40	6079.04	8.77
5.1	36.86	24.98	5.00	88.36	0.40	6079.04	8.79
6.2	20.43	14.66	6.07	51.86	0.49	7390.21	8.71
6.1	28.28	25.78	5.98	91.19	0.48	7271.01	10.50
6.1	37.09	29.63	5.98	104.81	0.48	7271.01	11.17
7.4	20.29	14.61	7.25	51.68	0.58	8820.57	11.02
7.4	28.25	25.95	7.25	91.80	0.58	8820.57	12.96
7.4	36.75	31.80	7.25	112.49	0.58	8820.57	13.24
8.1	20.55	16.25	7.94	57.48	0.63	9654.95	11.95
8.2	28.47	25.26	8.03	89.36	0.64	9774.14	13.33
8.1	36.71	31.91	7.94	112.88	0.63	9654.95	13.33

Figure 7 shows the variation of the net liquid mass flow rate carried by waves. As expected, the net liquid mass flux increases with the increase in gas and liquid velocities. As the gas velocity increases, more liquids are carried upwards due to the increasing interfacial shear stress.

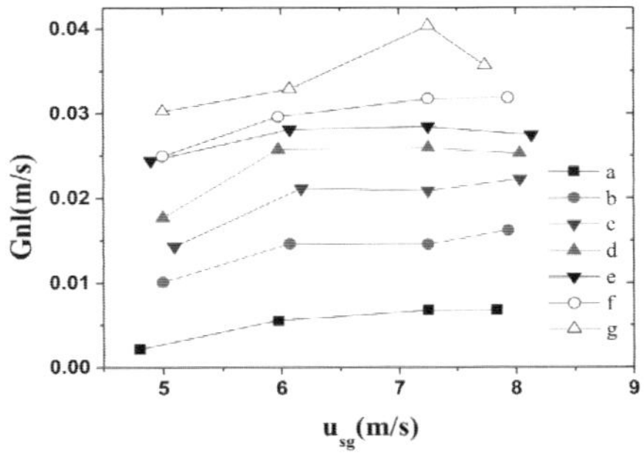

Figure 7. Variations of the net liquid mass flow rates.
(a=16.29×10^{-3} kg/s, b=20.45×10^{-3} kg/s, c=24.36×10^{-3} kg/s, d=28.38×10^{-3} kg/s, e=32.75×10^{-3} kg/s, f=36.85×10^{-3} kg/s, g=40.89×10^{-3} kg/s)

Figure 8 shows the variation of the wave frequency under different flow conditions. The result indicates that the wave frequency increases with increasing of gas and liquid velocities. Note that the wave frequency gradually approaches a constant value when the gas velocity increases, and the dependence on the liquid rate are fragile. It can be interpreted that when the wave amplitude reaches its maximum value, the collapse of the huge wave bursts more quickly at higher gas velocity before the liquid affects the frequency. Figure 9 compares the present study with the work of [16, 21]. Based on the current work, the wave frequency directly correlates with the Reynolds number of the gas phase, and there seems little dependency on liquid mass flow rate. Due to the limited experiment data from Barbosa et al. [21], little information can be inferred. However, the experiment data from Hewitt et al. [16] are quite different, i.e., the span of wave frequency is much larger, and the wave frequency has a strong dependency on liquid mass flow rate. The frequency increases with the increase in gas velocity at a lower net liquid mass flow rate but inverses at a higher net liquid mass flow rate. No doubt that the fluid properties and pipe diameter are suspected to be the main reason. The experiment of Hewitt et al. [16] was taken within a 10mm vertical tube, and the working fluids are trichloroethylene and air. Moreover, near the average cross-section of the same gas Reynolds number and net liquid mass flow, the smaller the diameter is, the higher the fluctuation frequency is. It is assumed that the smaller the gas circulation area is, the higher the gas velocity is and the higher the fluctuation frequency is.

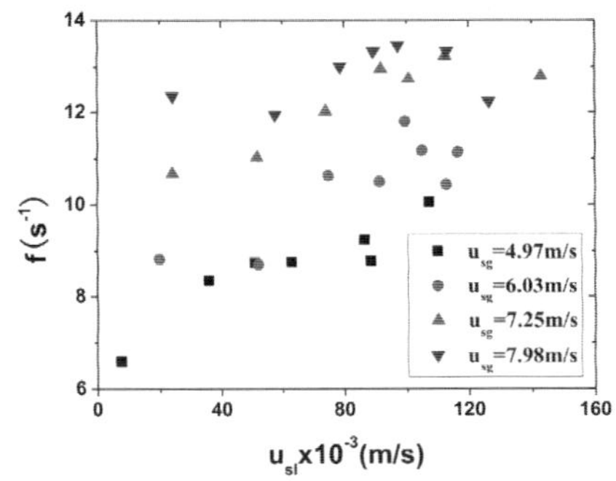

Figure 8. the frequency of huge waves at different flow conditions.

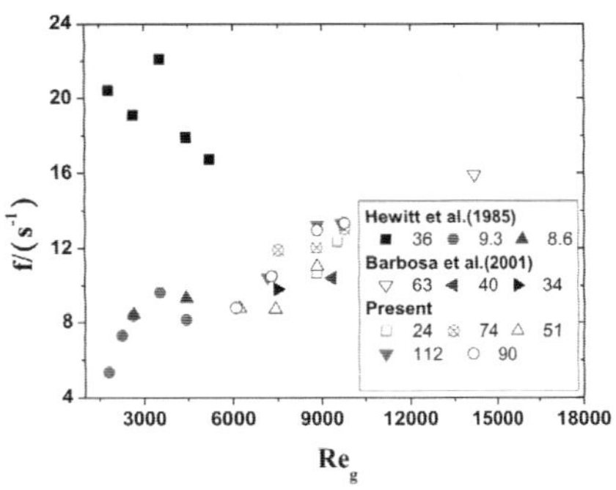

Figure 9. Comparison of wave frequency with the data available in literature.

4. Conclusions

A depth understanding of the properties of huge waves in churn flow is important to accurately predict the occurrence of CHF. The characteristics of churn flow are studied qualitatively by using high-speed visualization technology. It was experimentally shown that huge waves periodically appear at the water inlet section, travel upward, and eventually entrain into the gas phase. In addition, the liquid film performs more unstable at the smaller gas velocity due to the insufficient shear stress. The liquid non-uniformly distributes in the cross-section of the pipe, and the droplet size in churn flow is intuitively larger than that in the annular flow. Moreover, the experiment data analysis indicates that the wave frequency increases with the increasing gas and liquid flow rates, but the amplitude is contrary. Comparison of data with Hewitt et al. [16] and Barbosa et al. [21] shows a sensible difference, which was suspected that physical properties and tube diameter are the reason for this difference.

References

[1] Whalley P B, Hutchinson P and Hewitt G F 1974 Calculation of critical heat flux in forced convection boiling *Proceedings of the 5th International Heat Transfer Conference* Tokyo, Japan p 290-294.
[2] Saito T, Hughes E D and Carbon M W 1978 Multi-fluid modelling of annular two-phase flow *Nuclear Engineering and Design* **50**(2) 225-271.
[3] Hewitt G F and Govan A H 1990 Phenomenological modelling of non-equilibrium flows with phase change *International Journal of Heat and Mass Transfer* **33**(2) 229-242.
[4] Lee K W, Baik S J and Ro T S 2000 An utilization of liquid sublayer dryout mechanism in predicting critical heat flux under low pressure and low velocity conditions in round tubes *Nuclear Engineering and Design* **200**(1-2) 69-81.
[5] Okawa T, Kotani A, Kataoka I and Naito M 2003 Prediction of critical heat flux in annular flow using a film flow model *Journal of Nuclear Science and Technology* **40**(6) 388-396.
[6] Okawa T, Kotani A and Kataoka I 2004 Experiments for equilibrium entrainment fraction in a small vertical tube *Proceedings of the 5th International Conference Multiphase Flow* Yokohama p 224.
[7] Okawa T, Kotani A, Kataoka I and Naitoh M 2004 Pre-diction of the critical heat flux in annular regime in various vertical channels *Nuclear Engineering and Design* **229**(2-3) 223-236.
[8] Govan A H 1988 Phenomenological prediction of critical heat flux *Proceeding of the 2nd UK National Heat Transfer Conference* p 315-326.
[9] Sugawara S 1990 Droplet deposition and entrainment modelling based on the three-fluid model *Nuclear Engineering and Design* **122**(1-3) 67-84.
[10] Ahmad M, et al. 2010 Drop entrainment in churn flow *7th International Conference on Multiphase Flow* Tampa, FL USA.
[11] Zuber N and Findlay J A 1965 Average volumetric concentration in two-phase flow systems *Journal of Heat Transfer* 453-468.
[12] Taitel Y, Barnea D, Dukler A E 1980 Modelling of flow pattern transitions for steady upward gas-liquid flow in vertical tubes *A.I.Ch.E.J.* **26** 345-354.
[13] Hewitt G F and Hall-Taylor N 1970 *Annular Two-Phase Flow* (Oxford: Pergamon Press).
[14] Jayanti S and Hewitt G 1992 Prediction of the slug-to-churn flow transition in vertical two-phase flow *International Journal of Multiphase Flow* **18**(6) 847-860.

[15] Jayanti S, Hewitt G F, Low D E F and Hervieu E 1993 Observation of flooding in the Taylor bubble of co-current upwards slug flow *International Journal of Multiphase Flow* **19**(3) 531-534.

[16] Hewitt G F, Martin C J and Wilkes N S 1985 Experimental and modelling studies of annular flow in the region between flow reversal and the pressure drop minimum *Physico-Chemical Hydrodynamics* **6** 43-50.

[17] Hewitt G F, Martin C J and Wilkes N S 1985 Experimental and modelling studies of annular flow in the region between flow reversal and the pressure drop minimum *PCH Physicochemical Hydrodynamics* **6**(1/2) 69-86.

[18] Wallis G B 1969 *One-Dimensional Two-Phase Flow* (New York: McGraw-Hill).

[19] Owen D G 1986 *An Experimental and Theoretical Analysis of Equilibrium Annular Flows* (UK: University of Birmingham).

[20] Govan A H, Hewitt G F, Richter H J and Scott A 1991 Flooding and churn flow in vertical pipes *International Journal of Multiphase Flow* **17** 27-44.

[21] Barbosa J, Govan A H and Hewitt G F 2001 Visualization and modelling studies of churn flow in a vertical pipe *International Journal of Multiphase Flow* **27**(12) 2105-27.

[22] Da Riva E and Del Col D 2009 Numerical simulation of churn flow in a vertical pipe *Chemical Engineering Science* **64**(17) 3753-65.

[23] Azzopardi B J 1997 Drops in annular two-phase flow *International Journal of Multiphase Flow* **23** 1-53.

[24] Verbeek P H J, Miesen R and Schellenkens C J 1992 Liquid entrainment in annular dispersed upflow *8th Annual European Conference on Liquid Atomization and Spray Systems* Amsterdam, p 33-45.

[25] Azzopardi B J and Zaidi S H 2000 Determination of entrained fraction in vertical annular gas/liquid flow *Journal of Fluids Engineering* **122** 146-150.

[26] Barbosa J R, Hewitt G F, König G and Richardson S M 2002 Liquid entrainment, droplet concentration and pressure gradient at the onset of annular flow in a vertical pipe *International Journal of Multiphase Flow* **28**(6) 943-961.

[27] Wallis G B 1962 The onset of droplet entrainment in annular gas–liquid flows *General Electric Report* No. 62GL127.

[28] Azzopardi B J and Wren E 2004 What is entrainment in vertical two-phase churn flow *International Journal of Multiphase Flow* **30**(1) 89-103.

[29] Hewitt G F and Roberts D 1969 *Studies of Two-Phase Flow Patterns by Simultaneous X-Ray and Flash Photography* (Harwell, England: Atomic Energy Research Establishment).

[30] Ohba K and Nagae K 1993 Characteristics and behaviour of the interfacial wave on the liquid film in a vertically upward air-water two-phase annular flow *Nuclear Engineering and Design* **141**(1-2) 17-25.

Research on the Effect of Key Technological Parameters on Molten Pool during Selective Laser Melting Processing

Wentong Wang, Linfeng Tang, Congyan Chen*, Yu Li and Tao Zhou

School of Automation, Southeast University, Nanjing, China
Email: chency@seu.edu.cn

Abstract. Selective laser melting (SLM), as an emerging technology in additive manufacturing, often has various defects in the forming process. To ensure the consistency and stability of the parts forming quality, the effects of two typical technological parameters, laser power and scanning speed, on the temperature of molten pool are investigated in this paper. Firstly, the temperature field of Ti-6Al-4V is simulated theoretically via ANSYS software, and the effects of two typical technological parameters on the temperature field are studied. Then, in the experiment, using the designed radiation monitoring device and Ti-6Al-4V powder as forming material, the influence of these two typical factors on the state of molten pool is studied. The simulation and experimental results show that the temperature of molten pool shows positive correlation with the laser power and negative correlation with the scanning speed. This will provide a certain reference value for upgrading and optimizing SLM equipment.
Keywords. Selective laser melting, temperature field simulation, molten pool.

1. Introduction

As one of the most advanced and promising additive manufacturing technologies, selective laser melting (SLM) is widely used in medical devices, template manufacturing, vehicles, aerospace, architecture, military and many other fields. In SLM forming process, it is really essential to understand the influence of different technological parameters on the temperature of molten pool. As the smallest unit in the thermal evolution process of constituent materials, the temperature of molten pool determines the viscosity, affects the wetting and spheroidizing properties, and affects finally the forming quality of SLM. Therefore, to investigate the influence of typical technological parameters on the temperature of molten pool has a certain guiding role for the actual SLM Industrial production. Laser power and scanning speed are important parameters of input energy in SLM equipment, so it is necessary to study their influence on the temperature of molten pool. The experiments in work [1], by way of Ti-6Al-4V Alloy powder as material, show that the relative density, martensite lath size and microhardness of the workpieces increase with the input of laser energy. In [2], the influence of scanning speed on surface roughness is analyzed, and the parameter optimal solution set is established to improve the surface quality by additional milling. In [3], the temperature field and velocity field in SLM process are studied by numerical simulation and experimental test. Three states of molten pool and the threshold of scanning speed are given. In [4], the factors of typical technological parameters affecting porosity are explored, and the post-treatment methods to reduce porosity are also discussed.

In this paper, Ti-6Al-4V Alloy powder is used as the forming material to investigate the effects of the laser power and scanning speed on the temperature of molten pool, and the temperature field of the material is numerically simulated. The forming process of SLM is analyzed from theory, simulation and experiment, so as to help the actual SLM production practice set reasonably technological parameters and improve product quality.

2. Theoretical Simulation and Analysis

Finite element analysis is a common mean to research the SLM forming process theoretically. It can be used to theoretically analyze the influence of different parameters on the temperature of molten pool.

2.1. Basis of Simulation

The heat conduction evolution of each molten pool in SLM forming process can be expressed by heat transfer control equation as [5, 6]:

$$\frac{\partial}{\partial x}\left(k_x\frac{\partial T}{\partial x}\right)+\frac{\partial}{\partial y}\left(k_y\frac{\partial T}{\partial y}\right)+\frac{\partial}{\partial z}\left(k_z\frac{\partial T}{\partial z}\right)+q=\rho c_p\frac{\partial T}{\partial t} \tag{1}$$

where, x, y and z respectively represent the distance of heat conduction along the X, Y and Z axes in the spatial rectangular coordinate system; ρ Is the material density, c_p is the specific heat capacity of the material.

For the initial and boundary conditions of the specified temperature, the unique solution of the equation can be obtained, and then the temperature field distribution of the molten pool is getted. For the simulation analysis of temperature field, the classical Gaussian surface heat source is used as the heat source model. In the actual forming, there is latent heat of phase change in the material phase transformation process. In this paper, the thermal method is used to handle the latent heat of phase change. The relationship between enthalpy H(J) and temperature can be expressed as:

$$H=\int \rho c_p dT \tag{2}$$

For material properties, it is assumed that Ti-6Al-4V powder belongs to isotropic material, that is, $k_x=k_y=k_z=k$, the thermal conductivity k in all directions is taken as 6.7 $\mathrm{W/(m\cdot K)}$, and the laser absorptivity of Ti-6Al-4V powder α is taken as 0.68 .

In order to simulate the temperature field of Ti-6Al-4V powder, the finite element model is established in ANSYS and meshed. The model and meshing of powder layer and substrate are shown in figure 1. The parameter settings are listed in table 1.

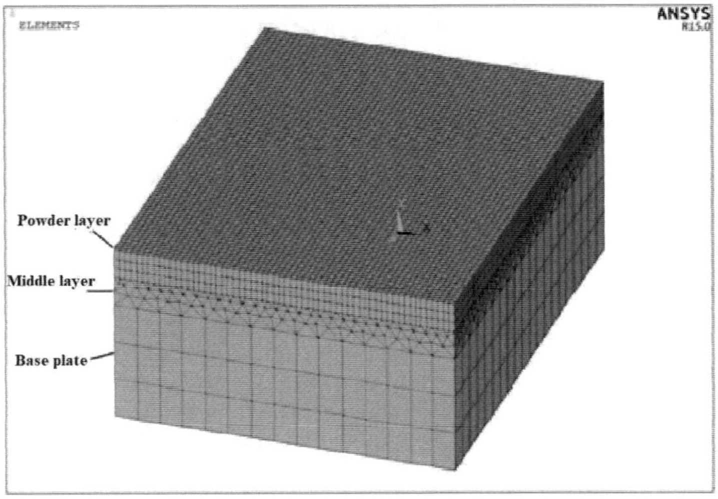

Figure 1. The finite element temperature field model and mesh generation of Ti-6Al-4V.

Table 1. Parameters in finite element simulation experiment.

Parameters	Ranges
Laser Power (W) (v=800 mm/s)	110, 120, 130, 140, 150
Laser Scanning Speed (mm/s) (P=130W)	400~1800 (200 mm/s interval)
Powder Layer Thickness (μm)	30
Laser Spot Diamater (μm)	50

2.2. Results of Simulation

As considering the factors such as birth and death element and heat source loading, the simulation results of temperature field can be obtained. For different power, the maximum temperature of molten pools after the heat source moves steps 17, 19 and 21 (corresponding time is 0.56 ms, 0.63 ms and 0.7ms) are compared. The summary of simulation results is listed in table 2 and the curve is given in figure 2.

Table 2. Maximum temperature of molten pool under different laser power.

Power (W)	Maximum temperature (°C)		
	0.56ms	0.63ms	0.7ms
110	2028.96	2028.74	2028.6
120	2103.79	2103.8	2103.75
130	2173.3	2173.23	2173.21
140	2250.73	2250.76	2250.75
150	2334.91	2334.83	2334.77

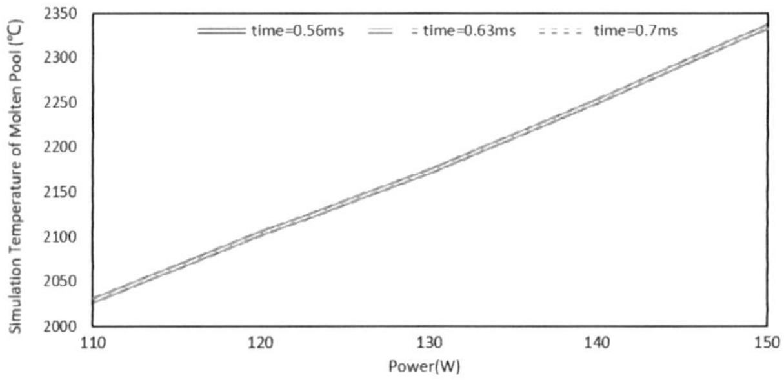

Figure 2. The relation curve between simulation temperature and laser power.

For different speeds, the maximum temperatures of molten pool after steps 17, 19 and 21 of heat source movement are also taken for comparison. The summary of simulation results is listed in table 3, and the curve is shown in figure 3. At different speeds, these three steps correspond to different moving times of the heat source, so they are only marked by steps in the charts for simplicity.

Table 3. Maximum temperature of molten pool at different scanning speeds.

Speed(mm/s)	Maximum temperature (°C)		
	Steps 17	Steps 19	Steps 21
400	2377.43	2378.02	2378.52
600	2270.5	2270.75	2270.87
800	2173.3	2173.23	2173.21
1000	2083.96	2083.91	2083.85
1200	2006.8	2006.77	2006.7
1400	1908.58	1908.7	1908.75
1600	1828.89	1828.9	1828.9
1800	1742.36	1742.36	1742.36

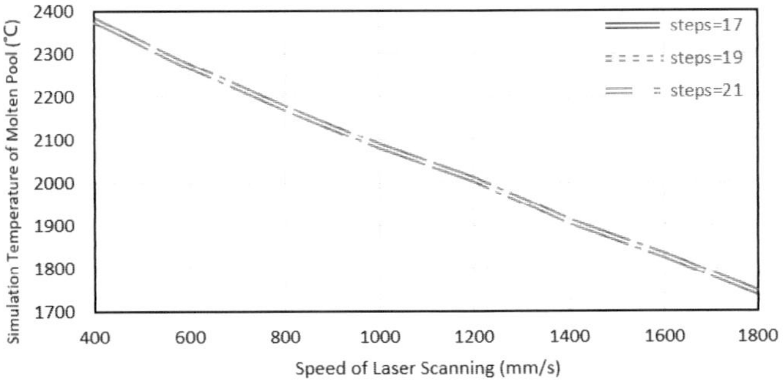

Figure 3. The relation curve between simulation temperature and laser scanning speed.

The simulation results show above that the temperature of molten pool is linearly related to the laser power. The stronger the power is, the higher the corresponding temperature of molten pool is; The higher the laser scanning speed, the less energy absorbed by the molten pool and the less energy radiated outward.

3. Experimental Results and Discussion

3.1. Experimental Setup and Design

The metal powder material used in this paper is TC4 (Ti-6AI-4V), which has low density, high strength and good corrosion resistance [7]. Our setup system has been developed in [8], as is shown in the figure 4. And the printed workpiece is shown in the figure 5.

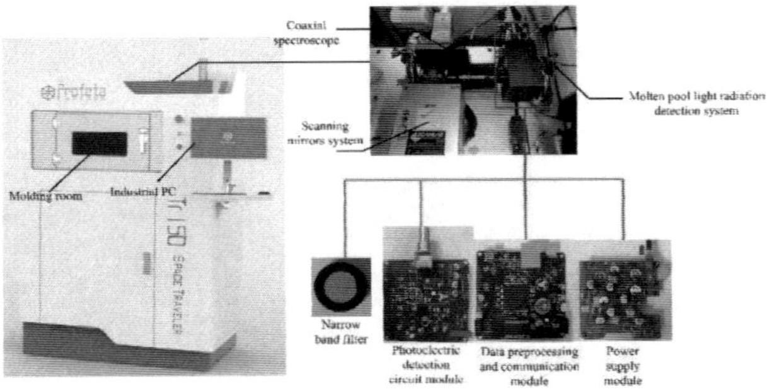

Figure 4. Picture of equipment and molten pool light radiation detection system.

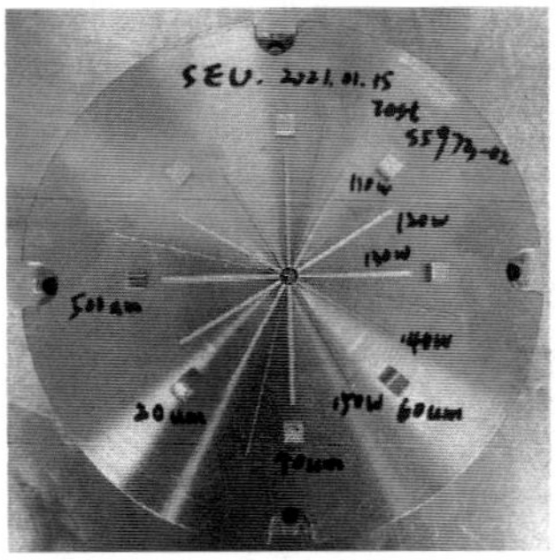

Figure 5. Picture of printed workpiece.

In order to study the radiation state of molten pool with different above parameters and their settings, two groups of experiments were designed. The scanning span between flutes is 80 μm. Other parameter settings are listed in table 4. Before the formal experiments, the simulation results are compared with the measured results when the laser power is 130 W, with the scanning speed 800 mm/s. The results of comparison show that the simulation results are consistent with the results of the experiments.

Table 4. Parameters of experiments.

Experimental Groups	The First Group	The Second Group
Number of Test Pieces	5	8
Laser Power (W)	110~150	130
Laser Scanning Speed (mm/s)	800	400~1800
Scanning Mode	Unidirectional Scanning	Unidirectional Scanning
Distance between Starting Point and Origin (mm)	2	2
Distance between End Point and Origin (mm)	35	35
Width (mm)	1	1
Number of Forming Layers	100	100

3.2. Experimental Results and Analysis

To explore the impact of power on radiation signal, five test pieces of the first group of experiments in the table 4 are taken. For each test piece, the signals of all melting lanes 5 ~ 30 mm from the origin in the same layer (with an interval of 5mm) are acquired to collect their mean value and standard deviation, and the error bar diagram of signal at each distance is shown in figure 6.

It can be clearly seen from the curves in figure 6 that when the molten pool is more than 10mm away from the origin, the radiation signal and the laser power show a good linearity. This is because the higher the laser power, the more energy the material absorbs. So the higher the temperature is, and the higher the radiation energy. When the molten pool is 5 mm away from the origin, the linearity between the radiation signal and the laser power decreases. And in the range of 110 ~ 130W, the radiation signal attenuates with the increase of power, which may be due to the remelting of the molten pool caused by too high laser energy at the origin, and the molten pool sinks downward.

It can also be seen from the figure 6 that the deviation of radiation signals between different flutes is small, indicating that the forming stability and consistency are good under the condition of maintaining constant power.

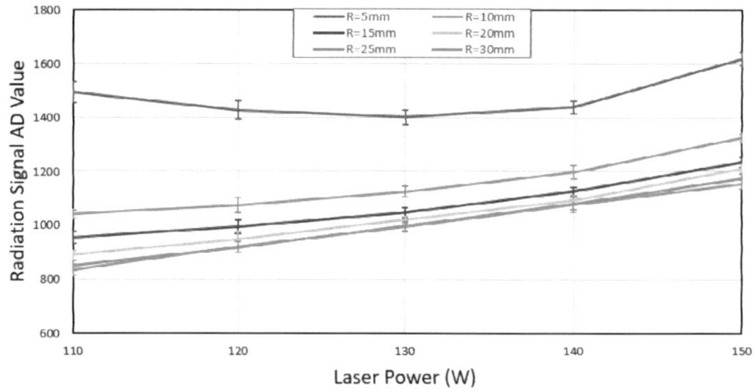

Figure 6. Relation curve between radiation signal and laser power.

In order to study the impact of different scanning speeds on the state of molten pool, eight test pieces of the second group of experiments in the table 4 are taken. For each test piece, the signals of all melting lanes 5 ~ 30mm away from the origin in the same layer (with an interval of 5mm) are measured. Their mean value and standard deviation are collected, and the error bar diagram of signal and scanning speed at each distance is shown in figure 7.

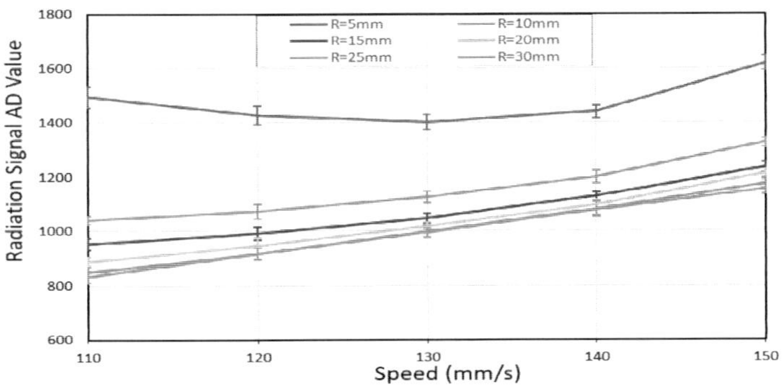

Figure 7. Relation curve between radiation signal and laser scanning speed.

It can be seen from the trend of each curve in figure 7 that no matter how far away from the molten pool to the origin, the radiation signal always decreases first and then increases with the enhancement of scanning speed. It is considered that the enhancement of laser scanning speed means that the energy of laser per unit distance is reduced, the energy absorbed by powder is also reduced, and the radiation energy released is also reduced. When the scanning speed continues to increase, the weld bead length corresponding to the sampling data in a sampling period also continues to increase. Because the coaxial optical path system actually observes a circular area with a radius of 1mm centered on the current molten pool, the number of molten pools actually observed increases and the induced radiation energy increases accordingly with the increase of the distance of the observed molten pool.

Additionally, figure 7 also shows that the farther the molten pool is from the origin, the less the influence of the scanning speed on the radiation signal. This may be due to the change of the shape of the laser spot when the distance becomes larger, which has a preheating effect on the subsequent powder and reduces the impact of the change of laser energy density on the energy absorbed by the molten pool.

4. Conclusion

For the effects of different laser power on molten pool temperature or radiation, figure 2 and figure 6 show that the molten pool temperature or radiation is linearly related to the laser power, both of the simulation results and the experimental results have the same trend.

For the influence of scanning speed on molten pool temperature or radiation, comparing figure 7 to figure 3, however, the experimental results are slightly different from the simulation results. It can be considered that this is because the actual detection is the total radiation energy of multiple molten pool areas in a sampling period, while the simulation selects the maximum temperature of a single molten pool. But the essence of the both trends is the same, that is, the faster the laser scanning speed, the less energy absorbed by the molten pool and the less energy radiated outward.

In conclusion, the two typical technological parameters studied in this paper have a direct impact on the state of molten pool. The experimental and simulation results verify each other, and this regulation is helpful to guide the reasonable setting and adjustment of parameters in the actual production processing.

Acknowledgements

This work was supported by Shenzhen Fundamental Research Program (grant project: Defect identification and forming quality intelligent management in SLM).

References

[1] Do D K and Li P 2016 The effect of laser energy input on the microstructure, physical and mechanical properties of Ti-6Al-4V Alloys by selective laser melting *Virtual and Physical Prototyping* **11(1)** 41-47 https://doi.org/10.1080/17452759.2016.1142215

[2] MATRAS A 2020 Research and optimization of surface roughness in milling of SLM semi-finished parts manufactured by using the different laser scanning speed *Materials* **13**(1) 9-10 https://doi.org/10.1002/sca.20274
[3] Yuan W, Chen H, Cheng T and Wei Q S 2020 Effects of laser scanning speeds on different states of the molten pool during selective laser melting: simulation and experiment *Materials & Design* **189** 21-26 https://doi.org/10.1016/j.matdes.2020.108542
[4] Ni X Q, Kong D C, Wen Y, Dong C F, Zhang L, Lu L, Song J and Wu W H 2019 Influencing factors and improvement methods of porosity in 3D printing metal materials *Powder Metallurgy Technology* **37(03)** 163-9.
[5] Sun C 2020 *Study on Molten Pool Structure in Laser Selective Melting Additive Manufacturing Process* (Zibo: Shandong University of Technology)
[6] Zhang Hui 2017 *Study on Solidification Microstructure Evolution of TC4 Alloy Melted by Selective Laser* (Harbin: Harbin Institute of Technology)
[7] Li H X, Huang B Y, Sun F and Gong S L 2013 Microstructure and tensile properties of Ti-6Al-4V alloys fabricated by selective laser melting *Rare Metal Materials & Engineering* **S2** 209-212
[8] Yuan J G, Li Y, Liu J N, Chen C Y, Pang E L, Tan S J and Jin Y M 2020 Online measurement of molten pool temperature in the forming process of selective laser melting *Chinese Journal of Lasers* **47(03)** 158-166.

Numerical Simulation of Internal Flow Field of Self-Designed Centrifugal Pump

Xin Wang[1], Jun Zhang [1*] and Zongshun Li [2]

[1] Department of Marine Equipment and Mechanical Engineering, Jimei University, Xiamen, China,
[2] Fujian Sea Knight Pump Co., Ltd., Zhangzhou, China.
E-mail: bull0202 @sina.com.

Abstract. The two self-designed of centrifugal pump with small vane and centrifugal pump without small vane were simulated numerically to select a centrifugal pump with higher efficiency. The internal flow characteristics of the centrifugal pump was simulated by using Reynolds time-averaged N-S equation and RNG turbulence model to obtain pressure and velocity distribution and the cavitation characteristics were simulated by using SST turbulence model and Schnerr-Sauer cavitation model to obtain the gas volume fraction distribution. The results show that at the same flow rate, the change of velocity in the pump cavity of centrifugal pump with vane is smoother and the gas volume is less, but the back-flow is aggravated near the small vane, especially when interacting with the tongue, and a large amount of gas is generated at the suction surface of the small vane. In addition, the efficiency of centrifugal pump without vane is higher than that of centrifugal pump with vane, which provides a basis for the structural optimization of centrifugal pump.
Keywords. Centrifugal pump, numerical simulation, internal flow field, cavitation.

1. Introduction

Centrifugal pump, as a widely used fluid machinery, plays a role in agriculture, water conservancy, industry and other fields. In addition to meeting the basic requirements of flow rate and head, the most important thing in the design of a pump is the high efficiency, which is closely related to the energy consumption and affects the production cost and product competitiveness [1]. The impeller is the core component of the centrifugal pump, and its performance affects the efficiency of the whole pump. Centrifugal pumps not only have a complex internal flow structure, with the computational domain divided into two sub-regions: the stator and the rotor, but also the flow is of a three-dimensional, viscous, non-constant form [1, 2]. The complex operating environment requires increasingly high performance of centrifugal pumps, and the limitations of traditional empirical methods make it particularly important to apply numerical simulations to predict the performance of centrifugal pumps [3, 4]. Siddiqi [5] used a multi-objective genetic algorithm (MOGA) to optimize a seven-vane impeller, and the results showed that the optimized impeller resulted in a more uniform pressure and velocity distribution on the meridional plane of the impeller and higher efficiency. Young-Joon An [6] used CFX to numerically simulate the three-dimensional incompressible flow of a centrifugal pump, and the impeller with a smooth curve along the front cover showed good performance. The inlet and outlet pressure difference decreases with increasing flow rate under the influence of flow variation; high head can be obtained at low flow rates. For hydraulic machinery, in addition to the basic performance parameters of head and efficiency, cavitation instability is also a crucial issue and is quite harmful to its operational safety [7]. Yang [8] studied the effect of cavitation on centrifugal pumps by changing the inlet pressure, and the results showed that when the inlet pressure is less than the saturated vapor pressure of the transported liquid, the liquid vaporizes and gas is generated, and efficiency decreases due to cavitation in the pump. With the increase of cavitation, the efficiency decreases obviously. Wei [9] based on the SST turbulence model, the pressure pulsation characteristics of centrifugal pumps at four typical flow rates were Numerical simulations were performed based on the SST turbulence model. The results showed that uneven flow distribution at the impeller inlet would produce fluctuations at the leading edge of the impeller and large size return vortices in the flow channel, leading to low pressure areas at the impeller and the spacer tongue, which may produce vaporization and thus affect cavitation. Deng [10] used the SST turbulence model and the Zwart-Gerber-Belamri cavitation model to numerically simulate the full flow channel of a centrifugal pump and obtained that the gas first appears in the suction surface at the leading edge of the vanes, as well as cavitation exacerbates the instability of the flow and affects the centrifugal pump's ability to do work.

Impeller and volute of the water impact will produce vortex, is one of the reasons for the decline in pump efficiency, but there is less research on how to change the impeller structure affects the reflux zone. In order to reduce this loss, this paper designed two kinds of impellers with and without small vane. The large vane of the two impellers adopts the forward vane type, and the first impeller adds the small vane of the same curvature on this structure, the purpose is to compare and analyze the influence of the two vane forms on the pressure and velocity distribution of the centrifugal pump, which can provide some reference basis for improving the efficiency of the centrifugal pump.

2. Computational Model

2.1. Physical Model

According to the given structural parameters, two different three-dimensional models of the impeller were designed as shown in figure 1. In order to balance part of the axial thrust on the impeller, the impeller was designed with back vanes. The physical model of the centrifugal pump basin was obtained by flow channel extraction using SCDM software. Due to the complex structure of the centrifugal pump model, the mesh was divided with unstructured mesh in ICEM as shown in figure 2. The number of grid cells of centrifugal pump A impeller is 296527, the number of grid cells of centrifugal pump B impeller is 295601, the number of grid cells of worm housing runner is 579347, the total number of grid cells of centrifugal pump A is 1275914, and the total number of grid cells of centrifugal pump B is 1274948.

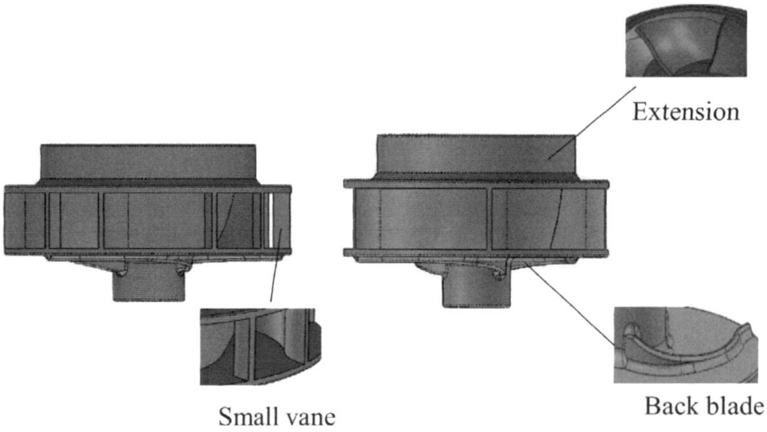

Figure 1. Impeller three-dimensional model.

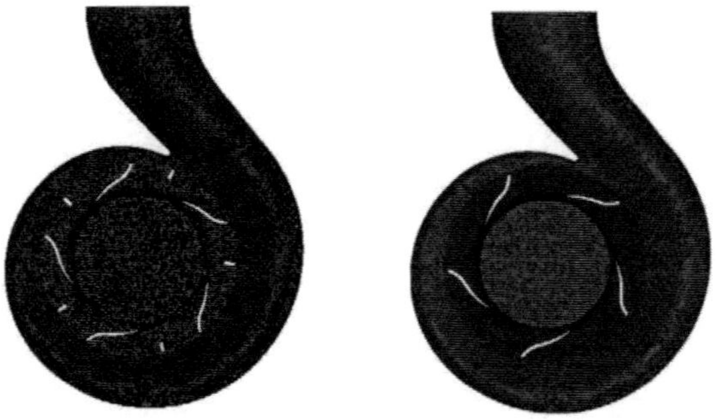

(a) (b)

Figure 2. Volute and impeller mesh division.

2.2. Numerical Simulation

2.2.1. Control Equations. The MRF model is applied to the internal flow of this centrifugal pump to convert the internal flow field non-constant problem into a constant problem to solve. The equations of motion in the rotating coordinate system can be expressed as [1]:

$$\frac{\partial \vec{v}_r}{\partial t} + \vec{v}_r \cdot \nabla \vec{v}_r + 2\vec{\Omega} \times \vec{v}_r + \vec{\Omega} \times \vec{\Omega} \times \vec{r} + \frac{\partial \vec{\Omega}}{\partial t} \times \vec{r} = -\frac{1}{\rho}\nabla p + \frac{1}{\rho}\nabla \cdot \tau + \vec{f} \tag{1}$$

where: $\vec{v}_r$ -relative velocity vector, $\vec{\Omega}$ -the rotational angular velocity of the rotating coordinate system, $\vec{r}$ -the position vector of the point in the rotating coordinate system, τ -viscous stress tensor, ρ -fluid density, $\vec{f}$ -force per unit mass, $2\vec{\Omega} \times \vec{v}_r$ -Gothic acceleration, $\vec{\Omega} \times \vec{\Omega} \times \vec{r}$ -centrifugal acceleration.

For this centrifugal pump the full flow channel three-dimensional constant numerical calculation, the medium is water and the N-S equation for the time homogenization can be expressed as [11]:

$$\frac{\partial u_i}{\partial t} + u_j \frac{\partial u_i}{\partial x_j} = f_i - \frac{1}{\rho}\frac{\partial p}{\partial x_i} + \frac{\partial}{\partial x_i}\left(V \frac{\partial u_i}{\partial x_j} - \overline{u_i' u_j'} \right) \tag{2}$$

where: u_i' , u_j' -the pulsation value of the instantaneous velocity

For constant, incompressible flow, neglecting the bulk force, the above equation can be rewritten as:

$$u_j \frac{\partial u_i}{\partial x_j} = -\frac{1}{\rho}\frac{\partial p}{\partial x_i} + \frac{\partial}{\partial x_i}\left(V \frac{\partial u_i}{\partial x_j} - \overline{u_i' u_j'} \right) \tag{3}$$

2.2.2. Turbulence Model. Rotational effects are considered in the model, so the RNG turbulence model is used to improve the accuracy of the rotational flow calculation. The RNG turbulence model expression is [12]:

$$\frac{\partial(\rho k)}{\partial t} + \frac{\partial(\rho k u_i)}{\partial x_i} = \frac{\partial}{\partial x_i}\left(\alpha_k \mu_{eff} \frac{\partial \varepsilon}{\partial x_j} \right)\frac{\partial k}{\partial x_j} + G_k + \rho\varepsilon \tag{4}$$

$$\frac{\partial(\rho\varepsilon)}{\partial t} + \frac{\partial(\rho\varepsilon u_i)}{\partial x_i} = \frac{\partial}{\partial x_j}(\alpha_\varepsilon \mu_{eff} \frac{\partial \varepsilon}{\partial x_j}) + \frac{\varepsilon}{k}(C_{1\varepsilon} G_k - C_{2\varepsilon}\rho\varepsilon) \tag{5}$$

which: $\mu_t = \rho C_\mu \frac{k^2}{\varepsilon}$, $\mu_{eff} = \mu + \mu_t$, $C^*_{1\varepsilon} = C_{1\varepsilon} - \frac{\eta(1-\eta/\eta_0)}{1+\beta\eta^3}$, $\eta=\sqrt{2}E_{ij}\frac{\varepsilon}{k}$, $E_{ij} = \frac{1}{2}(\frac{\partial u_i}{\partial x_j} + \frac{\partial u_j}{\partial x_i})$.

where: $\alpha_k = \alpha_\varepsilon = 1.39$, μ_{eff} - viscosity, $N{\bullet}s/m^2$, $C_\mu = 0.0845$, E_{ij} -time-averaged strain rate, $C^*_{1\varepsilon}$ - empirical constant correction term, η_0=4.377, $C_{1\varepsilon} = 1.42, C_{2\varepsilon} = 1.68$, $\beta = 0.012$.

2.2.3. Cavitation Model. In this paper, we use the Schnerr-Sauer cavitation model, whose basic equation can be expressed as [13]:

$$\dot{m} = \begin{cases} \dot{m}^+ = C_{evap}\dfrac{3\alpha_v(1-\alpha_v)}{R_B}\dfrac{\rho_1\rho_v}{\rho_m}{\bullet}\left(\dfrac{2}{3}\dfrac{(P_V-P)}{\rho_1}\right)^{\frac{1}{2}} & P < P_V \\ \dot{m}^- = C_{cond}\dfrac{3\alpha_v(1-\alpha_v)}{R_B}\dfrac{\rho_1\rho_v}{\rho_m}{\bullet}\left(\dfrac{2}{3}\dfrac{(P-P_v)}{\rho_1}\right)^{\frac{1}{2}} & P \geq P_V \end{cases} \quad (6)$$

where: $\dot{m}$ -mass transfer source term, when $\dot{m} > 0$,indicates evaporation process, $\dot{m} < 0$,indicates the condensation process, ρ_m, ρ_1 and ρ_v -mixed phase density, liquid phase density and gas phase density, α_v -volume fraction of gas phase, R_B -radius of bubble, P_B -surface pressure of bubbles, P - environmental pressure.

2.3. Boundary Conditions and Numerical Simulation Methods

The fluid inside the centrifugal pump has a complex flow form, and three pairs of cross interfaces are set at the junction of the inlet section and the impeller, the impeller and the worm gear flow path, and the worm gear outlet and the outlet section, which are used to exchange fluid parameters and ensure the continuity of the cross interfaces. The design flow rate of this centrifugal pump is known, so the inlet boundary condition of this centrifugal pump is set as velocity inlet, and the outlet boundary condition is free outflow. The medium is water, and gravity needs to be considered. The wall boundary condition is set to wall without slip condition, and the flow near the solid wall is determined by the standard wall function. The solver uses SIMPLE separated solver. The viscous term automatically adopts the second-order accuracy discrete format, and the discrete format of both turbulent kinetic energy and turbulent dissipation rate equations adopt the first-order windward format. If the residuals increase, the default values of the relaxation factors for pressure, momentum, turbulent kinetic energy and turbulent dissipation rate can be adjusted downward.

2.4. Fluid Physical Parameters and Pump Design Working Conditions Parameters

The transfer medium is water, the pump design working parameters and fluid physical parameters are shown in table 1.

Table1. Pump design working parameters and fluid physical parameters.

Design Parameters	Numerical value
Flow rate $Q/\left(m^3 \cdot h^{-1}\right)$	130,140,150
Head H/m	20
Media density $\rho/\left(kg \cdot m^{-3}\right)$	1000
Viscosity coefficient $kg/\left(m \cdot s\right)$	1.005×10^{-3}

3. Simulation Results and Analysis

3.1. Pressure Distribution Cloud

The cloud diagram of pressure distribution at the impeller inlet section is shown in figure 3. From figure 3 (a) and (b), it can be seen that the fluid pressure rises gradually in a step-like manner from the blade inlet to the outlet, and the pressure reaches the maximum at the outlet. It can be seen that the simulation results of both meet the performance requirements of pressurizing the fluid by impeller work. Low pressure zone appears in the blade leading edge of the suction surface, the pressure is negative, according to the conditions of cavitation occurrence, here easy to produce gas, and the actual situation.

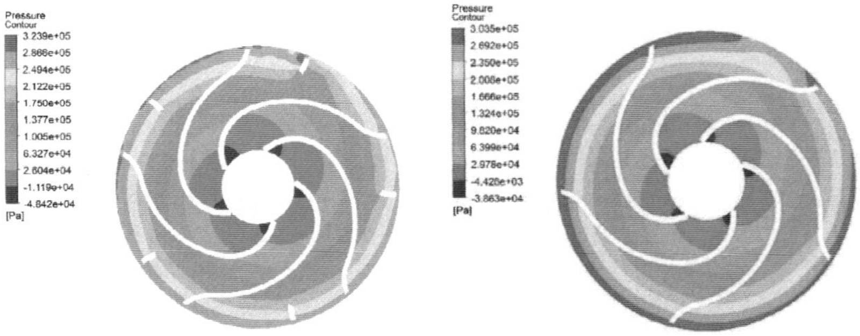

(a) (b)

Figure 3. Impeller inlet section pressure cloud distribution.

3.2. Velocity Vector Distribution Cloud

The cloud diagram of the velocity vector distribution of the center section of the impeller of the centrifugal pump A worm casing is shown in figure 4(a). As can be seen from the figure, the velocity of the vane at the inlet is higher and the flow is uniform, which indicates that the impeller structure is reasonably designed. From the blade inlet to the outlet, the speed gradually decreases, while the pressure distribution cloud chart can be seen, the pressure is gradually increasing, so to meet the law of conservation of energy. Fluid in the impeller rotation driven by work to obtain energy, high speed left the outer edge of the impeller into the worm gear flow channel, the role of the small blade worm gear flow channel velocity vector is roughly uniform state distribution. By figure 4 (b) 1, 2, 3 can be seen in the worm gear flow channel there are a few tiny reflux vortex; by figure 4 (b) 1, 3, 4 can be seen in the small blade suction surface near all produced a tiny reflux vortex, which is due to the different curvature of the blade and rotational movement caused; by figure 4 (b) 3 to see the speed distribution near the spacer tongue disorder and accompanied by reflux, which is because of the resistance of the spacer tongue and impeller driven by high speed rotation of fluid [2], and when the small blade rotation to the septum area reflux situation aggravated, in the worm shell reflux vortex and small blade reflux vortex between another obvious reflux vortex, this is the septum and small blade structure interaction results, the fault type small blade to increase the pressure drop, speed more turbulent, resulting in energy loss. That small blade structure design rationality needs to be improved.

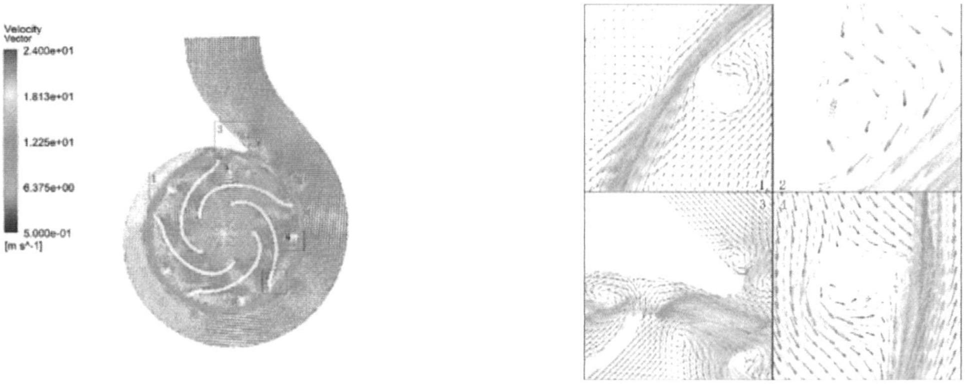

(a) (b)

Figure 4. Cloud diagram of velocity vector distribution in the center section of volute impeller of centrifugal pump A.

The cloud diagram of velocity vector distribution in the center section of the worm gear impeller of centrifugal pump B is shown in figure 5(a). From figure 5(b) 1, 2 and 4, it can be seen that there are several continuous tiny backflow vortices in the worm gear flow channel, and the velocity vector of the worm gear flow channel is not uniform between 0° and 135° cross-section; the velocity near the spacer tongue is turbulent and accompanied by tiny backflow, and from figure 5(b) 3, it can be seen that tiny backflow vortices are generated at the outlet of this blade. Overall, the internal flow field of this centrifugal pump B is in good condition, and there is no obvious backflow vortex.

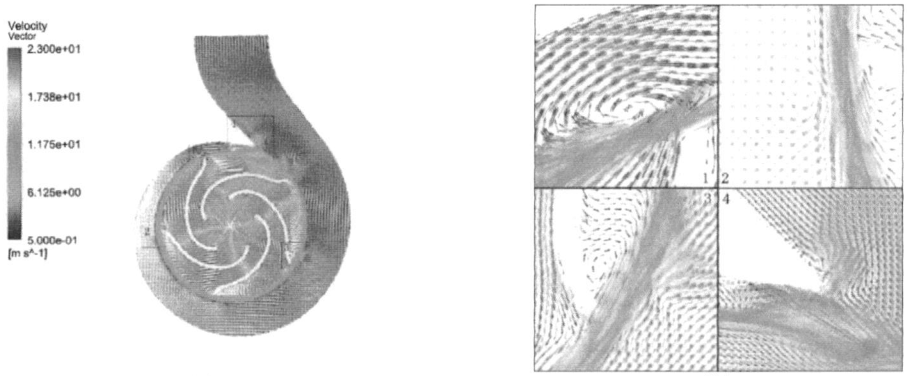

(a) (b)

Figure 5. Cloud diagram of velocity vector distribution in the center section of volute impeller of centrifugal pump.

3.3. Cavitation Analysis

Cavitation is closely related to the pressure instability of centrifugal pumps, which can reduce pump performance, generate vibration and noise, and even corrode the pump flow components [14]. To eradicate cavitation, it is necessary to prevent the generation of gas at the impeller inlet, which is the pressure at the inlet is too low. In this paper, cavitation is simulated using Mixture multiphase flow model, turbulence model, Schneer-Sauer model for cavitation, Coupled solver, and PRESTO for pressure discretization format [15]. The inlet and outlet boundary conditions of the centrifugal pump are set to be pressure boundaries, and the inlet pressure is taken to be the inflection point value of 3% head drop, the reference pressure is 0 Pa, the saturated vapor pressure is 3540 Pa under the current operating conditions, and other boundary conditions are kept as default.

The cloud diagram of the gas volume fraction distribution of the impeller of centrifugal pump A is shown in figure 6. As can be seen from the figure, the cavitation of the impeller is manifested as the attached gas developing up on both sides of the blade leading edge surface, and the gas volume fraction is not uniformly distributed on each blade surface, and gradually separates from the blade surface with the growth of the gas volume fraction and develops toward the center of the blade flow path.

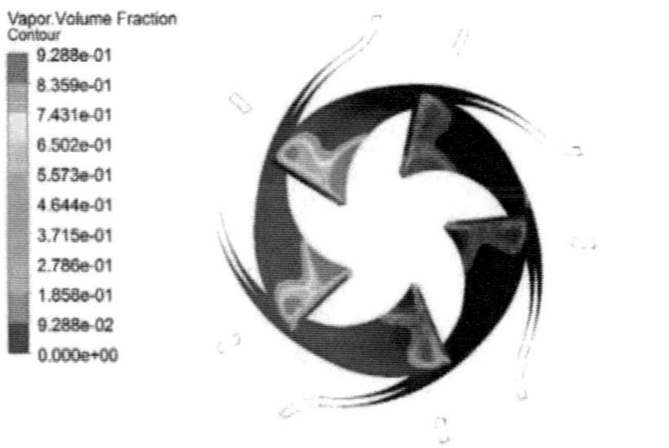

Figure 6. Impeller A gas volume fraction distribution.

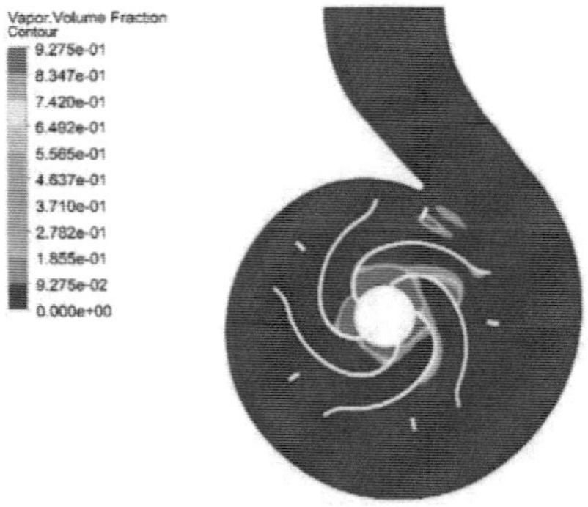

Figure 7. Impeller A inlet section gas volume fraction distribution.

The cloud plot of the gas volume fraction distribution in the inlet section of the impeller of the centrifugal pump A is shown in figure 7. As can be seen from the figure, there is also a small amount of adherent gas at the exit of some of the blades and it does not develop into the central flow channel [15]. From the velocity vector diagram, it can be seen that the flow in the septal tongue region is turbulent and the flow impact is large, which causes energy loss and the existence of low pressure area; in addition, when the small blade rotates to the vicinity of the septal tongue, the resulting multiple backflow vortices cause the pressure drop to become large, and the small blade suction surface also produces gas, which is more damaging to the small blade. On the whole, the low pressure area at the impeller inlet, the low pressure area near the spacer tongue and the suction surface of the small blade are easy to produce gas, and the cavitation performance needs to be improved.

Centrifugal pump B impeller and its inlet section gas volume fraction distribution cloud diagram as shown in figures 8 and 9. It can be seen that centrifugal pump B has similar cavitation characteristics with centrifugal pump A, but it generates more gas volume than centrifugal pump A, which is more likely to produce cavitation in the high-pressure area of the pump cavity, resulting in a decrease in pump efficiency.

Figure 8. Impeller B gas volume fraction distribution.

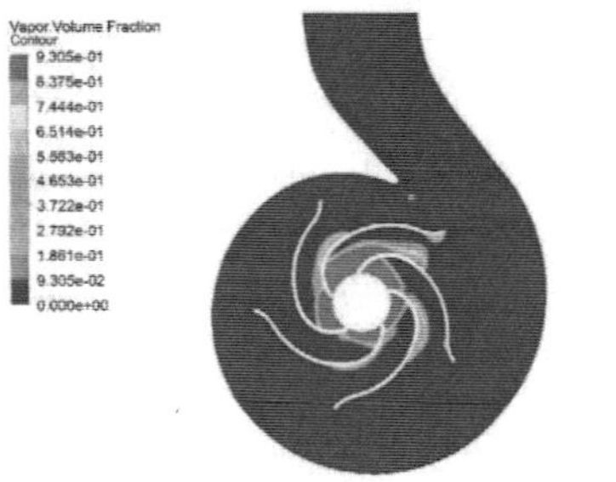

Figure 9. Impeller B inlet section gas volume fraction distribution.

4. Efficiency Comparison

With the FLUENT solution setup, parameters such as pressure, torque, and flow rate of the internal flow of the centrifugal pump can be obtained, which in turn leads to performance characteristics such as head, shaft power, and efficiency.

According to the definition of centrifugal pump head H, its equation expression is [16]:

$$H=\frac{P_{out}-P_{in}}{\rho g} \tag{7}$$

where: P_{out},P_{in} -total pressure of inlet and outlet, Pa , ρ -media density, kg/m^3 , g -gravitational acceleration, m/s^2 .

axis power: $P=M\omega, \omega=\frac{2\pi n}{60}$

Centrifugal pump efficiency:

$$\eta=\frac{\rho gQH}{3600M\omega} \tag{8}$$

where: Q -Volume flow rate of centrifugal pumps, m^3/h .

The comparison of the efficiency of centrifugal pumps A, B at different design flow rates is shown in figure 10. As can be seen from the figure, the efficiency of centrifugal pump A is lower than that of centrifugal pump B at the same flow rate.

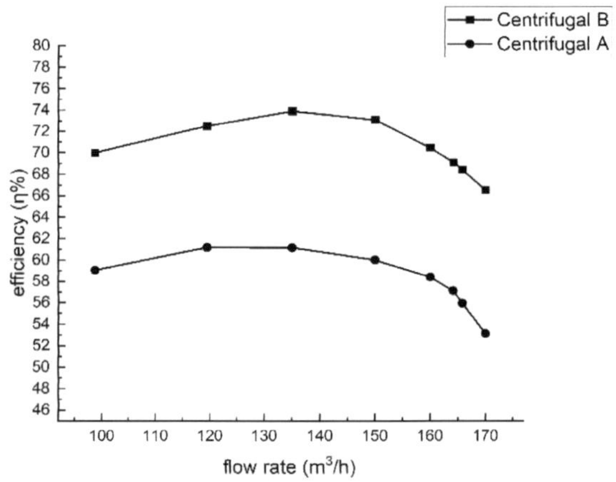

Figure 10. Comparison of efficiency of centrifugal pumps A,B at different design flow rates.

5. Conclusion

Based on CFD technology and MRF model, the three-dimensional turbulent flow inside the centrifugal pump is simulated numerically using Reynolds time-averaged N-S equation with RNG and SST turbulence models and Schnerr-Sauer cavitation model, and the pressure distribution, velocity vector distribution and cavitation characteristics of the internal flow field of the centrifugal pump are obtained. The results show that:

(1) The small vane can play a role in preventing the expansion of the counterflow area, reduce the convective impact of the fluid, so that the water flow evenly and the gas volume is reduced, but the small vane makes the pressure gradient in the partition tongue area increase, the speed is more turbulent, reflux aggravation, resulting in energy loss, reducing the efficiency of the centrifugal pump

A. The structure design rationality needs to be improved.

(2) Small vane can increase the head of the centrifugal pump, but it will increase the shaft power and reduce the efficiency.

Acknowledgements

This study is supported by Fujian Province Science and Technology Plan Funding Project (2017H0024); Fujian Province Science and Technology Commissioner Project (Min Ke Nong "2019" No. 11)

References

[1] Tang H and He F 2002 Numerical simulation of the flow field inside a centrifugal pump *Pump Technology* (03) 3-8+14.

[2] Wang Z J, Tong L and Zheng J S 2012 Numerical simulation and performance prediction of internal three-dimensional flow of centrifugal pumps based on CFD *Fluid Machinery* **40**(06) 14-18.

[3] Song D M, Liao G L and Liu X Y 2017 Performance prediction of double suction centrifugal pumps based on turbulence numerical simulation *Machinery* **44**(08) 18-21.

[4] Song S F and Liu Y T 2020 Numerical simulation of internal flow field of centrifugal pump based on CFD *Energy and Energy Conservation* (03) 99-100+156.

[5] Siddiqi M A, Boppa P and Strongin M P 2013 Multi-Parameters Pump Impeller Optimization *American Society of Mechanical Engineers Digital Collection* 387–394.

[6] Young-Joon An, Shin B R and Wahid M A 2010 Numerical Simulation of Centrifugal Pump with Double-Suction Impeller *Kuala Lumpur (Malaysia)* 456–464.

[7] Luo X W, Ji B and Yoshinobu Tsujimoto 2016 A review of cavitation in hydraulic machinery *Journal of Hydrodynamics* **28**(03) 335-358.

[8] Yang Z J, Wang K T and Nie Y L 2020 Experimental study of the effect of cavitation on the performance of centrifugal pumps *China Water Power & Electrification* (02) 50-54.

[9] Wei Z, Yang W and Xiao R 2019 Pressure Fluctuation and Flow Characteristics in a Two-Stage Double-Suction Centrifugal Pump *Symmetry* **11**(1) 65.

[10] Deng Z Q, Jiang J R and Liu X B 2021 Analysis of non-constant cavitation in the internal flow field of a centrifugal pump *Water Resources and Power* **39**(06) 166-170.

[11] Chen W 2007 Simulation of internal flow and structural optimization of low specific rpm centrifugal pumps *Zhejiang University*.

[12] Yao Z F, Wang F J and Xiao R F 2012 Experimental investigation of pressure instabilities affected by cavitation for a double-suction centrifugal pump *Earth and Environmental Science* **15**(6) 062040.

[13] Hu J, Hou X Y and Yu Y 2021 A Modified Schnerr-Sauer Cavitation Model with Local Flow Properties *Transactions of Beijing Institute of Technology* **41**(01) 9-15.

[14] Yang S S, Kong F Y and Wang W T 2010 Numerical calculation and analysis of cavitation performance of centrifugal pumps *Journal of Huazhong University of Science and Technology (Natural Science Edition)* **38**(10) 93-95.

[15] Hu K, Hu T T and Ma H F 2019 Ansys Fluent Example details *Beijing: China Machine Press*.

[16] Huang S and Sang D K 2010 Numerical simulation and performance prediction of 3D flow field of multi-stage centrifugal pump *Mechanical Science and Technology for Aerospace Engineering* **29**(06) 705-708.

Feasibility of Boosting Low-Energy-Quality Steam by Using Wave Rotor

Yuqiang Dai[1,2,a*], Zhipeng Tang[1], Mohan Li[1], Gang Hao[2,3], Luwei Zhang[1] and Jintao Wu[1,2,b*]

[1]Department of Chemical Machinery and Safety Engineering, Dalian University of Technology, Faculty of Chemical Engineering, room H515, Chemical Experiment Building, Dalian, Liaoning, China.
[2]Special Fluid Technology and Equipment Joint Research and Development Centre of Dalian University of Technology and Ackam (Jiangsu) Industrial Technology Co., Ltd. China.
[3]Ackam (Jiangsu) Industrial Technology Co., Ltd., No.280 Fumin Road, Economic development zone, Changzhou, Jiangsu, China.
Email: [a] daiyuqiang@dlut.edu.cn, [b] wujt75@dlut.edu.cn

Abstract. Owing to the difficult utilization of the low-pressure level in the process industry, the low-energy-quality steam is often condensed to recover the demineralized water or just discharged directly, causing a huge waste of thermal energy. A novel technology of enhancing the steam's energy quality by using the wave rotor based on the principle of moving shockwave compression is proposed. The supercharging ability of 3-port wave rotor is studied by meaning of 1-dimension unsteady theory and computational fluid dynamic. A practical thermodynamic flowsheet of boosting the low-pressure steam driven by high-pressure steam is also proposed and analysed in detail. As an example, to boost the saturated steam of pressure 1.0 MPa to 1.953 MPa, a three-stage wave rotor solution is proposed and is verified its feasibility. The high supercharging ratio and entrainment ratio of the wave rotor are much higher than the traditional steam ejector shows the feasibility of enhancing energy-quality of low-pressure steam.
Keywords. Steam boosting, shockwave, wave rotor, work exchanging network.

1. Introduction

Steam is an important part of utilities in process industries such as petroleum, chemical and energy. Energy power cost accounts for a large proportion of the total cost of production and is an important part of the price of products. It has always been the focus of energy conservation and emission reduction in factories. The energy quality of industrial steam is determined by the pressure level. The high pressure or high energy quality steam can be used flexibly in the process equipment. However, the-state-of-the-art situation is that after the integrated optimization of the process equipment's energy, a large amount of saturated steam by-products with the pressure level of one bar will be obtained. Their energy quality is low and often cannot be effectively utilized. Finally, low-energy-quality steam will be discharged or recycled to softened water by the external condenser, so lots of low-quality energy is often wasted.

Low-quality steam energy recycling has become an important issue for energy saving in various process industries. However, surveys of domestic and international research have found that typical solutions use the remaining heat to recycle condensed soft water and use superheaters to turn steam into superheated steam. The main technical disadvantage in such heat transfer technology of steam energy utilization is that the total heat transfer factor between superheated vapour and process stream is always very low, and the heat exchanger has very large volume or large investment.

The solution aiming at improving low-quality steam pressure is a natural choice. The conventional method uses a compressor to boost the steam to the desired high pressure. The technology also has economic and technical problems. On the one hand, electric energy has higher quality, so it is not an ideal choice to consume it. On the other hand, due to the low viscosity of steam and the large specific capacity, it is difficult to maintain the boundary layer of centrifugal steam compressor blades. It is hard for engineers to develop such a high-speed centrifugal compressor with a large volume flow rate.

In this article, a novel idea that uses moving shock waves to supercharge the low-quality steam to increase its energy quality, is proposed. The steam after enhancing quality can be used as process utility in equipment again. This low-energy-quality improving technology based on shock wave supercharging principle will become the key technology in energy conservation of process industry.

2. Gas-dynamic and Thermodynamic Model of Wave Rotor

2.1. Gas Dynamic Model of Wave Rotor

The shock boosting process is implemented in an advice called wave rotor which is an unsteady flow device where two streams exchange pressure energy directly due to movement of gas waves generated in channels [1]. It can get higher isentropic process efficiency than steady flow devices like diffusers [2-3]. Strong and weak discontinuities such as moving shock wave *S*, entropy waves (including rarefaction wave *E* and compression wave *C*) and contact face *J* are formed in wave rotor channels by using a series of circumferentially evenly arranged, double-ended open-ended shock wave tubes under periodic discontinuous boundaries (periodic on/off of inlet and outlet ports) to complete the shock wave supercharging and rarefaction wave self-cooling processes which is different from the steady expansion and compression processes of turbine technology. It is the unsteady moving shock supercharging and rarefaction wave expansion processes.

Wave rotors complete directly pressure energy exchange by gas wave which is different from kinetic energy transfer of turbine technology. This energy transfer pathway of the wave rotor is short (without kinetic energy transfer) and quick (at level of the sonic speed) [4]. Therefore, the rotors can be designed at a low revolution speed. It solves the problem of kinetic energy loss caused by high rotation speed and liquid hammer of turbine technology. And the structure of the bilateral opened channel of the wave rotor's drum makes the liquid drainage convenient and is suitable for operation with wet steam.

Both turbine technology and the wave rotor are the key technologies for efficient use of pressure, they can achieve compression with high expansion refrigeration efficiency. However, the supercharging expansion process of moving gas wave is efficient, and its application can reduce the volume of the equipment significantly due to its unsteady characteristics [5]. These are the fundamental reasons why the powerful nations of science and technology have tried to apply this kind of mechanical development to various fields.

The port number of the wave rotor can be three, and the working wave diagram is shown in figure 1. The through-flow (TF) type of such a wave rotor has the topical characteristics as follows:

- It can fully utilize the self-cooling characteristic of the periodic motion rarefaction wave to cool the channel and achieve a relatively small temperature rise when supercharging [6]. It undergoes multiple rarefaction wave expansion and shock wave compression in one working cycle at a certain axial position in the rotor channel.
- There are two advantages for unchanged direction of the steam velocity in the channel: Firstly, it is easy to discharge from the right port when there is liquid in the rotor channels. Secondly, both hot and cold streams flow through the channel from the left to the right, causing great reduction of the axial heat conduction of the channel wall, and reduction of the heat loss.

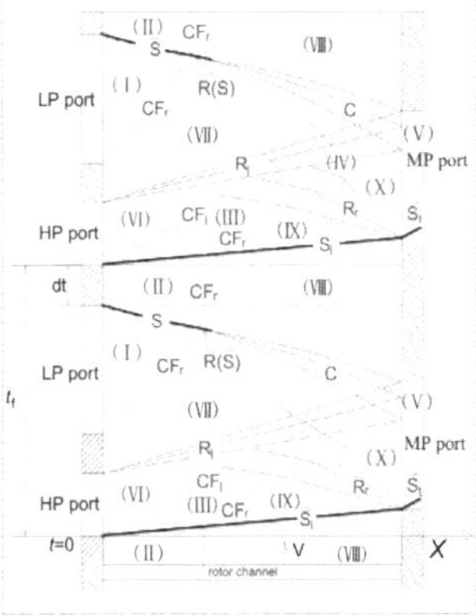

Figure 1. The typical working wave diagram of a three-port wave rotor supercharger.

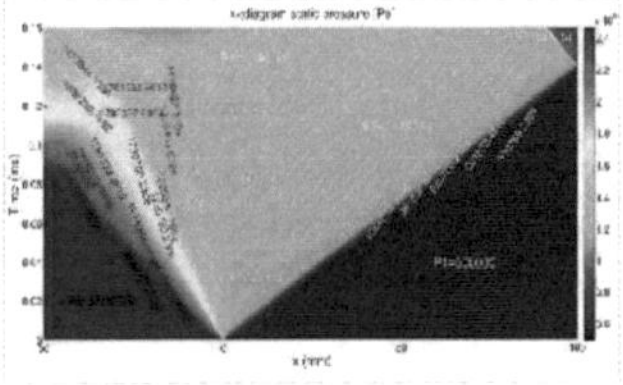

(a)

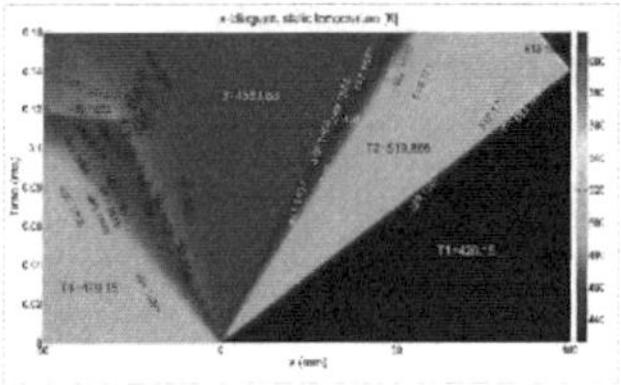

(b)

Figure 2. Wave diagrams illustrated by pressure (a) and temperature (b) in shocktubes.

Table 1. The results of the steam shock tube with p_4=2.5MPa and p_1=0.5MPa.

Zone No.	Pressure /kPa	Temperature /K	Enthalpy /(kJ.kg^{-1})	Velocity / (m.s^{-1})
1	500.00	426.15	2750.92	494.81
2	1126.69	520.67	2932.80	547.01
3	1126.69	458.28	2649.73	NaN[a]
4	2500.00	497.15	2802.10	504.73

[a] Vapor-liquid two phase exists.

The supercharging capacity of the wave rotor can be estimated by gas dynamics methods. A feasible way to evaluate the supercharging performance of the moving shockwave is to use the shock tube, where there is a high pressure zone 4 separated with the low pressure zone 1 by a diaphragm at t=0.0s. Table 1 shows the typical results of the supercharging ability of the moving shockwave with p_4=2.5MPa and p_1=0.5MPa in the shocktube filled with saturated steam on both sides of the diaphragm. The maximum theoretical pressure of the incident shock is about 1.1MPa, or the pressure ratio of the incident shock is $p_{21}=p_2/p_1=2.2$. These results can be verified by the computational fluid dynamics (CFD), and the wave diagram is shown in figure 2.

2.2. Thermodynamic Model of Wave Rotor

According to the wave rotor working wave diagram (figure 1), both driving high pressure steam and the driven low-energy-grade steam experience multiple times shockwave compression and rarefaction wave expansion processes. The driving side experiences 4 expansion and 1 compression processes crossing the zones of HP port → (VI) → (VII) → (VIII) → (IX) → (X) →MP port in *x-t* diagram, the driven side experiences 1 expansion and 2 compression processes crossing the zones of LP port→ (I) → (II)→ (III) →(IV)→ (V)→MP port. It follows that the expansion side (driving fluid side) and the compression side (driven fluid side) of the wave rotor are separated by the contact surface *J* which is just like a massless piston. Theoretically, the driving and driven fluids have no mass transfer, and the energy exchange between them is directly transmitted through the volumetric work w ($= -pdv$, if reversible). It is completely different from the velocity-type transfer shaft work way of the expansion-compression process of turbine and is a thermodynamic process without shaft loss. According to the unique energy transfer principle, a single thermodynamic model of the wave rotor can be established, as shown in figure 3.

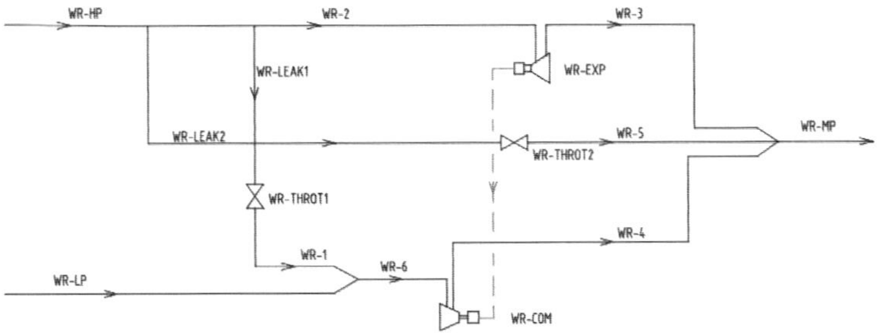

Figure 3. The thermodynamic model of wave rotor.

As the model shown in figure 3, because of the unsteady working process of the expansion work and compression work, the efficiency of wave rotor is always higher than that of the turbomachine. The thermodynamic model of wave rotor can also be modelled with the compression isentropic efficiency ε_{com} and expansion isentropic efficiency ε_{exp} [7-8]. At the same time, there are many leakage paths [9] at the driving or driven side in the wave rotor. All these leakage processes can be regarded as throttling processes. Besides, the driving and driven streams inevitably have the mixing and thermal diffusion phenomena caused by the distortion of the contact face in the practical working process [10-11]. Especially, for 3-port wave rotor, the driving and driven streams will mix in the MP port and have the mixing loss. All these mixing losses are also included in the factors of ε_{com} and ε_{exp}.

3. The Feasibility of Application Examples of Stream Supercharging Technology

Based on the boosting characteristic and the thermodynamic model of wave rotor, it is convenient for engineers to analyse a special process system, such as the automated-cascaded refrigeration cycle with wave rotor, the multiple effect evaporation system with gasdynamic wave vapour recompression, and so on. The following provides a novel solution to enhance the steam energy quality in the process industry. For example, steam with a flow rate of 7.1tons/h can be supercharged from 1.0 MPa to 1.953 MPa by using the wave rotor. The process flowsheet diaphragm (PFD) of such a system can be modelled as shown in figure 4.

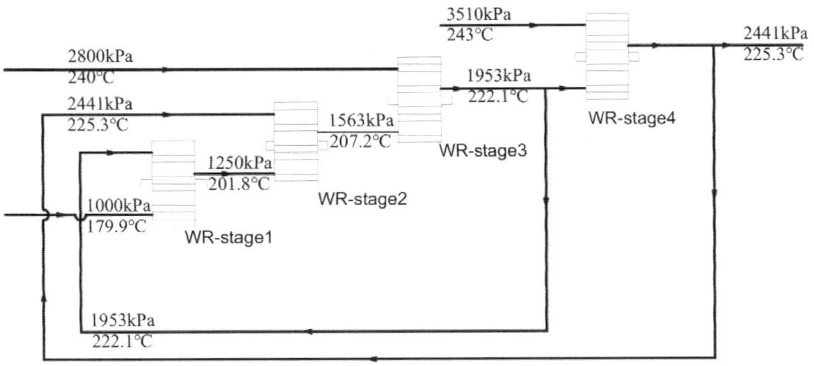

Figure 4. A practical flowsheet of boosting low-pressure steam by using wave rotor.

Accounts for the-stage-of-the-art wave rotor boosting technology, to get about the boost pressure ratio 2.0, one can use three stages of wave rotors with the entrainment ratio of about 0.30. The mass flowrates, entrainment ratio (*ER*), and pressure ratio (*PR*) of each stage of the wave rotors are shown in table 2. The *PR* of the first stage is PR_1=1.25MPa/1.0MPa=1.25, it will consume 22.19 tons/h superheated driving steam with the pressure of 1.953Mpa, which is part of the output of a 3-stage wave rotor rig. The PR_2=1.563MPa/1.25MPa=1.25 at the second stage, the 2.441 MPa, 225.3 °C slightly superheated steam with the flow rate of 97.65 tons/h is needed. The final supercharged steam of 1.953MPa is slightly over-heated with 222.1°C. In the first stage, the wave rotor can accomplish the process with $\varepsilon_{exp} > 44.73\%$, and $\varepsilon_{com} > 60.0\%$; in the second stage, the isentropic efficiencies must be $\varepsilon_{exp} > 44.34\%$ and $\varepsilon_{com} > 60.0\%$, and in the third stage, the isentropic efficiencies must be $\varepsilon_{exp} > 46.13\%$ and $\varepsilon_{com} > 70.0\%$.

Table 2. The mass flowrate of driving (HP), driven (LP) and supercharged (MP) streams of the wave rotor.

	HP port	LP port	MP port	*PR*	*ERs*
1st stage	22.19	7.10	29.29	1.25	0.32/0.32 [a]
2nd stage	97.65	29.29	126.94	1.25	0.30/0.33 [a]
3rd stage	426.10	126.94	550.14	1.25	0.30/0.37 [b]
4th stage	1760.00	527.90	2288.00	1.25	0.30/0.27 [b]

[a] The CFD results of optimal structure.
[b] No CFD is available, so we give the in-house-code results.

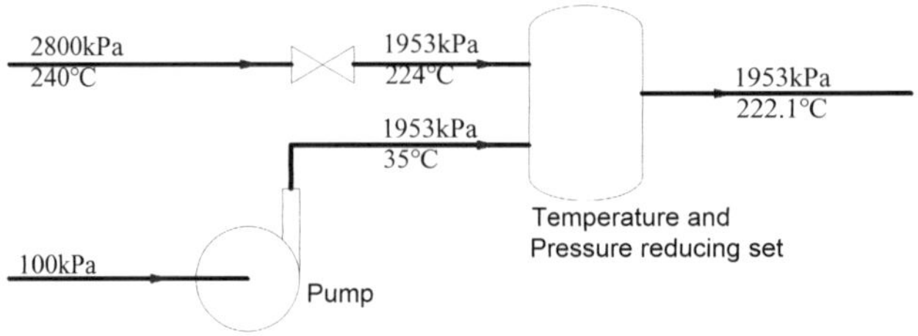

Figure 5. A practical industrial flowsheet of temperature and pressure reduction rig by using check valve set.

For contrast, the conventional steam ejector is also analysed at the same pressure and temperature boundary conditions. For getting the same *PR*= 1.25, at the same inlet process parameters, the first stage of the ejector has negative entrainment ratio. It follows that at a low expansion ratio, the boost performance of ejectors is much lower than that of the wave rotor. Therefore, the wave rotor can achieve a larger *ER* under the same *PR*, and it means more driven steam can be supercharged by using wave rotor. Figure 5 shows a practical industrial flowsheet of temperature and pressure reduction rig by using a check valve set. With the same driving steam of 2.8 MPa and 240 °C, we want to get the same object steam of 1.953 MPa and 222.1 °C. By the temperature and pressure reduction technology, only 0.864 tons/h additional steam is produced, which is much lower than the wave rotor boosting technology (7.1 tons/h).

4. Conclusion

Different from the working principle and energy conversion mode of turbine technology, wave rotor technology can realize unsteady shock supercharging and rarefaction wave expansion processes and direct energy exchange between driving steam and driven steam, and has a high energy conversion efficiency and compact size. Based on the thermodynamic model of the wave rotor, a novel solution to enhance the quality of low-energy steam is proposed and verified its feasibility by a typical case in process industry. It is a promising technology to use the wave rotor to exchange the work between the high-pressure steam and low-pressure-steam and will be widely used in work exchanger network [12] optimal integration system.

Acknowledgments
This present work is supported by the National Key Research and Development Program of China (2018YFA0704602), and the National Natural Science Foundation of China projects (51476015 51106017).

References

[1] Dai Y Q, Ding M X, Tan W H and Hu D P 2010 *J. Dalian Univ. Technol.* **50** 888-895
[2] Akbari P, Nalim R and Mueller N 2006 *J. Eng. Gas. Turb. Power*, **128** 717-735
[3] Akbari P, Kharazi A A and Müller N 2003 *Proc. of the ASME 2003 Inter. Mechanical Engineering Congress* IMECE2003-55222 p 1
[4] Weber H E 1986 *ASME 1986 Inter. Gas Turbine Conf. and Exhibit (Dusseldorf)* 86-GT-62 **1** p V001T01A025
[5] Piechna J, Akbari P, Iancu F and Müller N 2004 *Proc. of IMECE04 2004 ASME Inter. Mechanical Engineering Congress (Anaheim)* IMECE2004-59022 p 1
[6] Fatsi A and Ribaud Y 1999 *Aerosp. Sci. Technol.* **3** 293-299
[7] Chan S N, Liu H X and Xing F 2016 *J. Eng. Gas Turb. Power* **138** 112601
[8] Dai Y Q, Zou J P, Zhu C, Liu P Q, Zhao J Q, Zhang L M and Hu D P 2010 *J. Therm. Sci.*, **2** 021011
[9] Akbari P, Nalim R, Donovan E S and Snyder P H 2008 *J. Propuls. Power*, **24** 732-740
[10] Dai Y Q, Liu F X, Wu J T, Wei W, Hu D P Liu X W 2013 *Proc. of the ASME 2013 Inter. Mechanical Engineering Congress and Exposition (San Diego)* IMECE2013-63449 p 1
[11] Hu D, Li R, Liu P, Zhao J 2016 *Int. J. Heat Mass Transf.* **100** 497-507
[12] Razib M S, Hasan M M F and Karimi I A 2011 *Comput. Chem. Eng.* **37** 262-277

A Quantitative Study of Carmine Aqueous Solution Based on Drop-Coating Deposition Micro-Raman Spectroscopy (DCDR)

Wei Huang[1, 2]

[1]Department of Forensic Science, Fujian Police College, No.59, Shoushan Road, Cangshan District, Fuzhou, China
[2]Key Laboratory of OptoElectronic Science and Technology for Medicine of Ministry of Education, Fujian Provincial Key Laboratory of Photonics Technology, Fujian Normal University, No.8, Shangsan Road, Cangshan District, Fuzhou, China
Email: Huang84wei@163.com

Abstract. Carmine is a kind of colorant which is widely used in food, beverage, medicine, cosmetics and tobacco industry. However, excessive use of carmine may lead to the risks of carcinogenic, teratogenic and mutagenic, which seriously threaten the health and safety of consumers. In this paper, DCDR technology is utilized to develop a quantitative method for the detection of carmine, which requires only a small volume deposition of analyte solution (several μL) on a suitable hydrophobic substrate. The conventional Raman spectrum of carmine aqueous solution and corresponding Raman spectrum using DCDR method were compared, illustrating a much higher sensitivity for DCDR method. The Raman spectra of carmine aqueous solution with different concentrations of 100, 50, 10, 8, 4 and 2μg/mL are acquired from the spots on the "coffee-ring" with DCDR method. Using DCDR method, a good linear relationship has been observed between the intensities of the two characteristic peaks, $1364cm^{-1}$ as well as $1572cm^{-1}$, and the concentrations of the solution, with the linear correlation coefficient of $R^2>0.99$. The results illustrate that DCDR method has a good potential in the quantitative analysis of colorant like carmine, providing a promising technique for a rapid detection for food additives.
Keywords. Carmine, DCDR technology, micro-Raman spectroscopy, quantitative detection.

1. Introduction

Carmine is currently one of the most widely used monoazo synthetic colorants in China, which is used for colouring foods such as fruit juice drinks, mixed wines, carbonated beverage, candies, cakes, ice cream and yogurt [1]. Carmine is a compound (CAS Number: 1390-65-4) with a dark red colour in powder, which is 1,3-naphthalenedisulfonic acid, 7-hydroxy-8-[(4-sulfo-1-naphthalenyl) azo] trisodium salt in aqueous solution and solid state [2]. The molecular formula of Carmine is $C_{20}H_{11}N_2Na_3O_{10}S_3$, and the chemical structure is shown in figure 1. As a kind of azo compounds, carmine can be metabolized intoβ-naphthylamine and α - amino-1-naphthol in the body which are strong carcinogenic substances. The aromatic amines are further activated by metabolism and turn into electrophilic products which could combine with DNA and RNA to form adducts inducing mutations [3]. Also, carmine can be oxidized to free radicals, and then metabolizes with the substances in the body to produce reactive oxygen species (ROS), which could attack DNA and causing oxidative damage [4]. It has been proved to be certain carcinogenic and mutagenic effects in toxicology experiments [1, 5].

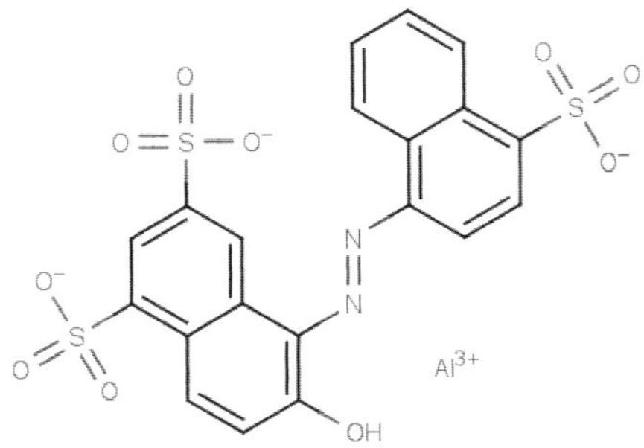

Figure 1. Chemical structure of carmine.

The addition of carmine is prohibited into food in the United States, Canada and Norway. There are strict regulations on the scope of use and maximum usage for carmine as an additive, even though the addition is allowed in food in China, the European Union, Japan and other countries. According to the provisions of the Chinese National Standard of GB/T 5009.35-2016, the detection limit of carmine an additive is 0.5mg/kg in foods. And carmine is not allowed to add to the in aquatic products or meat in China.

However, food safety incidents of excessive or illegal addition in food happen from time to time, seriously threatening the health and safety of consumers. With more and more attention to the excessive use of colorants in food safety, the study on the detection of carmine is increasingly a need.

Variety of modern analytical methods and instruments are used to detect additives, including thin layer chromatography, spectrophotometry, voltammetry, polarography, capillary electrophoresis, ion chromatography and high performance liquid chromatography (HPLC) etc. [6]. However, these methods either need complicated sample preparation process, or time-consuming, or need expensive analytical instruments, and are not suitable for large-scale rapid field screening requirements [7].

It is of great significance and urgency to develop a simple, fast, portable and method for detecting colorants, especially for azo compounds like carmine.

Drop coating deposition Raman (DCDR) spectroscopy is a technique based on the deposition of a small volume of analyte solution (several μL) on a suitable hydrophobic substrate. Due to the surface hydrophobicity, evaporation of the solvent may lead to preconcentration of studied analyte in ring-shaped drying pattern (known as "coffee-ring") from which Raman spectra are acquired. DCDR enables the measurement of solutions at very low concentrations and small volumes [8], and has been a widely used technique to study biological molecules and molecular mixtures like lipids, porphyrins, tears, amino acids, peptides, haemoglobin, oligosaccharides, liposomes in cholesterol and other molecules in recent years [9-13].

The aim of this paper is to develop a quantitative method for the detection of carmine based on DCDR technique, excepting to explore a rapid, easy quantitative detection method for azo dye additives.

2. Materials and Methods

2.1. Laboratory Equipment and Materials

Raman spectra of the samples were measured by confocal Raman microscope spectrometer (Invia, Renishaw Inc.), equipped with 1200 l/mm grating. Before the measurement, the wavenumbers were calibrated by using the excited line of 520 cm^{-1} from silicon. Measurements were carried out with monitoring by a 20× objective of microscope (N. A. =0.4, Leica). The 785 nm laser was used as the excitation source with a power of 50 mw and the final laser power on the sample was ~8 mW.

2.2. Standard Solution Preparation of Carmine
Carmine powder (99%, Sigma-Aldrich Chemicals, St. Louis, MO, USA) was dissolved in deionized water (Millipore-Q) and prepared as a stock solution with a concentration of 1000 μg/mL.
The stock solution was subsequently diluted into a series of concentrations of 100, 50, 10, 8, 4 and 2μg/mL for measurement.

2.3. Measurement of DCDR Spectra for Carmine Solutions
The preparation of samples for DCDR measurements could be divided into two steps as following: (1) Carmine solution with different concentrations were deposited by micropipette as 2μL droplets on an polished aluminium substrate (76×26cm); (2) Then the sample was dried left to dry naturally at ambient temperature of 20°C for 20min, where the "coffee-rings" are formed.

The prepared dried deposits on the substrate were placed on the stage of microscope Raman measurement. "Coffee-rings" of the deposits could be observed that most of the molecules of samples have aggregated in ring-shape to the edge of the deposits with the monitor of 20× objective. As shown in figure 2, the Raman spectra of samples from 400 to 2000 cm^{-1} were obtained with 3 spectral scans, and each with an accumulation time of 20s.

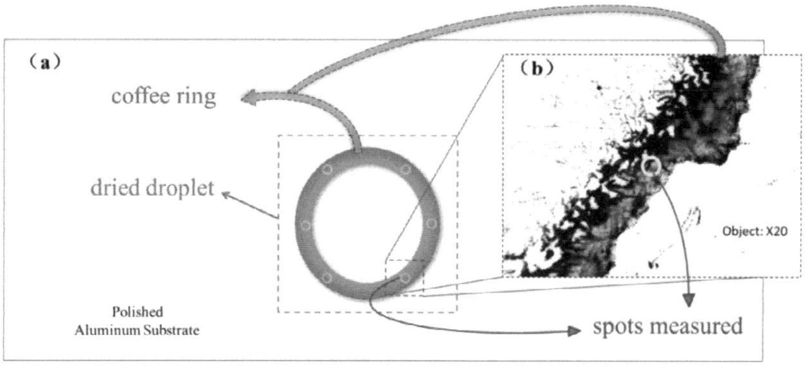

Figure 2. Schematic diagram of DCDR method for the detection of carmine: (a) Layout of the measuring spots on the coffee-ring; (b) White light image of carmine deposit under 20× objective.

For DCDR measurements, a minimum of 6 spectra from different spots with uniform distribution on the coffee-ring of the dried deposits were acquired in order to minimized the non-uniform distribution on the coffee-rings (shown in figure 2(a)). The white image of the measured spot on the "coffee-rings" is shown in figure 2(b) under 20× objective.

The 6 spectra from a coffee-ring were averaged as the DCDR spectrum of the dried deposit. All the samples were kept at ambient temperature of 25°C during the measurements.

3. Results and Discussion
The original spectra acquired were only subtracted by the baseline, without any other treatments, so as to reflecting the intrinsic characteristics of spectra from different sample such as intensities and wavenumbers of the peaks.

In order to investigate the advantages of DCDR technique in the detection of carmine solution samples, the conventional Raman measurement using droplet method and the DCDR method were carried out on the same sample droplet. As shown in figure 3, the DCDR method has a higher sensitivity than that of conventional Raman measurement.

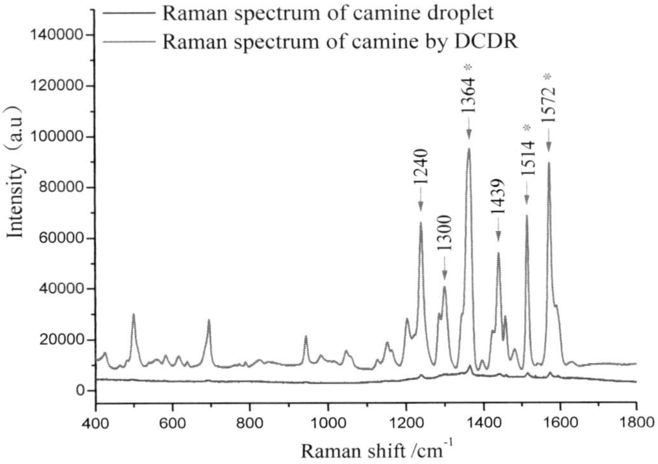

Figure 3. Comparison of conventional Raman spectrum of solution and corresponding Raman spectrum using DCDR method of carmine droplet aqueous.

The Raman peaks of carmine can be clearly observed in figure 3, including that of 1240 cm^{-1}, 1300 cm^{-1}, 1364 cm^{-1}, 1439 cm^{-1}, 1514 cm^{-1}, 1572 cm^{-1}. Tentative assignments of peaks from the DCDR spectra of carmine were shown in table 1 [2]. Specially, the peak at 1364 cm^{-1} assigned to stretching of CC (νCC), as well as the peak at 1572 cm^{-1} assigned to the bending of NH(δNH), deformation of CH(βCH) and stretching of CC (νCC), were characteristic peaks for the identity of carmine as an azo colorant.

Table 1. Characteristic Raman shifts and tentative assignments of peaks from the DCDR spectra of carmine [2].

Raman Shifts(cm^{-1})	Assignments
1572	[a]δNH, [b]βCH, [c]νCC
1514	δ NH, ν C=O, ν C=N, β CH
1439	β CH
1364	ν CC
1300	*δ* NH,*ν* N-N, β CH,*ν*CC
1240	*δ* NH,*ν* N-N, β CH,*ν*CC

[a] Bending.
[b] Deformation
[c] Stretching.

The averaged DCDR spectra were obtained from deposits of carmine solution with different concentrations of 100, 50, 10, 8, 4 and 2 μg/mL, which are shown in figure 4. It is illustrated clearly that the intensities of DCDR spectra showed a trend of decline with the decreasing concentration of the carmine solutions. That is, the intensities of DCDR are positively correlated with concentrations.

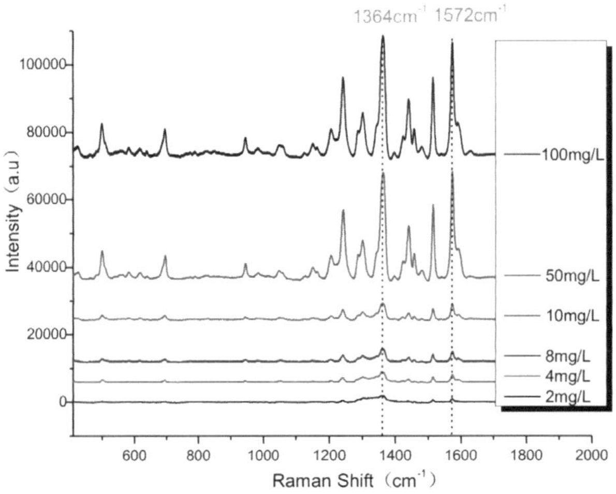

Figure 4. DCDR spectra of carmine from a dried drop in deposited concentration of 100, 50, 10, 8, 4 and 2μg/mL.

The two characteristic peaks of 1362 cm^{-1} and 1572 cm^{-1} can be also clearly observed when the concentration is as low as 2 μg/mL, indicating a promising sensitivity for the detection of carmine solution with a detection limit less than 2 μg/mL.

In order to verify the potential in the quantitative detection of carmine solution, the linearity of DCDR method using the intensities of the two characteristic peaks of 1362 cm^{-1} and 1572 cm^{-1} was further investigated.

It can be seen from figure 5 and figure 6, that there is a good linear relationship between the intensities of the two characteristic peaks of DCDR spectra and concentrations of carmine solutions in the range of 2-50μg/mL, with a linear correlation coefficient R^2 greater than 0.99.

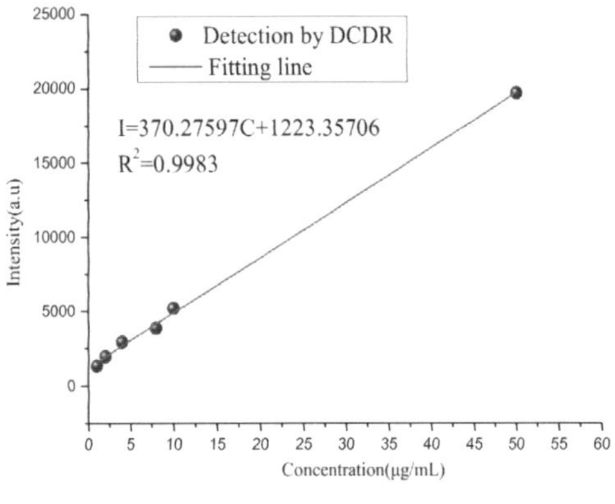

Figure 5. Fitting result of carmine DCDR spectra for intensities of 1365 cm^{-1} within concentrations of 2-50 μg/mL.

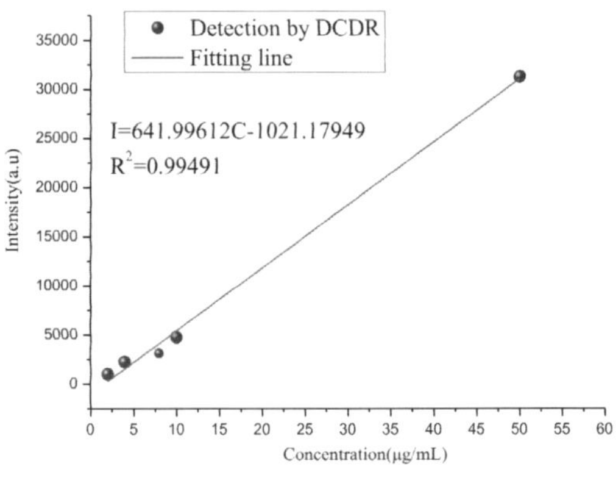

Figure 6. Fitting result of carmine DCDR spectra for intensities of 1572 cm^{-1} within concentrations of 2-50 μg/mL.

4. Conclusion

In this paper, DCDR technique was utilized as a quantitative study of the detection of azo colorant carmine. DCDR spectra from dried deposits of different carmine solution concentrations were obtained for a quantitative investigation.

Intensities of the two characteristic peaks of 1364 cm^{-1} and 1572 cm^{-1} showed a good linear relationship within concentrations of 2-50 μg/mL, with a correlation coefficient of $R^2 > 0.99$. The results show that DCDR technique has a good potential in the quantitative detection of colorant like carmine, providing a promising technique for a rapid detection for food additives.

Acknowledgments
This work is supported by the 2020 Research Project for middle-aged and Young teacher of Education Bureau of Fujian Province (Technology) (No. JAT200382) and supported by Open Project Program of Key Laboratory of OptoElectronic Science and Technology for Medicine of Ministry of Education, Fujian Provincial Key Laboratory of Photonics Technology, Fujian Normal University (No. JYG2104). The author thanks to Prof. Rong Chen's group in Fujian Normal University, for their great support and help in instruments and data analysis.

References

[1] Arif A, Ahmad A and Ahmad M 2020 Toxicity assessment of carmine and its interaction with calf thymus DNA *Journal of Biomolecular Structure and Dynamics* **39** 5861 - 71
[2] Almeida M R, Stephani R, Dos Santos H L F and Oliveira L 2010 Spectroscopic and Theoretical Study of the ``Azo"-Dye E124 in Condensate Phase: Evidence of a Dominant Hydrazo Form *Journal of Physical Chemistry A* **114** 526-34
[3] Talaska G and Al-Zoughool M 2003 Aromatic amines and biomarkers of human exposure *Journal of Environmental Science and Health, Part C* **21** 133-64
[4] Chung K T 2000 Mutagenicity and carcinogenicity of aromatic amines metabolically produced from azo dyes *Journal of Environmental Science & Health Part C* **18** 51-74
[5] Hu Q, Sun H, Liu L, Xiao L-x, Yang Z-q, Rao S, Gong X and Han J 2021 Development of an ultrasensitive spectrophotometric method for carmine determination based on fluorescent carbon dots *Food Additives & Contaminants: Part A* **38** 731 - 40
[6] Altunay N 2021 An optimization approach for fast, simple and accurate determination of indigo-carmine in food samples *Spectrochimica Acta Part A: Molecular and Biomolecular Spectroscopy* **257**
[7] Wu Y X, Liang P, Dong Q M, Bai Y, Yu Z, Huang J, Zhong Y, Dai Y C, Ni D and Shu H B 2017 Design of a silver nanoparticle for sensitive surface enhanced Raman spectroscopy detection of carmine dye *Food Chemistry* **237** 974-80
[8] Kuj~ová A b, QYikryl M, Procházka M and Kp iaová E 2021 Drop coating deposition Raman (DCDR) spectroscopy of contaminants *Spectrochimica acta. Part A, Molecular and biomolecular spectroscopy* **262** 120109
[9] Kočišová E and Procházka M 2015 Drop-coating deposition Raman spectroscopy of porphyrins *Journal of Raman Spectroscopy* **46** 280-2
[10] Ranc V and Chaloupková Z 2020 Perspectives of DCDR-GERS in the analysis of amino acids *Analyst* **145** 7701-8
[11] Kocišová E, Procházka M, Štepánek J and Mojzeš P 2010 Interaction of porphyrin/oligonucleotide complex with liposomes studied by drop coating deposition Raman spectroscopy *Spectroscopy: An International Journal*
[12] Šimáková P, Kočišová E and Procházka M 2013 Sensitive Raman spectroscopy of lipids based on drop deposition using DCDR and SERS *Journal of Raman Spectroscopy* **44** 1479-82
[13] Kocisova E, Petr M, ípová H, Kylian O and Prochazka M 2016 Drop coating deposition of a liposome suspension on surfaces with different wettabilities: "coffee ring" formation and suspension preconcentration *Physical Chemistry Chemical Physics Pccp* **19** 388

Density Functional Theory Study of Hydrogen Adsorption on Pt Pd/γ-Al_2O_3 Surface

Liang Zhang, Jia Li, Yong Chen, Cheng Zeng, Wu Kang, Panxing Yang, Ru Xiong and Jinhong Liu*

Nuclear Fuel Element and Material Sub-institute, Nuclear Power Institute of China, Chengdu, China
Email: zulin117@163.com

Abstract. At present passive hydrogen recombiners (PAR) are used to prevent hydrogen explosion. Hydrogen removal catalyst is the core component of PAR. The adsorption of hydrogen on the solid catalyst surface is the premise of catalytic hydrogen removal and is of great significance for deeper understanding of hydrogen removal mechanism. The adsorption behavior of H_2-Pt Pd/γ-Al_2O_3 system has been studied by using density functional theory and periodic slab model. The results of different adsorption sites indicate the adsorption energy of top site is highest, which is -1.2584eV. Higher adsorption energy means stronger interaction between H_2 and catalyst substrate, which elongates H-H bond and increases the negative charge on H_2. With increasing doping content of Pd, the adsorption energy of substrate decreases gradually. The adsorption energy absolute value of Pt_4/γ-Al_2O_3 is highest and its H-H bond is longest, arriving at 0.0967nm. After adsorbed on substrate, the energy gap of H_2 decreases drastically with the lowest energy gap of H_2-Pt_4/γ-Al_2O_3 that is 0.5197eV, and the peaks of density of state pattern move to lower energy level. This is because that the d orbital of Pt/Pd atoms interacts with the σ^* anti-bond orbital of H_2 strongly, transferring electrons to the σ^* anti-bond orbital of H_2. Doping Pd increases the energy gap of molecule orbital.

Keywords. Passive H_2 recombiner, catalyst carrier, adsorption, density functional theory, γ-Al_2O_3

1. Introduction

Nuclear safety is an important goal of nuclear power development. Nuclear reactors will produce hydrogen in accident conditions (due to the zirconium-water reaction, core melt-concrete reaction) and normal conditions (due to irradiation decomposition and adding hydrogen to suppress water oxidation). So there is a danger of hydrogen explosion [1, 2]. At present, nuclear power plants use igniters and passive autocatalytic hydrogen recombiners (PAR) to prevent hydrogen explosions. The hydrogen removal catalyst is the core part of the passive hydrogen recombiner, which is mainly composed of three parts: carrier, coating and catalytically active components. The adsorption of hydrogen on the surface of the solid catalyst is the prerequisite for catalytic hydrogen removal. Therefore, studying the adsorption of hydrogen on the solid surface is of great significance for understanding the mechanism of catalytic hydrogen removal. With the continuous development of computer technology and theoretical calculation methods, theoretical calculations have become more and more sophisticated, and density functional theory have been widely used to explore the adsorption laws of gas elements on the surface of crystals. Watson et al. [3] calculated the adsorption energies of H atoms at different adsorption sites on Pt(111) and Ni(111) surfaces and the surface mobility of H atoms. Okamot et al. [4] used density functional theory to calculate and compare the adsorption energy of H_2 molecules on the surface of Pt(111) and Pt13, Pt38, Pt55, and Pt147 clusters of FCC structure.

Ishimota et al. [5] studied the adsorption behavior of NO on Pt clusters and found that the adsorption energy of NO at the vertex is the largest, and the charge is transferred from metal to NO, which elongate N-O bond and increase activity. Meanwhile, other scholars utilized density functional theory to calculate the effect of the geometry parameters of H_2 molecular orbital [6], the adsorption behavior of O_2 and SO_2 on catalyst substrate [7, 8], effect of doping element on vacancy forming and adsorption energy [9-11], and charge distribution [12, 13], revealing the doped elements can increase the vacancy formation energy and adsorption energy, and alter the charge distribution of the adsorption system. However, there are few reports on the research of the γAl_2O_3 supported Pt and Pd multi-element catalytic system. The hydrogen removal catalyst is used in the harsh working environment of the reactor containment, which puts forward higher requirements on the performance of the catalyst carrier. It is necessary to study the mechanism of multi-element catalytic material removing hydrogen.

This paper intends to use density functional theory combined with the periodic slab model to simulate the adsorption behavior of H_2 on the surface of the γAl_2O_3/Pt Pd catalyst, and calculate the adsorption energy, bond length and charge distribution of the different adsorption sites of the $Pt_4/\gamma Al_2O_3$ model, as well as the effect of doping Pd on adsorption performance, molecular orbital energy gap and density of state. It provides a theoretical basis for further research on the interaction between hydrogen and solid catalyst surface and the improvement of catalyst performance.

2. Model and Calculation Method

The Inorganic Crystal Structure Database (ICSD) number of γAl2O3 is 173014, the spatial point group is P1, and the unit cell parameters are a=3.4009, b=2.7895, c=7.0762, α=β=γ=90°. On this basis, a supercell of 16 Al atoms and 24 O atoms was established, and the γAl_2O_3 (001) crystal plane was cut, and a vacuum layer was established along the c-axis for research.

Density functional theory combined with periodic slab model was used to simulate the adsorption behavior of H_2 on the surface of γAl_2O_3/Pt and Pd catalysts. The model was established by Material Studio, and the numerical simulation calculation was completed by the CASTEP software package. In the calculation, generalized gradient approximation (GGA) with Perdew-Burke-Ernzerhof (PBE) is selected for the exchange-related functional, the cutoff energy is 300eV, and spin polarization is used. K-piont and convergence accuracy are set to mediate. The thickness between two adjacent slabs is 1 nm to ensure that the interaction between the plates is small enough.

The adsorption energy is defined as the change in the energy of each substance before and after adsorption, and its magnitude can indicate the stability of the adsorption system. The calculation formula is: $E_{ads}=E_{H2-Pt4/\gamma-Al2O3} - E_{H2} - E_{Pt4/\gamma-Al2O3}$. In the formula, E_{ads} is the adsorption energy, E_{H2} and $E_{Pt4/\gamma-Al2O3}$ are the energy of the H_2 molecule and the substrate before adsorption, respectively, and $E_{H2-Pt4/\gamma-Al2O3}$ is the total energy of the system after adsorption.

3. Results and Discussion

3.1. The Activity of Different Adsorption Sites of Pt4/γ-Al_2O_3 Model

The adsorption of H_2 molecules at three different positions of the Pt4/γ-Al_2O_3 model was observed. The three adsorption positions were lateral position, bridge position and vertex position, as shown in figure 1. The adsorption energy, H-H bond length and charge distribution of H_2 adsorbed at three different positions are calculated, as shown in table 1. The results show that the absolute value of the adsorption energy at the vertex site is the largest, followed by the adsorption energy at the bridge and the smallest at the lateral site. The greater the adsorption energy, the stronger the interaction of H_2 with the catalyst matrix and the more stable the combination.

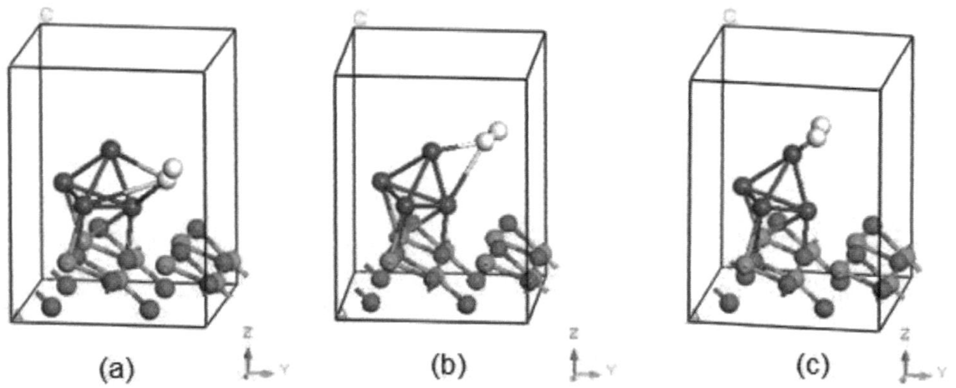

Figure 1. Three different adsorption sites of Pt_4/γ-Al_2O_3 model (a) is lateral site, (b) is bridge site and (c) is vertex site.

Table 1. Adsorption energy, bond length and electronic population of H_2 at different site.

Adsorption site	Adsorption energy eV	H-H bond length I-nm	q(H_2)
Vertex	-1.2584	0.0967	-0.19
Bridge	-0.0644	0.0768	-0.18
Lateral	-0.0318	0.0758	-0.09

The calculation results of the bond length show that the H-H bond length of each adsorption site has different degrees of elongation comparing with the bond length free H_2 molecule(0.074nm) [6], and the vertex site has the largest elongation of the H-H bond. This is consistent with the analysis result of adsorption energy. The greater the adsorption energy, the stronger the interaction between H_2 and the matrix, which will extend and weaken the H-H bond, then being conducive to the further dissociation of H_2 molecules. This is consistent with the calculation results in literature [12]. Liu studied the adsorption of SO_2 by different adsorbents and showed that the S-O bond length was elongated with the increase of adsorption energy.

The results of Mulliken charge distribution analysis show that the charge transferred from the matrix to H_2 at the vertex is at most -0.19, and the charge transferred at the side is at least -0.09. The more charge transfer, the stronger the interaction between H_2 and the matrix. Later, the vertex position with the largest adsorption energy is selected for the research and analysis of Pd doping. This corresponds with the results of references [7] and [13]. When the adsorption energy increases, the charge transferred from the substrate to the adsorbed gas molecules increases accordingly.

3.2. Influence of Pd Doping on Adsorption Performance

The adsorption energy, chemical bond and charge distribution of the catalyst substrate under different Pd doping ratios were calculated. The results are shown in table 2. The ratio of Pt/Pd is Pt4, Pt3Pd, Pt2Pd2, Pt1Pd3, Pd4, respectively. The results show that the absolute value of the adsorption energy of Pt4/γ-Al_2O_3 is the largest. With the increase of Pd doping, the adsorption energy gradually decreases. The greater the adsorption energy, the stronger the adsorption bond between H_2 and the catalyst matrix, and the more stable the system.

Table 2. adsorption energy, bond length and electronic population of matrix by doping different content of Pd.

Pt/Pd ratio	Adsorption energy eV	H-H bond length nm	H-Pt bond length nm	q(H_2)	q(Pt)
Pt_4/γ-Al_2O_3	-1.2584	0.0967	0.167	-0.21	0.18
Pt_3Pd/γ-Al_2O_3	-1.2173	0.0931	0.1701	-0.17	0.17
Pt_2Pd_2/γ-Al_2O_3	-0.9400	0.0954	0.1677	-0.2	0.21
Pt_1Pd_3/γ-Al_2O_3	-1.0187	0.0913	0.1719	-0.2	0.23
Pd_4/γ-Al_2O_3	-0.7148	0.0847	0.1835	-0.19	0.26

The calculation results of the bond length show that the H-H bond length is extended at different degree, after H_2 molecules are adsorbed on catalyst substrate with different Pt/Pd ratios. Among them, the maximum extension of the H-H bond at the vertex is 0.0967nm, and the H-H bond length decreases with the increase of doping Pd. At the same time, the H-Pt bond length increases with the increase of Pd doping, that is, the interaction between the matrix and the hydrogen molecules is weakened. This is consistent with the analysis results of adsorption energy. The greater the adsorption energy, the stronger the interaction between H_2 and the matrix, which will extend the H-H bond and weaken its binding strength, which is conducive to the further dissociation of H_2 molecules. This is consistent with the research results of Zhang [10], and the experiment of Lee [11] shows that Pt has higher catalytic performance than Pd and can dissociate oil fume more effectively.

The Mulliken charge distribution results show that the H_2 molecule gets electrons, and its charge is negative, between -0.17 and -0.21. It is not sensitive to the amount of Pd doping; while the Pt atoms in contact with H2 molecules lose electrons and are positively charged, and their charge decreases with the increase of Pd doping. This may be due to the fact that the electronegativity of Pd is less than that of Pt, which transfers electrons to Pt to reduce its positive charge. This corresponds with the calculation results in references [10] and [13]. Zhang and Xie studied the adsorption properties of black phosphorene and iron oxide respectively, and both showed that with the increase of adsorption energy, more charge was transferred from the substrate to the adsorbed molecules.

3.3. Effect of Doping Pd on Energy Gap and Density of States of Molecular Orbital

The energy gap Eg between the highest occupied molecular orbital and the lowest unoccupied molecular orbital with different doping Pd content was calculated, and the results are shown in table 3.

The energy range of free hydrogen molecules is 12.56eV, and the molecular orbital form is $\sigma_{1s}^{2}\sigma_{1s}^{*0}$, whose energy gap sharply decreases to between 0.52 and 0.70eV after being adsorbed on the catalyst substrate. This is because the d orbital of the Pt/Pd atom interacts strongly with the σ* antibonding orbital of the H_2 molecule, which transfers electrons to the σ* antibonding orbital of H_2, causing it to move to a lower energy range, thereby reducing the energy gap. Through further analysis of the energy gap of the molecular orbital with different doping Pd content, it is found that the minimum molecular orbital energy gap of H_2-Pt4/γ-Al_2O_3 is 0.52. With the increase of doping Pd content, the molecular orbital energy gap increases, and the energy gap of the H_2-Pt1Pd3/γ-Al_2O_3 molecular orbital is largest, reaching 0.70eV. The reduction of the energy gap makes it easier for the adsorbed molecule to form a bond with the matrix, and the bond is more stable, and it is helpful to weaken the H-H bond and further dissociate the H_2 molecule. This is consistent with the previous analysis results of adsorption energy and chemical bonds.

Table 3. Energy gap of molecule orbit by doping different content of Pd.

Pt/Pd ratio	HOMO eV	LUMO eV	Eg eV
H_2	-10.1684	2.3890	12.5574
H_2-Pt_4/γ-Al_2O_3	-4.2339	-3.7142	0.5197
H_2-Pt_3Pd/γ-Al_2O_3	-4.2312	-3.5700	0.6612
H_2-Pt_2Pd_2/γ-Al_2O_3	-4.1958	-3.6380	0.5578
H_2-Pt_1Pd_3/γ-Al_2O_3	-4.2856	-3.5836	0.7020
H_2-Pd_4/γ-Al_2O_3	-4.2393	-3.5618	0.6775

The analysis results of the density of states are shown in figure 2. The results show that the density of states of free H_2 has a peak at the Fermi level, which corresponds to the low-level orbital of the H_2 molecule, and the electron density of the high-level orbital is dispersed between 9~12eV. When H_2 is adsorbed on the catalyst matrix, the peak of the density of states diagram shifts to a lower energy level, and the energy range of the high-level orbital is further widened. This is mainly due to the interaction between the d orbital of the Pt/Pd atom and the orbital of the H_2 molecule to transfer electrons to the σ* antibonding orbital of H_2, causing it to move to a lower energy range. Among them, the peak density of states of H_2-Pt4/γ-Al_2O_3 shifts the most, and the peak shift of H_2-Pd4/γ-Al_2O_3 is the smallest. This is consistent with the analysis results of energy gap and bond length.

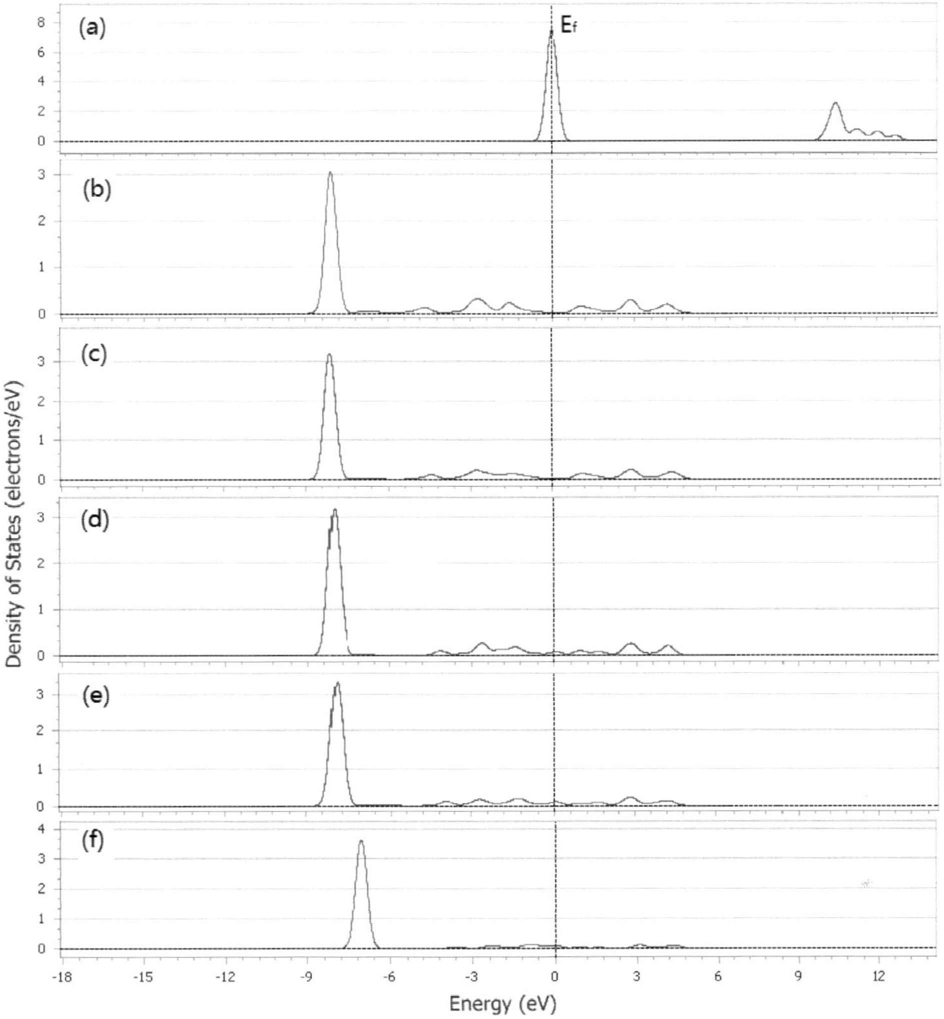

Figure 2. DOS of H_2 adsorbed on substrate with different content of Pd: (a) H_2, (b) H_2-Pt_4/γ-Al_2O_3, (c) H_2-Pt_3Pd/γ-Al_2O_3, (d) H_2-Pt_2Pd_2/γ-Al_2O_3, (e) H_2-Pt_1Pd_3/γ-Al_2O_3, (f) H_2-Pd_4/γ-Al_2O_3.

4. Conclusions

The adsorption behavior of the H_2-PtPd/γ-Al_2O_3 system is calculated by density functional theory combined with the periodic slab model. The results show that the absolute value of the adsorption energy at the vertex site is the largest. The greater the adsorption energy is, the stronger the interaction between H_2 and the catalyst substrate. Thus the H-H bond is elongated and the negative charge of H_2 increases, which is conducive to the further dissociation of the H_2 molecule. With the increase of doping Pd, the adsorption energy of the catalyst substrate gradually decreases, and the absolute value of the adsorption energy of Pt_4/γ-Al_2O_3 is the largest. After H_2 is adsorbed on the catalyst substrate, the energy gap of system decreases sharply, and the peak of the density of states shifts to a lower energy level. This is because the Pt and Pd atom's d orbital interacts strongly with the σ* anti-bonding orbital of H_2 molecule, transferring electrons to the σ* anti-bonding orbital of H_2. The increase of doping Pd makes the orbital energy gap increase.

References

[1] Bachellerie E, Arnould F, Auglaire M, et al. 2003 Generic approach for designing and implementing a passive autocatalytic recombiner PAR-system in nuclear power plant containments *Nuclear Engineering and Design* **221** 151-165.

[2] Zhu M and Wang S B 2006 Development and design of passive hydrogen compound device Ship *Science and Technology* **4**(2) 38-40.

[3] Watson G W, Wells R P K, Willock D J, et al. 2001 A comparison of the adsorption and Diffusion of Hydrogen on the (111) Surface of Ni, Pd, and Pt from Density Functional Theory Calculations *Journal of Physical Chemistry B* **105** 4889-4894.

[4] Okamoto Y 2006 Comparison of hydrogen atom adsorption on Pt clusters with that on Pt surfaces: A study from density-functional calculations *Chemical physics letters* **429** 209-213.

[5] Ishimoto R, Jung C H, Tsuboi H, et al. Periodic density functional and tight-binding quantum chemical moleculardynamics study of catalytic properties on g-Al2O3 supported Pt catalysts *Applied Catalysis A: General* **305** (2006) 64–69.

[6] He S and Hou X F 2018 Study on geometric parameters of hydrogen molecule and hydrogen molecular ion orbit *Shandong Chemical Industry* **47**(02) 3-4.

[7] Wang X, Chen W K, Lu C H, et al. 2007 Density functional theory study of oxygen adsorption on CuCl(111) surface *Chinese Journal of Catalysis* **28**(08) 696-702.

[8] Zuo H W, Lu C H, Ren Y R, et al. 2016 The first theoretical study on the adsorption of O_2 by Pt_4 clusters supported on monolayer graphite phase carbon nitride *Chinese Journal of Physical Chemistry* **32**(05) 1183-1190.

[9] Zhu L G, Zhou J and Sun Z M 2016 Effect of W doping on the formation energy of oxygen vacancies in TiO2 *Journal of Beijing University of Aeronautics and Astronautics* **42**(01) 54-58

[10] Zhang G Y, Jiao X Q and Liu G L 2020 Adsorption of CH2O molecules on Al-doped BP by density functional theory *Journal of Shenyang Normal University (Natural Science Edition)* **38**(01) 23-27.

[11] Lee J H, Trimm D L and Cant N W 1999 The catalytic combustion of methane and hydrogen sulphide *Catalysis Today* **47** 353-357.

[12] Liu Y P, Zhou Q, Wang J X, et al. 2021 Cr doped MN (M=In,Ga) monolayer: A promising candidate to detectand scavenge SF6 decomposition components *Sensors and Actuators A* **330** 112854.

[13] Xie C Y, Sun Y L and Zhu B Z 2021 The promoting mechanism of doping Mn, Co, and Ce on gas adsorption property and anti-SO_2 oxidation over γ-Fe_2O_3 (001) surface: A density functional theory study *Colloids and Surfaces A: Physicochemical and Engineering Aspects* **628** 127218.

Numerical Comparison of Transpiration Cooling Designs Using Real Gas Effects and Approximate Gas Model

Meng Wang[1], Jianhua Wang[1, a*], Fei He[1, b*], Kang Qian[1], Yadong Wu[2] and Guangsheng Zhu[2]

[1] CAS Key Laboratory of Mechanical Behaviour and Design of Materials, Department of Thermal Science and Energy Engineering, University of Science and Technology of China, Hefei 230027, Anhui, China.
[2] Beijing Institute of Astronautical System Engineering, Beijing 100076, China.
Email: [a]jhwang@ustc.edu.cn, [b]hefeihe@ustc.edu.cn

Abstract. In the severe high-temperature environment caused by aerodynamic heating, the vibrational excitation, dissociation and ionization of gas may successively occur, which are known as real gas effects. Under the real gas effects, the thermodynamic properties of gas vary drastically and significantly influence the performances of the active thermal protection system of hypersonic vehicles, especially in the case with coolant outflow, for example transpiration cooling. This paper numerically investigates the transpiration cooling performance with the consideration of the interaction between coolant outflow and hypersonic flow under the real gas effects. The mathematical models and coupled numerical strategy are firstly validated by experimental data, then the influences of real gas effects on the transpiration cooling of a wedged leading edge (WLE) are studied under a flight Mach number range from 8 to 12 and a flight height of 40 km. The analysis and discussions of the numerical results reveal some important phenomena and demonstrate the need to consider real gas effects.
Keywords. Transpiration cooling, real gas effects, porous media, hypersonic vehicles, high-temperature environment.

1. Introduction

Hypersonic vehicles are generally designed to fly or cruise in the near space with the Mach numbers higher than 5 [1], and the superiorities of hypersonic vehicles in future military field and civilian traffic have raised many concerns [2]. However, the hypersonic flight will cause a severe aerodynamic heating effect [3], and therefore a series of active thermal protection systems (ATPS) have been researched for the hypersonic vehicles [4].

It is well known that the air around hypersonic vehicles is significantly heated. In high-temperature environment, the phenomena of gaseous vibrational excitation, dissociation and ionization alleged "real gas effects" (RGE) may successively occur. The air composition will change and the thermal properties may deviate from the behavior of ideal gases and traditional approximate models [5]. RGE is one of the important issues in the design of ATPS for hypersonic vehicles, because when the Mach number exceeds 4, strong RGE are not negligible [6]. Using the numerical method, Emelyanov et al. [7] compared the results calculated by RGE and the ideal gas model. Their comparison indicated that using the ideal gas model, the calculated temperatures may exceed the real values.

In the designs of the hypersonic vehicles with long endurance, the development of effective ATPS is very important, because the aerodynamic heat may exceed the limits of current materials [8]. In recent years, transpiration cooling as a high-efficiency cooling technique attracts more and more attention. Due to the interaction between coolant flow and hypersonic flow, the accurate predictions of transpiration cooling performances need to consider RGE in the flight with a high Mach number. Overall, the comprehensive study of transpiration cooling with the interaction between coolant flow and hypersonic flow under the real gas effects is still lacking. In practice, due to open pores to external, the aerodynamic force and real gas effects of the hypersonic flow inevitably affect the transport characteristics of the coolant within and outside the transpiration cooling structure, which would cause unexpected changes to the temperature and the driving force of the system. Therefore, it is necessary and worthwhile to study the actual transpiration cooling process, which couples the hypersonic flow under the real gas effects.

This paper exhibits a numerical investigation on the transpiration cooling performances of WLE under the conditions of a Mach number range between 8 and 12, and flight height of 40 km, traditional approximate gas models (AGM) and RGE are used respectively, the numerical results are analyzed and compared. This work aims to provide the investigators and designers of the transpiration cooling system of the hypersonic vehicles with a relatively comprehensive reference.

2. Geometry Model

The geometrical model of the WLE used in this work is illustrated in figure 1. The model consists of a porous matrix and an impermeable part with a constant thickness of 2 mm, a total length of 114 mm and an inner cone angle of 14°. The porous matrix has an external radius of 3 mm, an average sphere diameter of 5×10^{-4} m, a porosity of 0.3, that correspond to a permeability of 1×10^{-13} m^2. At the altitude of 40 km, the static temperature of the air is 227 K and the static pressure is 1200 Pa. The gaseous coolant is injected from the inner surface, and exchanges heat with the porous matrix adequately. The outflow covers the hot surface of the porous and impermeable regions to prevent them from direct contact with high-temperature airflow.

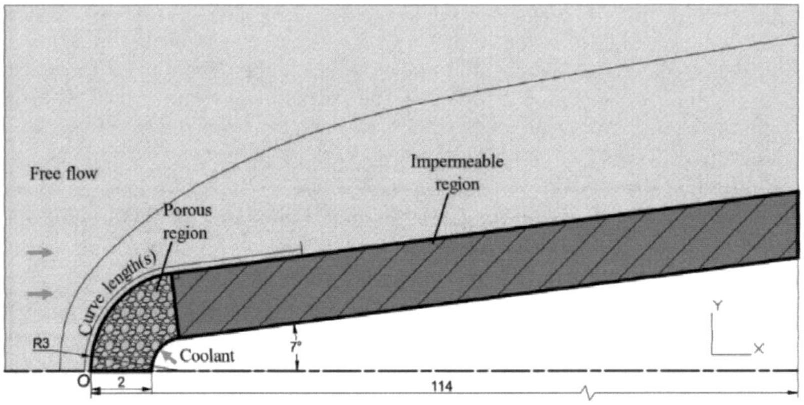

Figure 1. Geometry model of the wedge leading edge (WLE).

3. Mathematical Model

3.1. Conservation Equations

The entire computational domain consists of free flow, porous and impermeable regions. In the free flow region, Reynolds-averaged Navier-Stokes (RANS) equations are solved, and SST $k-\omega$ turbulent model is employed to close RANS equations.

Continuity equation:

$$\nabla \cdot (\rho \vec{u}) = 0 \tag{1}$$

Momentum equation:

$$\nabla \cdot (\rho\vec{u}\vec{u}) = -\nabla p + \nabla \cdot \bar{\tau} + \frac{\partial}{\partial x_j}\left(-\rho\overline{u_i' u_j'}\right) \tag{2}$$

where $\bar{\tau}$ represents shear stress, and the Reynolds stresses $-\rho\overline{u_i' u_j'}$ is calculated by SST $k-\omega$ model.
Energy equation:

$$\nabla \cdot \left(\vec{u}(\rho E + p)\right) = \nabla \cdot \left(k_f \nabla T + \bar{\tau} \cdot \vec{u}\right) \tag{3}$$

In the porous region, the Darcy–Forchheimer equation is used to describe the coolant flow:
Continuity equation:

$$\nabla \cdot (\varepsilon\rho\vec{u}) = 0 \tag{4}$$

Momentum equation:

$$\nabla \cdot (\varepsilon\rho\vec{u}\vec{u}) = -\nabla(\varepsilon p) + \nabla \cdot (\varepsilon\bar{\tau}) - \left(\frac{\mu}{K}\vec{u} + C_2\frac{1}{2}\rho|\vec{u}|\vec{u}\right) \tag{5}$$

where K and C_2 respectively represent the permeability of the porous matrix and the inertial resistance factor.

Energy equation

$$\nabla \cdot \left[\vec{u}\left(\rho_f E_f + p\right)\right] = \nabla \cdot \left(k_{eff} \nabla T + \bar{\tau} \cdot \vec{u}\right) \tag{6}$$

3.2. Thermodynamic and Transport Properties

In hypersonic flight, the non-equilibrium of chemical reaction in air occurs only when the vehicles fly above an altitude of 40km and at a velocity higher than 4km/s [9]. Therefore, considering the flight conditions, the properties of the free stream and coolant are derived from the equilibrium state of chemical reactions. The differences in the properties between the two calculation methods (AGM in dotted red lines and RGE in solid black lines) are shown in figure 2 especially above Ma 8.

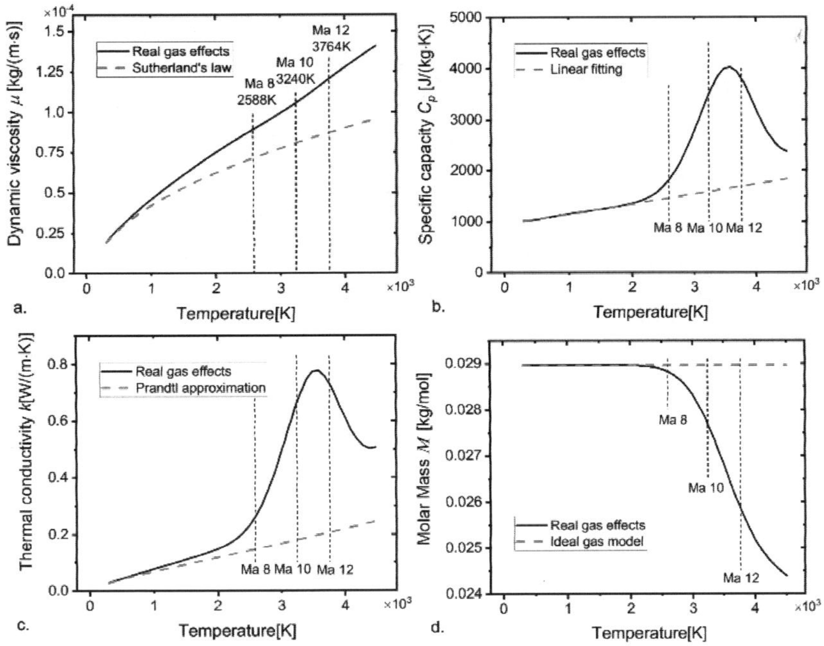

Figure 2. Transport properties calculated by approximate model and real effects.

4. Numerical Strategy

4.1. Calculation Grid and Independence of Numerical Results

As shown in figure 3, structured mesh with refined grids near the leading wedge is adopted. Table 1 displays the stagnation temperatures at Mach 12 calculated by the three mesh with different node numbers. The relative deviations are very small, therefore to balance the computational accuracy and load, the mesh with 76814 nodes is chosen in the following simulations.

Table 1. Temperature at stagnation point.

Node number	Stagnation temperature/K
49229	3752.26
76814	3754.02
109812	3754.49

4.2. Numerical Method

To overcome the convergence difficulty due to the coupled relation of the hypersonic flow in the free stream and the low-speed flow in the porous matrix, a three-step numerical solving strategy is applied:

1. The free stream region is calculated, the properties of air are seen as constant, here the entire outer surface of WLE is set as the adiabatic and impermeable wall.
2. The numerical results of constant properties are input as the initial values, the characteristics of the free-flow region are predicted again with AGM and RGE.
3. Enabling the porous and impermeable wall regions, the converged results in step 2 are used as the initial values of the free stream field, the boundaries among the three regions are set as interior and coupled wall.

4.3. Validation of Numerical Strategy Using Experimental Data

Experimental data obtained in the arc heating wind tunnel FD04 of China Academy of Aerospace Aerodynamic by Shen et al. [10] are used to validate the numerical strategy. The experiment was carried out at a Mach number of 4.2 and a total enthalpy of 2300 kJ/kg. Figure 4 indicates that the numerical results of RGE are closer to the experimental data than of AGM, especially at the stagnation point, which means the validity of the numerical strategy.

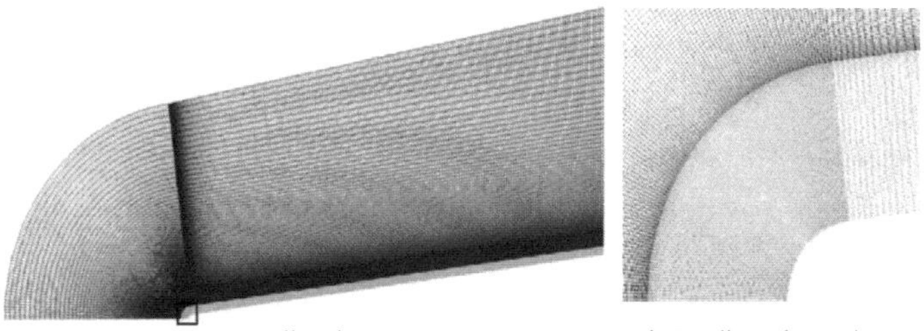

a. All region b. Leading edge region

Figure 3. Structured mesh of computational domain.

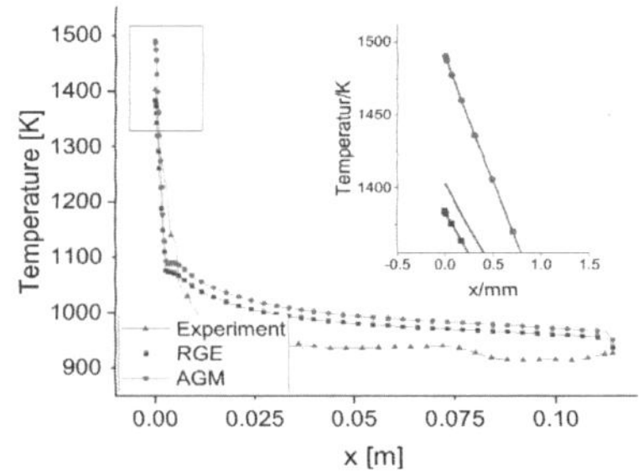

Figure 4. Comparisons of experiment data with numerical simulations.

5. Result and Discussions

Using the validated strategy, a series of numerical simulations are conducted at a flight altitude of 40km and three Mach numbers of 8, 10 and 12. The important design parameters of transpiration cooling systems are analysed and compared between the numerical results obtained by AGM and RGE.

5.1. Aerodynamic Characteristics Analysis

Due to the higher specific heat capacity of RGE, the temperatures around WLE shown in figure 5 are much lower than that of AGM, especially near the leading edge. At the stagnation point, the temperature predicted by RGE is lower than 3800 K, but by AGM achieves 6800 K. Then influenced by the differences in temperature and properties, the shock wave is closer to WLE using RGE. Figure 6 exhibits the pressure distributions predicted by AGM and RGE are quite similar. But more violently compressed after the shock wave, high-pressure region near the stagnation point occupies a larger area.

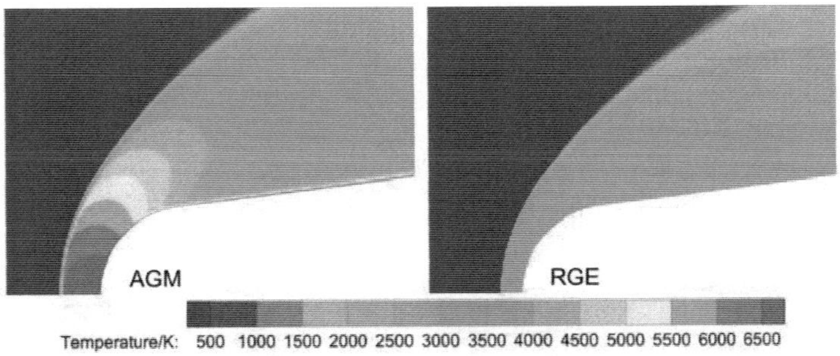

Figure 5. Aerodynamic heat environments around WLE without cooling air injection.

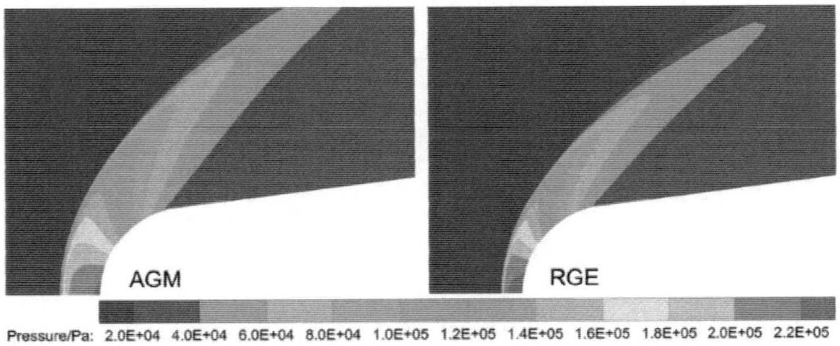

Figure 6. Pressure fields around WLE in two cases.

5.2. Transpiration Cooling Performances

Figure 7 illustrates the cooled wall temperature distributions predicted by AGM and RGE at different Mach numbers when cooling air is injected into the pores with the same velocity 6m/s. The temperatures predicted by REG are lower than that by AGM. This difference can be explained by the lower temperatures around WLE and more cooling air injected into the porous matrix using REG. Figure 8 illustrates the chamber pressure of the cooling air passing through the pores. From figure 8, a significant excess can be observed using REG due to the higher viscosity resistance. With Mach number increasing, these phenomena become more obvious, because of differences in the thermal properties between RGE and AGM are more remarkable, as shown in figure 2.

Figure 9 shows the cooling air mass flow rate distributions over the outer surface of the porous matrix in the two cases, and the total amounts are respectively calculated and listed in table 2. When Mach number increases, one can find the following three trends: 1) More mass flowrate enters the porous matrix in the cases of AGM and RGE. 2) The non-uniformity of the cooling air on the hot surface of the porous matrix is more obvious. 3) In comparison with the case of AGM, the consumption and non-uniformity of the cooling air of RGE increase significantly. These phenomena can be explained by the higher density of the cooling air due to the higher driving pressure, and the more obviously changing of the viscosity leads to a larger non-uniformity of the surface cooling air distribution.

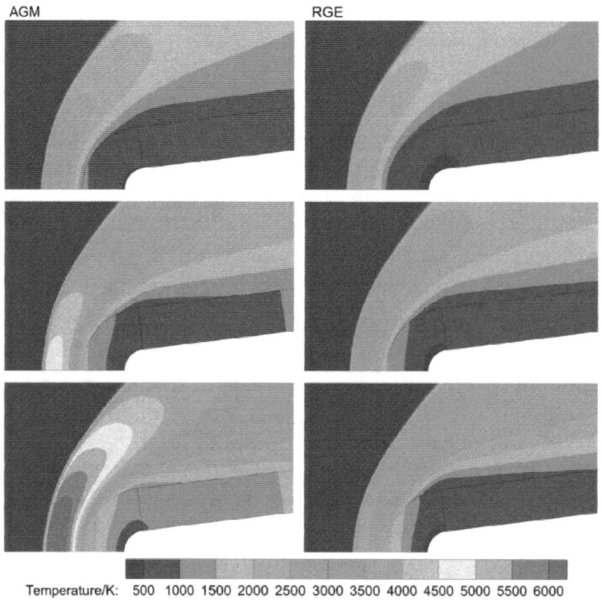

Figure 7. Temperature fields around and within WLE under different Mach numbers in two cases

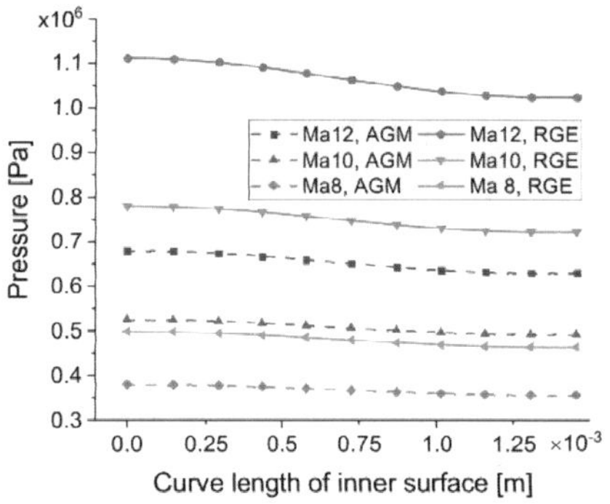

Figure 8. Cooling air injection pressure under different Mach numbers in two cases.

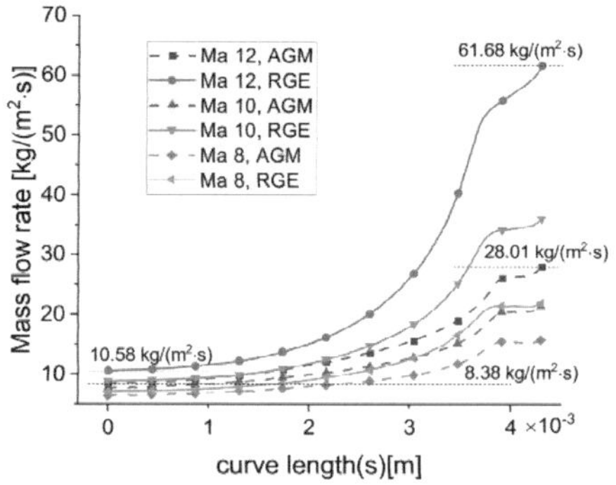

Figure 9. Mass flow rate distributions over porous surface in three Mach numbers in two cases.

Table 2. Mass flow rate at different Mach numbers with the same velocity 6 m/s.

Mach Number	Mass Flow Rate [kg/(m · s)]	
	AGM	RGE
8	0.04015	0.04935
10	0.05096	0.07159
12	0.06113	0.1047

5.3. Comparison of Cooling Air Consumption and Driving Force

At the same cooled stagnation point temperature, the cooling air consumption and the corresponding driving force are compared between the cases of RGE and AGM. In the case of AGM, the velocity of the cooling air is fixed at 6 m/s. Adjusting the velocity of the cooling air using RGE, the same stagnation point temperatures with AGM, corresponding chamber pressures and mass flow rates are listed in table 3. From table 3, the following important phenomena can be found: 1) The injection velocity in the case of RGE is smaller than that in the case of AGM, and with an increase in Mach number, the cooling air velocity reduces further. 2) In the case of RGE, the corresponding mass flow rates of the cooling air are always smaller than that in the AGM case. And at the higher Mach number, the consumptions of different cases become closer due to the more obvious non-uniformity of mass flowrate distributions. 3) Although the injection velocities and cooling air consumption are lower in the case of RGE, the driving forces of the cooling air or the chamber pressures required are higher than that in the case of AGM, especially at a high Mach number.

Table 3. Coolant injection velocity and mass flow rate at different Mach numbers.

Mach number	Stagnation point temperature [K]	AGM			RGE		
		Injection velocity [m/s]	Mass flow rate [kg/(m · s)]	Driving pressure [Pa]	Injection velocity [m/s]	Mass flow rate [kg/(m · s)]	Driving pressure [Pa]
8	1090		0.04015	367668	5.0	0.03619	481829
10	1734	6	0.05096	507092	4.5	0.05028	765729
12	2652		0.06113	651907	3.78	0.06071	1089126

6. Conclusion

The thermal properties of gases including freestream and coolant will significantly change in high temperature, and deviate from the traditional approximate gas model (AGM). Therefore RGE should be used in the cooling system designs of hypersonic vehicles. This paper exhibits a numerical analysis of the transpiration cooling system design of a wedge leading edge under three hypersonic conditions, and compares the differences between using RGE and AGM. Through analysis and comparisons, the following conclusions can be drawn:

1. The thermal environment estimated with RGE is much milder, although the aerodynamic pressures on the hot surface show small differences between the two models.
2. At the same cooling air injection velocity, using RGE, the wall temperatures are much lower, but the cooling air mass flow rates and the driving force is much larger than using AGM, which means that to get the same cooled temperature at the stagnation point, the injection velocity required by RGE is always lower, corresponding driving force is still much higher than that by AGM.
3. The distributions of the temperature and cooling air mass flow rate in the stagnation region are more non-uniform, and these trends become more remarkable when the Mach number increases.

Acknowledgements
This work is supported by the National Natural Science Foundation of China (contract no. 51806206).

Reference

[1] Vinh N X, Busemann A and Culp R D 1980 Hypersonic and planetary entry flight mechanics *NASA STI/Recon Technical Report A* 81.
[2] Schmisseur J D 2015 Hypersonics into the 21st century: A perspective on AFOSR-sponsored research in aerothermodynamics *Progress in Aerospace Sciences* **72** 3–16, doi:10.1016/j.paerosci.2014.09.009.
[3] Sziroczak D and Smith H 2016 A review of design issues specific to hypersonic flight vehicles *Progress in Aerospace Sciences* **84** 1–28 doi:10.1016/j.paerosci.2016.04.001.
[4] Zhu Y H, Peng W, Xu R N and Jiang P X 2018 Review on active thermal protection and its heat transfer for airbreathing hypersonic vehicles *Chinese Journal of Aeronautics* **31** 1929–1953, doi:10.1016/j.cja.2018.06.011.
[5] D'Angola A, Colonna G, Gorse C and Capitelli M 2008 Thermodynamic and transport properties in equilibrium air plasmas in a wide pressure and temperature range *Eur. Phys. J. D* **46** 129–150, doi:10.1140/epjd/e2007-00305-4.

[6] Bertin J J and Cummings R M 2006 Critical hypersonic aerothermodynamic phenomena *Annual Review of Fluid Mechanics* **38** 129–157 doi:10.1146/annurev.fluid.38.050304.092041.
[7] Emelyanov V, Karpenko A and Volkov K 2019 Simulation of hypersonic flows with equilibrium chemical reactions on graphics processor units *Acta Astronautica* S0094576518320435 doi:10.1016/j.actaastro.2019.01.010.
[8] Glass D 2008 Ceramic Matrix Composite (CMC) Thermal Protection Systems (TPS) and Hot Structures for Hypersonic Vehicles *In Proceedings of the 15th AIAA International Space Planes and Hypersonic Systems and Technologies Conference* (Dayton, Ohio American Institute of Aeronautics and Astronautics).
[9] Bertin J J 1994 *Hypersonic Aerothermodynamics* American Institute of Aeronautics and Astronautics, Inc.
[10] Shen L, Wang J H, Dong W J, Pu J, Peng J L, Qu D J and Chen L Z 2016 An experimental investigation on transpiration cooling with phase change under supersonic condition *Applied Thermal Engineering* **105** 549–556 doi:10.1016/j.applthermaleng.2016.03.039.

Progress of Development and Application of Drag Reduction Agents for Slick-Water Fracturing

Ke Xu*, Yongjun Lu, Jin Chang, Dingwei Weng and Yang Li

PetroChina Research Institute of Petroleum Exploration & Development, Beijing, 100083, China.
Email: xuke69@petrochina.com.cn

Abstract. Fracturing technology is the key technology of shale oil and gas exploitation in the United States. The key of hydraulic fracturing lies in the formulation of fracturing fluid, which can improve the permeability of shale gas layer, reduce pumping resistance, optimize production conditions, reduce strata damage and other purposes. Slick-water is widely used in water-based fracturing for unconventional oil and gas exploitation, which can reduce the friction resistance. As the core of the slickwater fracturing fluid, the friction reducer determine the fluid's performances. Combining with the related literature at home and abroad, this paper analyzes the mechanism of friction reducer, introduces the research and application progress of natural polysaccharide, surfactant and polyacrylamide. It is considered that both the instant powder polymer and W/W dispersion polymer have great application potential. The friction reducer with high-efficiency sand-carrying and salt resistance is the focus of future research.

Keywords. Unconventional reservoirs, slickwater, anti-blocking agents, polymers, resistance reduction rates.

1. Introduction

With the development of conventional energy extraction and fracking technology, the target of oil and gas exploration at home and abroad has been shifted from conventional to unconventional. Unconventional reservoir materiality is poor, porosity is small, penetration rate is low, no natural production capacity, oil and gas resources include oil sands, heavy oil, shale oil, dense sandstone gas, coal seam gas, shale gas, etc., rich in world content, need to carry out reservoir transformation in order to obtain industrial output. Slippery water fracturing is one of the most effective ways to realize its economic exploitation. Slippery water is a common water-based fracturing fluid, in clear water to add anti-blocking agents, filter-reducing agents, support agents, surfactants, fungicides, clay stabilizers, etc. formed, 96%-99% by water composition, also known as water fracturing liquid or water fracturing solution. Belongs to the non-Newtonian fluid, shows the sticky elasticity, the resistance effect is significant, can use large displacement, large-scale, high pump pressure operation mode to communicate the natural cracks of the formation to achieve reservoir transformation, with easy return can be reused, not easy to form filter cake, small damage to the formation and so on. As one of the most important additives, anti-resisting agent has become a hot research topic of fracturing liquid in modern times [1-2].

Based on the relevant literature at home and abroad, this paper summarizes the development course, classification, resistance reduction mechanism and research and development application, and looks forward to the development direction.

2. The Development Course of the Anti-Blocking Agent

In 1948, Toms [3] observed a decrease in chlorobenzene flow resistance with a small amount of polymethyl acrylates (PMMA), and the polymer's resistance reduction, also known as the Toms effect, was first observed. Savins in 1963 [4] defined the resistance reduction agent as a small amount added to the fluid to increase transport capacity, energy consumption reduced substances, referred to as DRA, the resistance reduction effect is generally characterized by the resistance reduction rate. Adding one millionth of a millionth of a long-chain flexible resistance reduction agent to the liquid can reduce turbulence friction loss by up to 80%, because the realization of resistance reduction can bring great economic benefits, resistance reduction agent has become a common concern, widely used in oil and gas, biomedicine, fire protection, sewage treatment, regional heating and other fields [5].

Fracturing fluid with reducing inhibitors first appeared in the late 1950s and early 1960s [6], and has since been mainly used in cross-linking gel systems for reservoir modification. With the transition from conventional to unconventional oil and gas exploration, anti-fracturing fluid has received renewed attention and has shown great superiority in the fracking operations of shale gas layers. In 1997, Mitchell first adopted a water-skiing fracturing system to carry out a large-scale volume modification of the Barnett shale, the success of which provided an effective means of unconventional oil and gas exploration [4]. Large-scale applications began in the field in 1998 and accounted for 30 per cent of total fracking fluid use in the United States in 2004. It has been rolled out worldwide [7-9].

3. Reduce the Resistance Classification

The advantages and disadvantages of anti-blocking agent determine the performance of fracturing fluid, which is the core factor restricting reservoir mining [10-12]. After the success of the application of water-skiing fracturing fluid, there has been an increase in research reports on anti-blocking agents at home and abroad. At present, the most used water-based anti-blocking agent is formulated for water-based water-based anti-obstruction agent, according to molecular structure and source, can be divided into the following types.

3.1. Polyacrylamide (PAM)

Polyacrylamide is a synthetic polymer and is most widely used in oil and gas fields. There are sol-like, solid powder-like and emulsion-like forms. The production cost of sol-like anti-resistance agent is low, but the resistance reduction rate is low, and it is less used in fracturing fluid. Powdered polyacrylamide is easy to store and transport, the synthesis process is mature, the degradation rate is high, but the dissolution rate is slow, in the water first dissolved and then dissolved, the dissolution will not be sufficient fisheye caused reservoir damage, site construction can be overcome by means of instruments and equipment to overcome its dissolution problem. The United States developed a hybrid equipment power consumption is low, all aspects of better performance, the technical level in a leading position. China's current use of the most car-mounted hybrid car CSGT-240, the use of jet mixer for water powder mixing, the maximum dispensing capacity of 10m3/min, the general use of two vehicles at the same time dispensing to meet the construction needs. The powder-like anti-resistor CFZ-1, produced by Yanfei, et al., can be completely dissolved within 5min and has a certain viscosity to meet the requirements of continuous mixing on-site. The combination of instant powder-type anti-resistors and continuous mixing devices can greatly reduce costs and operational complexity. Emulsion-like polymers are micron-grade and can generally dissolve quickly in water within 3 min, including the following two types [13-15].

3.1.1. W/O Inverted Polymer Damper. Through inverting emulsion polymerization, the monomer water solution is added to the oil phase step by drop, emulsified and dispersed in the oil with the aid of oil-coated water emulsifier, the polymerization caused by the addition of the trigger agent, conforming to the free radical reaction mechanism, chain triggering, chain growth and chain termination of the three element reactions, and finally the water-soluble polymer particles are evenly dispersed in the oil phase emulsion. The distribution state of the oil and water was shown in figure 1. W/O-type inverting polymer relative to the molecular mass is large, solid content is about 25%-40%, the inner phase is water-soluble polymer, the outer phase is petroleum alkanes, dilution, the polymer in water can be quickly hydrated release. With good solubility and thickening ability, the synthetic path is easy to operate, the raw material source is wide, the field application is the most. Inverted emulsion polymerization when the amount of surfactant is large, polymers need to be phased before they can be dispersed in water to play a role, and the emulsion contains a large number of oil phases, will bring about environmental pollution problems, for which W/W dispersion polymer anti-blocking agent is proposed.

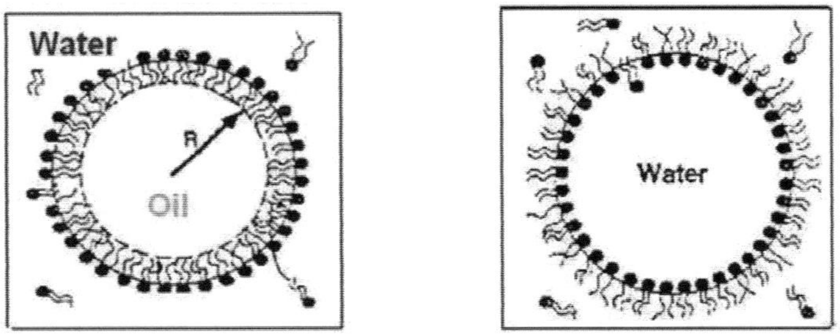

Figure 1. The distribution state of the oil and water.

3.1.2. W/W Dispersion Polymer Anti-Resistor. Through the water dispersion polymerization method, a high concentration of electrolyte solution is added to the water solution containing monomer as a phase separator, isolation and electrostatic action with the polymer chain, the polymer formed after adding the promoter can be dissolved directly in water by dispersion mechanism, forming a uniform and stable W/W solution. The particle size is usually 10um. The synthesis process is simple, the dissolving rate is fast, the safety is high, and the environmental impact is small. W/W-type anti-resistant than W/O type, does not contain organic solvents and surfactants, dissolves quickly, low cost, small environmental pollution. But most of them are still in the research and testing stage, and they are not being put into use on a large scale. The field application effect should be further studied so that it can replace the traditional W/O-type anti-resisting agent [16-17].

Changing the composition of polymer monomers can give the anti-resistance agent better performance. Monomers containing sulfonate roots, which can form strong hydrogen bonds with water and have greater stability, include vinyl acrylates, acrylic sulfonate, 2-acrylamide-2 methyl sulfonate (AMPS), etc., with AMPS being the most widely used. Ring monomers can increase the rigidity of polymers, including chitosan, vinyl glycoside, acrylamide, etc., can increase thermal stability. Adding a small amount (molar fraction <2%) hydrophobic group to the molecular main chain, usually 12-18 alkyls, can obtain hydrophobic polyacrylamide (HAPAM), which can be physically cross-linked to form a reversible spatial mesh structure, which in turn can increase the ability to resist shearing. When the concentration is lower than CAC (critical polymerization concentration), the polymer chain mainly takes place in-molecule action, the higher the content of hydrophobic monomer, the more curly the polymer, is not conducive to reducing resistance. When CAC is reached, intermolecular bonding occurs, DR increases, and when CAC is exceeded, viscosity decreases and DR decreases slightly.

3.2. Natural Polymers

Natural polymer inhibitors are biodegradable, the most representative of the natural macromolecules xenon gum (XG) and polysaccharide gum (GG), the price of the two are equivalent, the gum depends on imports, and xylitol can be produced for their own use.

Xyrogen is an efficient thickener that quickly dissolves in water to form a stable rigid rod-like double helix structure. Small influence by acid alkaline, with false plasticity, and compared with polymers with molecular weight, the ability to resist shear is stronger, but the resistance performance is generally, low temperature is difficult to break the glue, high temperature is easily oxidation degradation and loss of effect. The high salt resistance makes it show great application potential in the field of fracturing fluid formulation. Minghua, Lu Congjun, etc. [18] in 2016 using mineralization of 28074mg/L of sea water formulation quality score of 0.4% modified xylogen fracturing liquid, with low resistance, good stability, sand carrying capacity, safety and environmental protection, can meet the compact reservoir fracturing needs at 60 to 130 degrees C, and simple formulation, low cost, with broad application.

Linear gel and its derivatives in high concentration as a thickener to increase adhesion, low concentration can be used as a resistance reduction agent, in polyacrylamide-type anti-blocking agent before the epidemic, widely used in the formulation of fracturing fluid, low cost, is the earliest use of water-skiing anti-blocking agent. Usually when the mass concentration is 10mg/L, the effect of reducing resistance begins to appear, the mass concentration increases, the resistance reduction rate increases, at 300 mg/L is 30%, the resistance reduction rate is low, a small amount of crosslinking agent is added, can simultaneously improve the resistance reduction and sand carrying capacity of the acolyte [19-24]. However, when the glue is returned, the residue content is high, the damage to the reservoir is large, and it is easily decomposed by microorganisms, and the storage time is limited. Some researchers combine flexible polymers with gel to improve performance, R. Deshmukh [25] grafted 7 kinds of polyacrylamide onto the gum, found that the shear resistance and resistance of the gum have been improved, the longer the polymer chain, the better the performance, all grafted gum in ten days did not appear microbial degradation phenomenon.

3.3. Surfactant

According to the charged properties of hydrophilic groups, surfactant-type anti-resisting agents can be divided into four types, respectively, cation, anion, non-ion and gender ion surfactant as the main agent.

Among them, the anti-ion surfactant is poor anti-salt, will react with calcium and magnesium ions in the water to produce precipitation, non-ion surfactants have turbidity point, poor temperature resistance, high price of both sexes surfactants, and the preparation process is cumbersome. The application potential of cation surfactant in the reduction of resistance has been widely studied, even at low concentrations, can occur a stable resistance reduction effect. Among them, seasonal ammonium salt has better anti-temperature anti-shearing properties and is currently the most commonly used surfactant type anti-resisting agent. Xing L and others [26] prepared a solid cation surfactant CFZ-1, the mass score of 0.15%-0.2%, can reduce the resistance 68%-72 Eight wells have been successfully applied in ultra-low seepage dense reservoirs, and Tammano S [27] have a 70% resistance reduction rate when the mass score is 0.6%, which is synthesized by the main agent of 1dane dimethyl oxide.

When the amount of surfactant added is greater than the critical beam concentration, different forms of glue bundle can be formed to make the solution sticky elasticity and drop resistance, if the shear rate is too large, the beam structure is destroyed, the reduction resistance rate is reduced. The high dose of surfactant required to achieve resistance reduction, viscosity can only reach 10%-40% of the same polymer concentration, limiting the transport of proppants, coupled with the higher price of surfactants, resulting in increased fracking costs, rarely used as a water-skiing anti-blocking agent. Added to fracturing fluid is mainly used to change rock wetting and reduce surface tension, improve the water lock phenomenon caused by hair tube force, promote liquid re-discharge and thus reduce strata damage.

4. The Mechanism of Action of the Anti-Blocking Agent

Slippery water fracturing uses high displacement, large amount of liquid operation. In some oil fields, the injection rate reaches 25 to 30 m^3/min, the water injection of a single well is 2× 105 m^3, turbulence is formed in the tube, a large number of vortex is produced, and the polymer solution with the addition of a resistance reduction agent can play an effective role. The following image shows the curve of

Journal of Physics: Conference Series **2097** (2021) 012022 doi:10.1088/1742-6596/2097/1/012022

frictional resistance of polymer solutions (non-Newtonian fluids) and Newtonian fluids that produce a drop-resistance effect with Reynolds.

In the small Reynolds number, the shear rate is low, the viscous fluid is layered motion, at this time the anti-resistor molecules are disorderly, with the increase of Reynolds, the polymer solution curve deviates from the Newtonian fluid curve, began to occur drop resistance, the fluid interface within the unit length of the pipeline in the turbulence state is distorted and deformed, the curly anti-blocking molecule fully dispersed, the long chain is naturally stretched along the direction of flow, and is linearly arranged, the curve continues to deviate until the drop resistance effect reaches its maximum value, when the Reynolds number is too large, the polymer molecular structure is degraded by a larger shear, the two curves are close to each other until coincident, the drop rate gradually reduced to zero. Therefore, the drop effect exists only in turbulent areas within the limited Renault numerical range [28].

The earliest theory of drop resistance is the pseudo-plastic hypothesis put forward by TOMS, which holds that the viscosity of the fluid decreases with the increase of the shear rate, so the resistance decreases with the flow of the fluid. An in-depth study of non-Newtonian fluids found that polymethylene acrylic, which also has a damping effect, was a typical shear-thickened liquid, a theory overturned.

Since then, a large number of scholars have carried out in-depth study of the mechanism of descent, the complexity of near-wall turbulent flow mechanics and polymer dynamics, so that there is no definite model to explain. There are two more accepted mechanisms.

4.1. Turbulence Inhibits Drop Resistance

The polymer stretches in the near-wall flow area, and the stress generated by the stretch is opposite to the vortex force, resulting in vortex suppression. Zhang [29] used three-dimensional direct numerical simulation to find that the positive and negative torque of the polymer can inhibit the frequency and intensity of the vortex burst, reduce the vortex regeneration, thereby reducing turbulence pulsation, so that the flow in the quasi-stratospheric flow state. Using numerical simulation, Toonder [30] used particle image velocity measurement technology to study the flow velocity and turbulence distribution of liquid after adding the anti-blocking polymer, and found that radial turbulence intensity decreased and axial turbulence intensity increased. Polymer molecules mainly act on the area near the pipe wall, also known as buffer zones, resisting some radial forces of fluid microns and converting them into axial forces that flow in a smooth direction, reducing the exhaustive consumption and reducing the coefficient of friction of the flow by changing the turbulence structure and flow rate distribution.

4.2. Sticky Elastic Resistance Reduction

All polymers with a resistance reduction effect have a sticky elasticity at higher concentrations. Maxwell's equation shows that the vortex of the fluid is obviously inhibited with the increase of adhesive elasticity. Sticky elasticity gives it the ability to store energy, absorbs kinetic energy near the wall and converts it into its own elastic energy, giving the fluid a certain buffering effect, when the polymer molecule relaxes long enough, the energy is released into the buffer layer to weaken the near-wall turbulence. The anti-resisting agent can also form an elastic bottom layer in the pipe, with the increase of concentration, the bottom layer thickens, creating isolation between the pipe wall and the liquid, reducing the pipe wall resistance, thus achieving the drop effect. Some researchers have found that at low flow rate, the resistance reduction performance is mainly affected by viscosity, while at high flow rate, it is mainly controlled by elasticity. Liu [31] used a transmission electron microscope to observe the micro-resistance mechanism of the anti-blocking agent, indicating that the dispersed particles and continuous mesh-like structure of the anti-blocking agent have a good resistance reduction effect at low and high flow rates, respectively, as shown in figure 2.

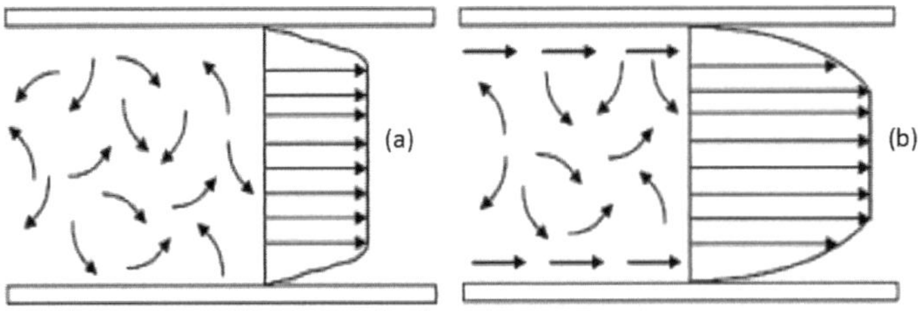

Figure 2. (a) Pure liquids; (b) Liquids with anti-blocking agents added.

The mechanism of drop resistance is not completely clear at present, but it is certainly related to the tensile properties of polymer flexible long chain molecules.

5. Summary and Outlook

Reducing mor resistance has always been the most important goal in the process of oil and gas extraction, based on the above-mentioned research on water-skiing fracturing anti-blocking agent, anti-blocking agent can be extended through the molecular chain to achieve resistance reduction. Natural polymer de-inhibitors come from a wide range of sources and are biodegradable, but have limited resistance reduction capacity. The amount of surfactant is large and the cost is high. Polyacrylamide is the most widely used, there are different kinds, the effect of resistance reduction is obvious, can meet the requirements of different site construction. Both at home and abroad have produced related products, on-site applications have achieved certain results. A wide range of foreign products, BJ Service produced the most widely used anti-blocking agents, limited domestic product categories, most of which are in the laboratory research stage, can reach 70% resistance reduction rate of products less.

Although the water-skiing fracturing liquid can achieve a large reduction resistance, but the sand carrying capacity is weak, not suitable for long-term effective exploitation of oil and gas. It is necessary to increase the adhesive elasticity to increase the production capacity under the condition of meeting the drop resistance, to use slippery water for volume modification, to improve the refive rate of fracturing fluid, to reduce the damage of the reservoir water lock, and to increase recycling. In view of environmental protection problems and water pressure, it is necessary to develop an economic and environmentally friendly anti-resisting agent with excellent de-resistance performance at high mineralization, low damage to the formation and high efficiency with sand. Dispersed polymerized W/W anti-resistant does not contain surfactants and oil phases, dissolves fast and low cost, and has broad application prospects.

China should learn from foreign experience, when formulating fracking fluid, for reservoir characteristics, water characteristics, fracking equipment and other chemical additives and other comprehensive considerations, the combination of resistance reduction and production capacity, the introduction of different properties of the monomer polymer inhibitor molecular structure design and improvement, or add other material interaction, the optimal site construction design, so as to achieve the lowest cost of maximum recovery results.

Acknowledgments
The project is supported by National Science and Technology major projects (2017ZX05023003).

References

[1] Wang L, Wang D and Shen Y 2016 Study on properties of hydrophobic associating polymer as drag reduction agent for fracturing fluid *Journal of Polymer Research* **23**(11) 235.
[2] Abubakar A, Al-Wahaibi T and Al-Wahaibi Y 2014 Roles of drag reducing polymers in single- and multi-phase flows *Chemical Engineering Research & Design* **92**(11) 2153-2181.

[3] Toms B A 1948 Some observations on the flow of linear polymer solutions through straight tubes at large reynolds numbers *First International Congress on Rheology*.

[4] Deshmukh S R and Singh R P 1987 Drag reduction effectiveness, shear stability and biodegradation resistance of guargum - based graft copolymers *Journal of Applied Polymer Science* **33**(6) 1963-1975.

[5] Palisch T T, Vincent M C and Handren P J 2021 Slickwater fracturing: food for thought *SPE Production & Operations* **25**(3) 327-344.

[6] Nechaev A I, Gorbunova M N, Lebedeva I I, et al. 2017 Synthesis by radical polymerization and structure of drag reducing terpolymers based on acrylamide, acrylonitrile, and 2-acrylamido-2-methylpropanesulfonic acid *Russian Journal of Applied Chemistry* **90**(9) 1524-1531.

[7] Zhao B Y and Mao J 2019 Review of friction reducers used in slickwater fracturing fluids for shale gas reservoirs *Journal of Natural Gas Science and Engineering* **62** 302-313.

[8] Asidin M A, Suali E, Jusnukin T, et al. 2019 Review on the applications and developments of drag reducing polymer in turbulent pipe flow *Chinese Journal of Chemical Engineering* **27**(8)

[9] Michael S, Kyle F and Steve T 2016 Dry polyacrylamide friction reducer: not just for slick water *SPE Hydraulic Fracturing Technology Conference* (Texas: SPE).

[10] Hong C H , Zhang K and Choi H J 2010 Mechanical degradation of polysaccharide guar gum under turbulent flow *Journal of Industrial and Engineering Chemistry* **16**(2) 178-180.

[11] Deshmukh S R and Singh R P 1987 Drag reduction effectiveness, shear stability and biodegradation resistance of guargum-based graft copolymers *Applied Polymer* **33** 1963-1975.

[12] Liang X, Yang K and Xu H 2019 Preparation and properties of amphoteric polyacrylamide/modified montmorillonite nanocomposites and its drag reduction performance *Colloids and Surfaces A: Physicochemical and Engineering Aspects* **574** 94-104.

[13] Feng C, Yasuo K and Bo Y 2008 Experimental study of drag-reduction mechanism for a dilute surfactant solution flow *International Journal of Heat and Mass Transfer* **51**(3) 835-843.

[14] Broniarz L, Rozanski S and Rozanska Z 2007 Drag reduction effect in pipe systems and liquid falling film flow *Reviews in Chemical Engineering* **23**(3-4) 149-245.

[15] Ronald R, Harsha S and Friction L 2013 Reducers as water management aids in hydraulic fracturing *SPE Production and Operations Symposium* Oklahoma City, Oklahoma, USA, SPE.

[16] Robert G and Eric F 2018 Geochemical formulary and analysis of hydraulic fracturing water quality to determine the optimal fracturing fluid design *SPE/AAPG Eastern Regional Meeting* Pittsburgh, Pennsylvania, USA, SPE.

[17] Shen L, Leonid V and David H 2018 Can friction reducers transport sand during fracturing treatment *SPE/AAPG/SEG Unconventional Resources Technology Conference* Houston, Texas, USA, URTEC.

[18] Leonid V, Andrey B and Shen L 2018 Far-field diversion system designed for slickwater fracturing *SPE Liquids-Rich Basins Conference - North America* Midland, Texas, USA, SPE.

[19] Arjun S, Martin J and Imad Z 2014 Successful slickwater fracturing in ultrahigh TDS produced water by novel environmentally preferred friction reducer *International Petroleum Technology Conference* Kuala Lumpur, Malaysia, IPTC.

[20] Alexandra M, Nikolay S and Ahmed F 2019 Completing ultra extended-reach wells: overcoming the torque and drag constraints of brine *Abu Dhabi International Petroleum Exhibition & Conference* Abu Dhabi, UAE, SPE.

[21] Mehdi H and Clark P E 2017 Drag reduction behavior of hydrolyzed polyacrylamide/xanthan gum mixed polymer solutions *Petroleum Science* **14**(02) 412-423.

[22] Mortazavi S M M 2016 Correlation of polymerization conditions with drag reduction efficiency of poly(1-hexene) in oil pipelines *Iranian Polymer Journal* **25**(8) 731-737.

[23] Den Toonder J M J, Hulsen M A, Kuiken G D C, et al. 1997 Drag reduction by polymer additives in a turbulent pipe flow: numerical and laboratory experiments *Journal of Fluid Mechanics* **337** 193-231.

[24] Lawrence C, Edomwonyi O and Panagiota A 2019 Separated oil-water flows with drag reducing polymers *Experimental Thermal and Fluid Science* **102** 467-478.

[25] Deshmukh S R and Singh R P 1987 Drag reduction effectiveness, shear stability and biodegradation resistance of guargum - based graft copolymers *Journal of Applied Polymer Science* 33(6) 1963-1975.

[26] Xing L, Ke Y C, Hu X, et al. 2019 Preparation and properties of amphoteric polyacrylamide/modified montmorillonite nanocomposites and its drag reduction performance *Colloids and Surfaces A: Physicochemical and Engineering Aspects* **574** 94-104.

[27] Tamano S, Kitao T and Morinishi Y 2014 Turbulent drag reduction of boundary layer flow with non-ionic surfactant injection *Journal of Fluid Mechanics* **749** 367-403.

[28] Nesyn G, Yu M and Suleimanova V 2012 Polymer drag-reducing agents for transportation of hydrocarbon liquids: Mechanism of action, estimation of efficiency, and features of production *Polymer Science* **54**(1) 61-67.

[29] Zhang L J, Huang X D, Fan X M, et al. 2019 Rapid fingerprinting technology of heavy oil spill by mid-infrared spectroscopy *Environmental Technology* **42**(2) 270-278.

[30] Toonder J D, Hulsen M A, Kuiken G, et al. 1997 Drag reduction by polymer additives in a turbulent pipe flow: numerical and laboratory experiments *Journal of Fluid Mechanics* **337** 193-231.

[31] Liu Z Y, Zhou F J and Qu H Y, et al. 2017 Impact of the microstructure of polymer drag reducer on slick-water fracturing *Geofluids* **2017** 9080325,

Preparation and Application of CQDs/Ni(OH)$_2$0.75H$_2$O as Electrodes in Supercapacitors

Yunlong Zhou[1,2], Hanxiu Fu[1], Junwen Peng[3], Weize Chen[1], Kunfeng Cen[1], Jingru Lv[1*] and Xiaofen Cao[1]

[1] College of Chemistry and Materials Science, Longyan University, Longyan, Fujian, 364000, China
[2] Fujian Provincial Key Laboratory of Clean Energy Materials, Longyan, Fujian, 364000, China
[3] Fujian Weidong New Energy Co., Ltd., Yongding, Fujian, 364101, China

Abstract. In this paper, a hydrothermal synthesis method for preparation Carbon Quantum Dots (CQDs)/Ni(OH)$_2$0.75H$_2$O composite electrode material, and loaded on Nickel Foam(NF) by rolling method. The crystal phase, structure, morphology and the reaction process were characterized by X-ray diffractometer (XRD) and scanning electron microscopy (SEM). The electrochemical properties of the composite were measured by cyclic voltammetry impedance spectroscopy and couple method charge-discharge techniques. The results showed that the composites material was evenly adhered to nickel foam. When the current density is 1 A g$^{-1,}$ the specific capacity of CQDs / Ni (OH) $_2$0.75H$_2$O composite can reach 2115.4 Fg^{-1}, which is close to 5 times of Ni (OH) $_2$0.75 H$_2$O electrode material.
Keywords. Ni(OH)$_2$0.75H$_2$O. carbon quantum dots. hydrothermal synthesis method. supercapacitor.

1. Introduction

With the emergence of environmental pollution problems caused by the large-scale use of non-renewable energy, green and sustainable energy such as solar, wind and tidal energy are needed to replace non-renewable energy. However, these sustainable energy resources are unstable and have many uncontrollable risk factors, which are inconvenient in their practical applications. Therefore, it is required to develop various energy storage devices in order to collect the random fluctuations of renewable energy [1-3]. As an energy storage device, supercapacitor has become one of the hot research areas in energy field, because it combines the characteristics of large capacity, high power density, long-term recycling, short charging time, and so on [4-5].

There are two types of supercapacitors and one is double-layer capacitor, in which ions are adsorbed on the surface of the electrode material during charging and desorbed during discharging, which is a reversible process. And the working process of double-layer supercapacitor is a pure physical adsorption process, in which charge is stored and released continuously for the energy storage [6]. The other one is a pseudo-capacitor, in which the charge is transferred during the redox process and the chemical energy and electrical energy are converted to achieve energy storage. Pseudo-capacitors can store more energy under the same conditions [7-8]. Therefore, the selection of electrode materials in this experiment is based on pseudo-capacitor capacitor.

In practical applications, pseudo-capacitive electrode materials usually require at least the following conditions: multivalent, low cost and redox reactions taking place easily [9]. In the transition metals, nickel not only accords with the above characteristics, but also has a higher theoretical capacitance and is environmentally friendly, too. Therefore, nickel is suitable as the electrode material of pseudo-capacitor [10].

Carbon quantum dots (CQDs) are zero-dimensional nano-materials with strong adsorption capacity, large surface area, many active sites and good crystallization behavior. Additionally, CQDs can promote transport of electrolyte ions and increase the rate of charge transport in the electrode materials, which have great development potential as supercapacitors [11-12].

In this work, CQDs/$Ni(OH)_2 0.75H_2O$ was prepared by the hydrothermal method and the conductivity property of $Ni(OH)_2 0.75H_2O$ was highly improved.

2. Experimental

2.1. Pretreatment of Electrodes

Nickel foam foils were cut into the size of (10 × 30) mm^2, then were ultrasonically cleaned in acetone for 15 min, then ultrasonically cleaned in 3 mol•L^{-1} HCl for 20 min, finally ultrasonically cleaned in ethanol and distilled water alternately for several times, and then put it into the vacuum drying oven for drying at 60 °C.

2.2. Preparation of Electrode Materials

$Ni(NO_3)_2 \cdot 6H_2O$ (0.7270 g, 0.025 moL•L^{-1}), NH_4F(0.1480 g, 0.04 moL•L^{-1}), CH_4N_2O(0.6426 g, 0.107 moL•L^{-1}) and distilled water (90 mL) were added to a 250mL beaker. CQDs and 10 mL ethanol were added to 50 mL beaker and ultrasound was applied for 10 min to make CQDs completely soluble in ethanol. The above two solutions were mixed and stirred using a magnetic stirrer for 120 minutes, then the mixed solution was transferred to a high temperature reactor, which was further heated to 120 °C and sealed for 8 h. After filtration, the electrode materials were obtained and washed with distilled water, and the final products were transferred to a constant temperature drying chamber at 60 °C for 4 hours.

2.3. Structural Analysis of Electrode Materials

The morphology of the samples was characterized by a transmission electron microscope (TEM). The crystalline structure of the samples was characterized by X-ray diffraction (XRD).

2.4. Preparation of the Working Electrode

10 mg of the prepared electrode material and 80 mg of acetylene black were added into the mortar, which were grinded with 1 ~ 2 drops anhydrous ethanol until the mixture became dry, and anhydrous ethanol was added repeatedly. After grinding for 2 hours, 5% PTFE (the mass ratio of acetylene black: sample: PTFE is 8:1:1) was added to grind the whole mixture into a ball of viscous material, which was transferred to the glass plate and rolled. Then, some of the rolled samples were transferred to the nickel foam foil, which were further pressed with the tablet press machine to stick well on the surface of the nickel foam foil to form the working electrode. Finally, the working electrode was dried at 60°C in the constant temperature oven for 2 hours before measuring its electrochemical properties.

2.5. Electrochemical Measurement

The electrochemical properties were measured by electrochemical workstation performing the following steps:

(1) Cyclic voltammetry: the voltage range was 0 ~ 0.6 V, and the scanning speed was 10 MV S^{-1} ~ 50 MV s^{-1} with 10mV as an interval

(2) Charge and discharge under constant current: and the current density was 1 a · g^{-1} ~ 5 a · g^{-1} with 1 A as an interval.

(3) AC impedance: the frequency range was 10-2 Hz ~ 105 Hz, and the amplitude was 5 mV.

3. Results and Discussion

3.1. Morphology and Structure Characterization

In order to further study the micro state of electrode materials, nickel hydroxide Ni(OH) $_2$0.75H_2O, carbon quantum dots CQDS and composite CQDS / Ni(OH) $_2$0.75H_2O were characterized by SEM. The SEM images of Ni(OH) $_2$0.75H_2O and CQDS were shown in figure 1(a) and figure 1(b) respectively. It could be seen from the figures that Ni(OH) $_2$0.75H_2O was a stacked lamellar structure and CQDS was a spherical structure. Figure 1(c) shew the SEM of CQDS / Ni(OH) $_2$0.75H_2O composite. It could be seen from the figure that Ni(OH) $_2$0.75H_2O was dispersed and attached to the surface of CQDS, which could increase the surface area of the composite and improve its chemical properties.

Figure 1. SEM images of Ni(OH) $_2$0.75H_2O, CQDS and CQDS / Ni(OH) $_2$0.75H_2O.

In order to further analyze the final form of the experimental products, the products doped by carbon quantum dots were characterized by XRD and compared with the samples without doping.

Firstly, the characterization of the hydrothermal experimental products of nickel was obtained and analyzed. As a result, it was found that the diffraction peak of the product was basically consistent with the JCPDS no.38-0715 card, so the main component of the product was Ni(OH) $_2$0.75H_2O.

Then, by comparing the XRD characterization of CQDS / Ni(OH)$_2$0.75H_2O with Ni(OH) $_2$0.75H_2O, it was also found that the diffraction peak of the product doped by carbon quantum dots existed a weaker diffraction peak near 27 degrees ,which was not found for the undoped product. Since the characteristic diffraction peak of carbon quantum dots is 27.5 degrees [13], it could be identified that this peak was the diffraction peak of carbon quantum dots. However, the diffraction peak intensity was weak, which may be because the diffraction peak intensity after doping was small. Therefore, when measuring the diffraction peak, the diffraction peak of carbon quantum dots could not be clearly shown on figure 2 due to the high intensity of $Ni(OH)_20.75H_2O$ diffraction peak.

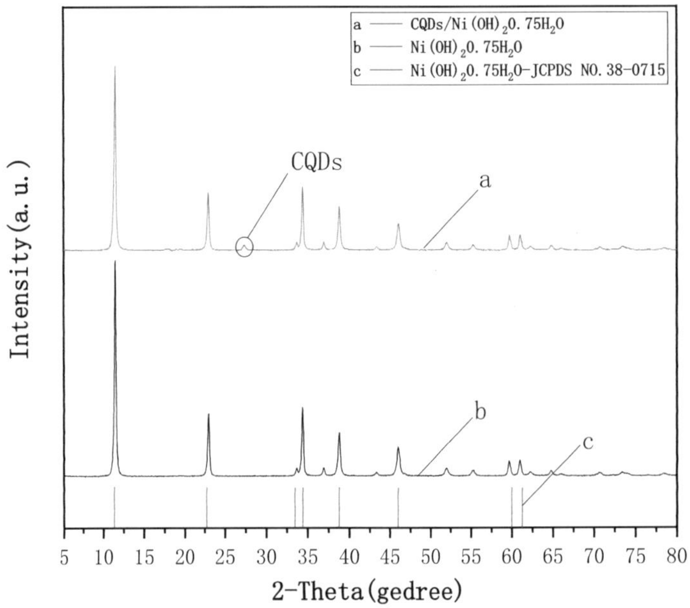

Figure 2. X-ray diffraction (XRD) of CQDs/ $Ni(OH)_20.75H_2O$ and $Ni(OH)_20.75H_2O$.

3.2. Electrochemical Measurement

3.2.1. Cyclic Voltammetry. Since the pseudo-capacitor stores energy by redox reaction, two obvious peaks could be observed from the cyclic voltammetry curves of the two substances prepared in the experiment, and the left peak was the reduction peak while the right peak was the oxidation peak. The corresponding electrochemical reaction equation was [14]:

$$Ni(OH)_2 + OH^- \rightleftharpoons NiOOH + H_2O + e^- \quad (1)$$

As shown in figure 3, it could be observed that the closed area of each curve gradually increased when the scanning rate of $Ni(OH)_20.75H_2O$ increased from $10mV \cdot s^{-1}$ to $50mV \cdot s^{-1}$, but the oxidation peak potential gradually shifted to the right and the reduction peak potential gradually shifted to the left, due to the electrode impedance. However, the oxidation-reduction peak of CQDs/ $Ni(OH)_20.75H_2O$ composite electrode as shown in figure 4 changed little due to the chemical interaction between oxygen-containing functional groups in carbon quantum dots and nickel hydroxide or the van der Waals force between carbon quantum dots and nickel hydroxide. As shown in figure 5. It could be concluded that CQDs/ $Ni(OH)_20.75H_2O$ composite electrode shew the characteristics of low electrode impedance, fast and stable electron and ion transport rate [15].

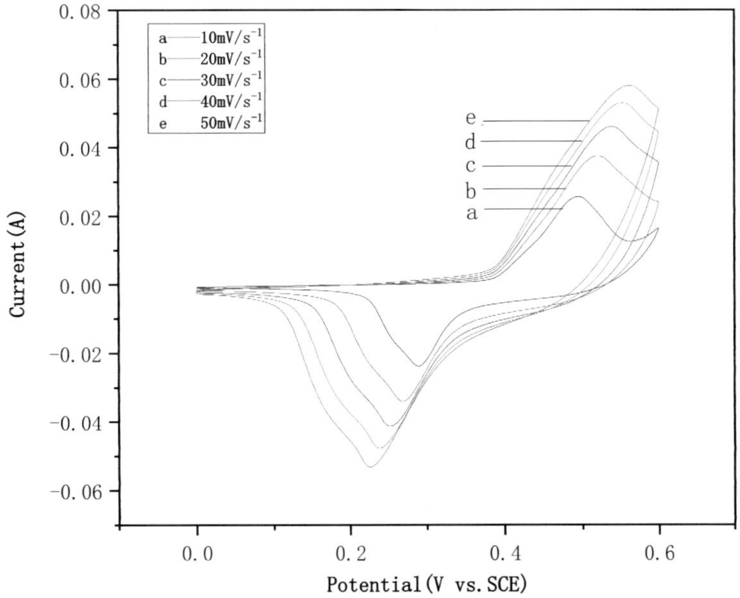

Figure 3. CV curves ofNi (OH) $_2$0.75 H2O with scanning rate of 10 ~ 50 mV· s^{-1}.

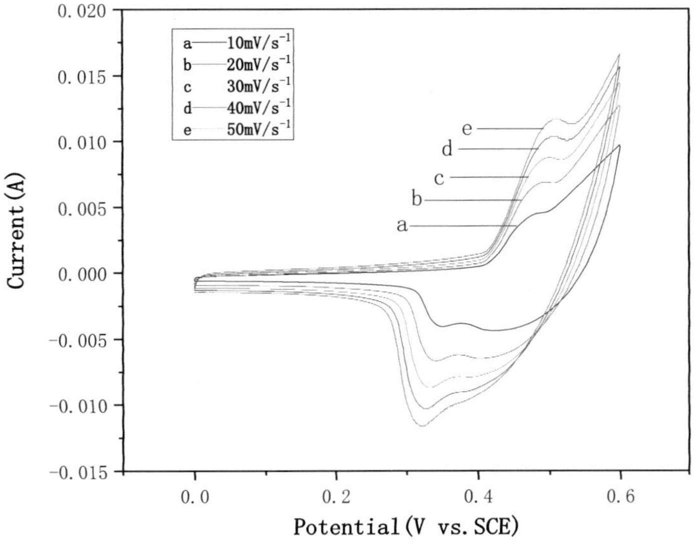

Figure 4. CV curves of CQDs/Ni(OH)$_2$0.75H_2O with scanning rate of 10 ~ 50 mV· s^{-1}.

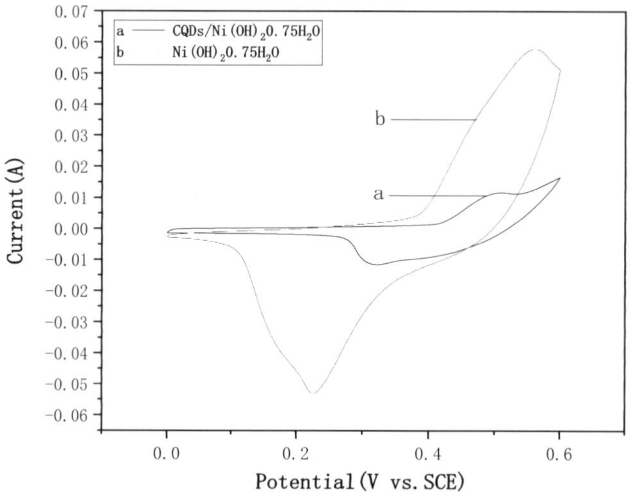

Figure 5. CV curves of $CQDs/Ni(OH)_20.75H_2O$ and $Ni(OH)_20.75H_2O$ with scanning rate of 50 $mV\cdot s^{-1}$.

3.2.2. Charge and Discharge under Constant Current. As shown in figure 6 and figure 7, the voltage of the two electrode materials gradually rose to 0.52 V during charging under the conditions of voltage range of 0 ~ 0.52 V and current density of 1 ~ 5 a · g^{-1}, and there were obvious flat periods in the charging curves. In the discharge process, the potential gradually decreased to 0 V with time, and the plateau period was obvious, which indicated that the material was a pseudo-capacitor.

According to the specific capacitance equation:

$$C_S = \frac{I \times \Delta t}{m \times \Delta V} \tag{2}$$

Where, C_S ($F\cdot g^{-1}$) is the specific capacitance; I (A) is the charge discharge current; Δt (s) is the discharge time; m (g) is the mass of the active substance involved in charge and discharge in the working electrode; ΔV (V) is the total voltage drop.

For instance, the specific capacitance of Ni (OH) $_20.75$ H2O could be calculated as 492.3 $F\cdot g^{-1}$ according to equation (2) under the current density of 1 $a\cdot g^{-1}$, as shown in figure 6. The proton transfer rate could accelerate as the increasing of the current density, so the protons could be transferred before involving in the reaction. Therefore, the specific capacitance would decrease with the increase of current density [16].

In the same way, the specific capacity of $CQDs/Ni(OH)_20.75H_2O$ in figure 7 was calculated to be 2115.4 $F\cdot g^{-1}$ at 1 $a\cdot g^{-1}$ .With the increase of current density, the specific capacitance would also decrease.

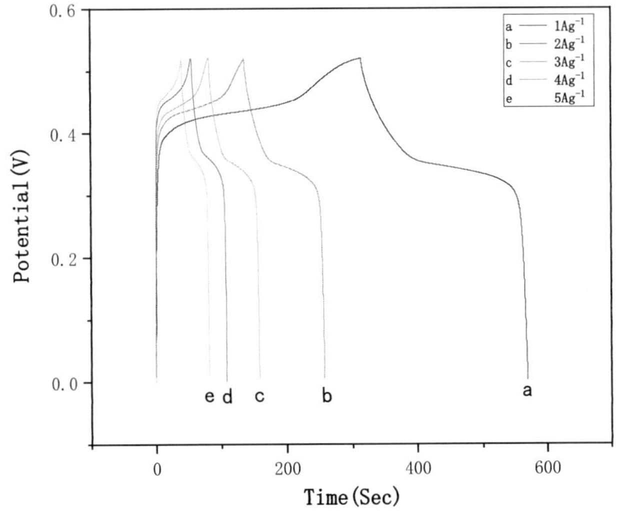

Figure 6. Charge and discharge curves of Ni (OH) $_2$0.75 H_2O under the current density of 1 ~ 5 $A{\cdot}g^{-1}$.

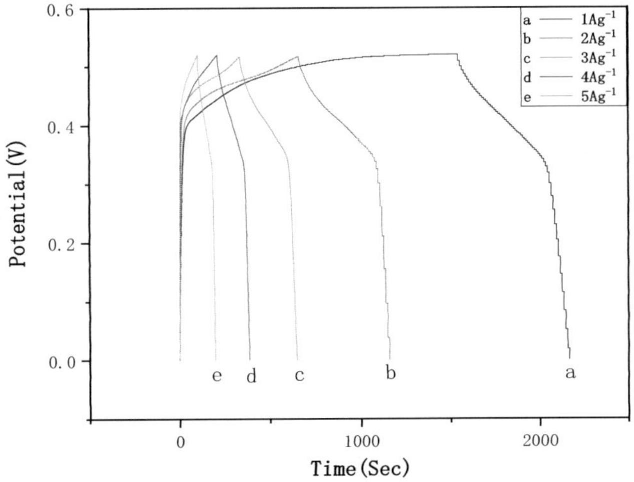

Figure 7. Charge and discharge curves of CQDs/Ni(OH)$_2$0.75H_2O under the current density of 1 ~ 5 $A{\cdot}g^{-1}$.

The comparison of the discharge curves at 1 $A{\cdot} g^{-1}$ and 0.52 V of CQDs/Ni(OH)$_2$0.75H_2O and Ni (OH)$_2$0.75 H_2O were shown in figure 8 . Consequently, the conductivity of the composite was increased by the doping of CQDS, and the specific surface area of CQDs/Ni(OH)$_2$0.75H_2O and the number of active sites were also increased, which would make the redox reaction could take place more completely, which was reflected in the increase of specific capacitance in the data. According to the comparison of the figure, the specific capacitance ratio of CQDs/Ni(OH)$_2$0.75H_2O and Ni(OH)$_2$0.75 H_2O was close to 5:1, which indicated that the electrochemical performance of CQDs/Ni(OH)$_2$0.75H_2O was significantly improved.

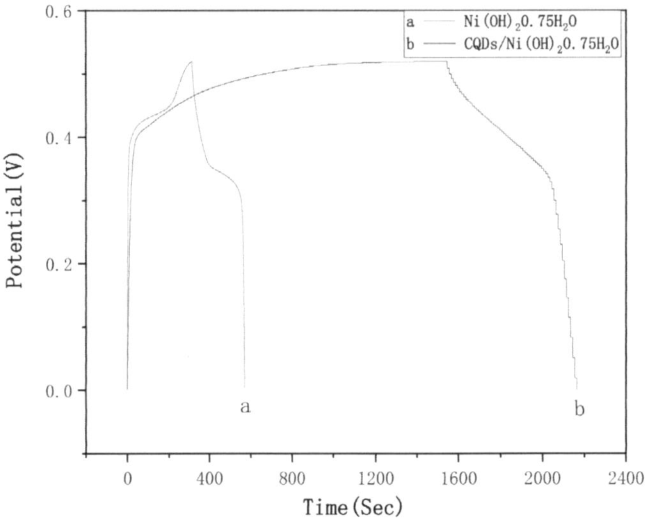

Figure 8. Charge and discharge curves of CQDs/Ni$(OH)_2$0.75H_2O and Ni (OH) $_2$0.75 H_2O under the current density of 1A· g^{-1}.

The AC impedance curves included two parts of high-frequency region and low-frequency region, which represented load transfer process control and diffusion control respectively. The resistance of charge transfer inside the electrode was shown in the high frequency region, while the rate of ion diffusion in the solution was indicated in the low frequency region. The high frequency region was represented by the semicircle radius in the AC impedance curve, and the low frequency region was represented by the linear slope in the AC impedance curve [17]. As shown in figure 9. the radius of CQDs/Ni$(OH)_2$0.75H_2O in the high frequency region was small and its slope in the low frequency region was high, which indicated that the internal charge transfer resistance of CQDs/Ni$(OH)_2$0.75H_2O composite was small and the ion diffusion rate of electrolyte was better.

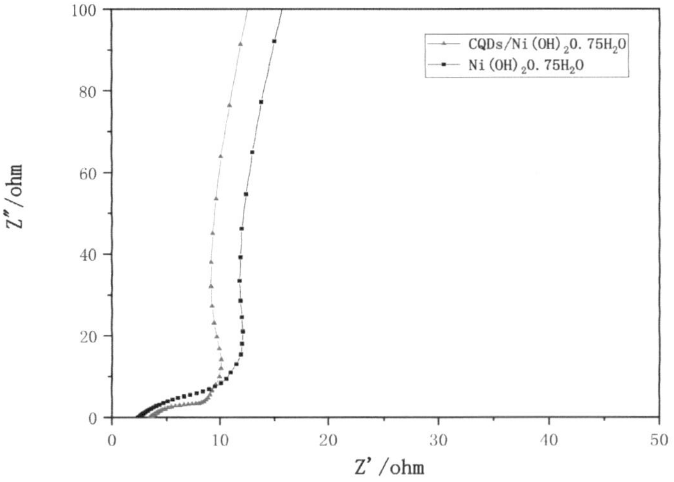

Figure 9. AC impedance curves of CQDs/Ni$(OH)_2$0.75H_2O andNi$(OH)_2$0.75 H_2O.

4. Conclusions

(1) Carbon quantum dots / nickel hydroxide composite electrode materials were prepared by hydrothermal method. Characterizations of XRD and SEM identified that the final product was $CQDs/Ni(OH)_2 0.75H_2O$, which was attached to the CQDS ball in a lamellar structure.

(2) The analysis of cyclic voltammetry (CV), AC impedance (EIS) and constant current charge and discharge (CCD) of the samples indicated that $CQDs/Ni(OH)_2 0.75H_2O$ composite electrode material owned the characteristics of lower impedance, better ion diffusivity, fast and stable electron and ion transport rate. And the specific capacity could reach 2115.4 f · g^{-1} when the current density was 1 a · g^{-1}.

Acknowledgments

The study was funded by the Key Laboratory of Clean Energy Materials of Fujian Province and the Department of Science and Technology of Fujian Province, under Grant (2019J01 799, 2019J01800).

References

[1] Ji X W, Wang Q, Yu M M, et al. 2020 Gold nanoparticles doped Mo—Ni bimetallic hydroxide electrode material *Electronic Components and Materials* **39**(11) 1-8.
[2] Xu X, Yang D Z, Fan Z J, et al. 2019 Reduced graphene oxide/carbon nanotube hybrid fibers with narrowly distributed mesopores for flexible supercapacitors with high volumetric capacitances and satisfactory duvability *Carbon* 152 134-143.
[3] Qiao S M, Huang N B, Gao Z Y, et al. 2019 Nickel-Manganese binary metal oxide as electrode materials for supercapacitors *Progress in Chemistry* **31**(8) 1177-1186.
[4] Yu X H, Luo Q L, Pan J, et al. 2019 Preparation and properties of flexible supercapacitor based on biochar and solid gel-electrolyte *CIESC Journal* **70**(9) 3590-3600.
[5] Qin H M, Yue L C and Mi J 2018 Research processes of electrode for supercapacitors *Journal of Taiyuan University of Science and Technology* **39**(6) 435-442.
[6] Xiao M 2019 Research progress of supercapacitors *Electronic Components and Materials* **38**(9):1-12.
[7] Qi P 2019 Talking about research progress of electrode materials for pseudocapacitor *Chinese Journal of Power Sources* **48**(8) 1235-1238.
[8] Li J S, Hu Y P, Zhao X D, Wang X, et al. 2019 Study on preparation and electrochemical properties of Co_3O_4 supercapacitor electrode materials *Journal of Science of Teachers'College and University* **39**(8) 48-53.
[9] Lv Z J, Zhong Q and Bu Y F 2016 Controllable synthesis of Ni-Co nanosheets covered hollow box via altering the concentration of nitrate for high performance supercapacitor *Electrochimica Acta* **215** 500-505.
[10] Li C Z, Zhang C, Zhu K Y, et al. 2017 Study on the supercapacitor properties of nano-flake $Co(OH)_2$/Ni foam composites *Advanced Ceramics* **37**(5) 357-362.
[11] Gao X 2017 Carbon quantum dots: synthesis, properties and applications *Chemical Industry and Engineering Progress* **36**(5) 1734-1742.
[12] Wang S Y, Pang H C and Ning G L 2019 Preparation of carbon quantum dots @ MnO_2 composite and its properties of flexible supercapacitors *Engineering Plastics Application* **47**(6) 102-106, 111.
[13] Qi F 2020 Carbon quantum dots: Preparation, properties and applications *Chemical Research* **31**(3) 270-277
[14] Li Z 2019 Study on Properties of Nickel Hydroxide Supercapacitors *Journal of Shenyang Ligong University* **38**(4) 70-74.
[15] Wei G G, Xu X R, Liu J X, et al. 2017 Carbon quantum dots decorated hierarchical $Na(OH)_2$ With lamellar structure for outstanding supercapacitor *Materials Letters* **186** 131-134.
[16] Yi C, Xiong X B, Zou Z B, et al. 2015 Fabrication of nickel-based composite film electrode for supercapacitors by a New Method of Anodization/GCD *Acta Phys. -January Chim. Sin.* **31**(1)

99-104.
[17] Zhong Y, Wang S G, Chen R, et al. 2017 Preparation and electrochemical performance of cobalt hydroxide and nickel hydroxide composite materials *Vacuum & Cryogenics* **23**(4) 235-240.

Journal of Physics: Conference Series **2097** (2021) 012024 doi:10.1088/1742-6596/2097/1/012024

Numerical Simulation of Adsorption and Heat Transfer in Porous Media Using Lattice Boltzmann Method

Jianhu Wang, Zhongdi Duan*, Cheng Cheng and Wenyong Tang

State Key Laboratory of Ocean Engineering, Shanghai Jiao Tong University, Shanghai 200240, China

Email: duanzhongdi@sjtu.edu.cn

Abstract. An adsorption model for fluid flow, heat, and mass transfer of the adsorbent bed was established. Based on the single relaxation time lattice Boltzmann method, a dual-distributed lattice Boltzmann model of density and concentration was established to solve the fluid flow and mass transfer process in the surface area of the adsorbent bed. The adsorption and heat transfer process on the surface of the adsorbent bed was incorporated into the dual-distributed lattice Boltzmann model by the fourth-order Runge-Kutta finite difference method. The multiphysics fields under Poiseuille flow were simulated by the presented model, and the adsorption capacity and temperature distribution during the adsorption process were investigated.

Keyword. Adsorption, heat transfer, lattice Boltzmann method, porous media.

1. Introduction

As a storage medium for gases such as H_2 and CH_4, porous media adsorbents have been widely used in the engineering industry [1]. The adsorption of gases such as methane occurs on the surface of porous media while giving off heat, and it is necessary to study the heat transfer process in the area near the adsorption bed during the adsorption process and the effect on the adsorption amount.

As an emerging and effective numerical simulation method at the mesoscopic level, LBM has been widely used in fluid flow, heat, and mass transfer problems, such as porous media flow, due to its advantages of simple programming, inherent parallelism, and easy handling of complex boundaries [2]. The heat of adsorption effect, however, was few incorporated in LB simulations for adsorption.

In this paper, a two-dimensional adsorption and heat transfer model is established based on LBM, and a Maxsorb adsorption bed of a specific size is selected for numerical simulation. The kinetic adsorption characteristics are investigated combining the heat of adsorption effect.

2. Numerical model

2.1. Dual-Distributed Lattice Boltzmann Model

The single relaxation time lattice Boltzmann method is used to solve the density field and concentration field in the fluid flow and mass transfer governing equations, in which the D2Q9 and D2Q5 models are selected respectively [3].

$$f_i\left(\mathbf{r}+\Delta t\mathbf{e}_i, t+\Delta t\right)-f_i(\mathbf{r},t)=-\frac{1}{\tau_f}\left[f_i(\mathbf{r},t)-f_i^{eq}(\mathbf{r},t)\right] \tag{1}$$

$$g_i\left(\mathbf{r}+\Delta t\mathbf{e}_i, t+\Delta t\right)-g_i(\mathbf{r},t)=-\frac{1}{\tau_d}\left[g_i(\mathbf{r},t)-g_i^{eq}(\mathbf{r},t)\right] \tag{2}$$

where f_i is the density distribution function; f_i^{eq} is the density equilibrium distribution function; g_i is the concentration distribution function; g_i^{eq} is the concentration equilibrium distribution function; τ_f and τ_d are the dimensionless relaxation times of the density distribution function and concentration distribution function, respectively.

2.2. Adsorption Kinetic Model and Heat Transfer Model

Adsorption occurs on the surface of porous media. Langmuir adsorption kinetics is generally used to describe the adsorption process on the surface of porous media.

$$D_f\frac{\partial C}{\partial \mathbf{n}}=\frac{\partial N}{\partial t}=k_1C\left(N_m-N\right)-k_{-1}N \tag{3}$$

where k_1 and k_{-1} are the constant adsorption and desorption rates; N is the adsorption capacity at that moment; N_m is saturated adsorption capacity.

The crystallite pore diffusion of methane in porous media [4] is also considered, and its governing equation is described as

$$\frac{\partial N}{\partial t}=D_s\left(\frac{\partial^2 N}{\partial x^2}+\frac{\partial^2 N}{\partial y^2}\right) \tag{4}$$

where D_s is the solid diffusion coefficient in porous media.

The heat transfer governing equation includes three regions: fluid domain, porous medium interface, and porous medium interior [5], which are respectively described as

$$\text{fluid domain: } \left(\rho c_p\right)_f\left(\frac{\partial \mathrm{T}_f}{\partial t}+\mathbf{u}\nabla \mathrm{T}_f\right)=\nabla\cdot\left(\lambda_f\nabla \mathrm{T}_f\right) \tag{5}$$

$$\text{interface: } \left(\rho c_p\right)_s\frac{\partial \mathrm{T}_s}{\partial t}=\nabla\cdot\left(\lambda_s\nabla \mathrm{T}_s\right)+\frac{\partial N}{\partial t}H_{ads} \tag{6}$$

$$\text{interior: } \left(\rho c_p\right)_s\frac{\partial \mathrm{T}_s}{\partial t}=\nabla\cdot\left(\lambda_s\nabla \mathrm{T}_s\right) \tag{7}$$

where c_p is specific heat capacity; λ is coefficient of thermal conductivity; H_{ads} is the heat of adsorption.

The adsorption capacity and temperature field are calculated iteratively using the fourth-order Runge-Kutta finite difference method [6], and the calculation variance is as follows

$$\frac{\partial T}{\partial t}\equiv K_1(T),\ \frac{\partial N}{\partial t}\equiv K_2(N) \tag{8}$$

$$h_1=K(\chi^n),h_2=K(\chi^n+h_1\frac{\Delta t}{2}),h_3=K(\chi^n+h_2\frac{\Delta t}{2}),h_4=K(\chi^n+h_3\cdot\Delta t) \tag{9}$$

$$\chi^{n+1}=\chi^n+\frac{\Delta t}{6}(h_1+h_2+h_3+h_4) \tag{10}$$

3. Adsorption Process Simulation

3.1. Adsorption Isotherm

In this paper, the Tóth adsorption isotherm model is selected to calculate the saturated adsorption capacity, which can be represented by

$$\frac{N_m}{N_0}=\frac{k_0\exp\left(H_{ads}/RT\right)p}{\left\{1+\left[k_0\exp\left(H_{ads}/RT\right)p\right]^s\right\}^{\frac{1}{s}}} \tag{11}$$

where s is adsorbent structural heterogeneity parameter, k_0 and N_0 are fitting parameters. The result of fitting experimentally measured adsorption uptake of methane onto Maxsorb III [7] based on the Tóth adsorption isotherm model is shown in figure 1, meanwhile fitting parameters and fitting errors are shown in table 1.

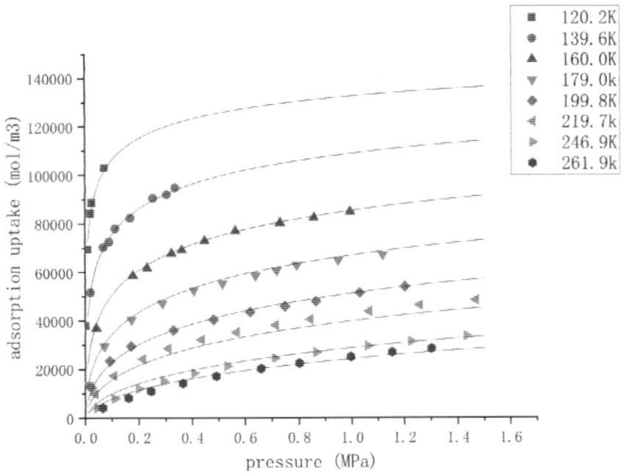

Figure 1. Adsorption isotherms of methane onto Maxsorb III, solid lines are the fitting curve from the Tóth adsorption isotherm model.

Table 1. fitting parameters for the Tóth adsorption isotherm model.

Fitting parameters	Value
N_0(mol/m^3)	174915.1
H_{ads}(J/mol)	15078.3
k_0	0.0036
s	0.2705
R^2	0.9969

3.2. The Simulation Process of LBM

The surface of porous media is rough and there are a lot of pores. Therefore, the structure diagram of maxsorb III adsorption bed with a lattice size of 500*500 is drawn in this paper, as shown in figure 2. The black part in the figure is Marsorb III particles, and the red part is the fluid domain.

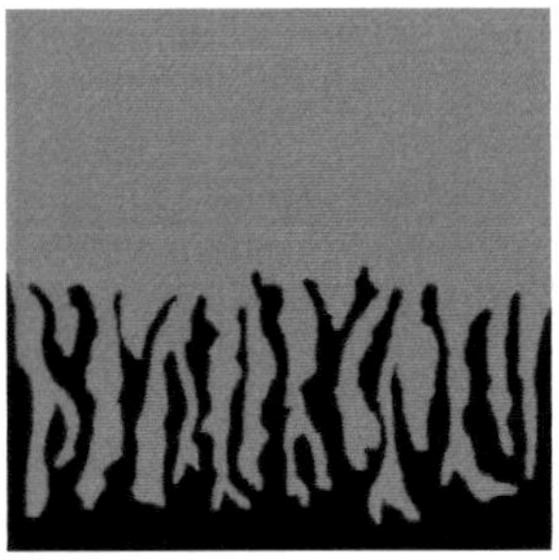

Figure 2. Schematic diagram of Maxsorb III adsorption bed structure.

In each lattice time step, the simulation process is calculated according to the time sequence of fluid flow, mass transfer, adsorption, and heat transfer. The fluid flow and mass transfer process are subjected to obtain information such as density, concentration, the velocity of each lattice point after collision and migration according to equation (1)(2). According to equation (3)-(10), the fourth-order Runge-Kutta finite difference method is used to calculate the adsorption capacity and heat transfer process of the Maxsorb III adsorption bed at the same time.

The variables in the LBM simulation process are all dimensionless lattice units, so the relevant physical parameters need to be converted to dimensionless lattice units by dimensionless correlation numbers, as shown in table 2.

Table 2. Conversion of dimensionless lattice units and physical parameters.

Physical parameters	Units	Physical symbol	Physical value	Lattice value
Sound velocity	m/s	c_s	340.40	0.5774
Domain length	m	L	1×10^{-4}	500
Time Step	s	Δt	3.40×10^{-10}	1
Gas density	kg/m^3	ρ_f	1.6550	1
Kinematic viscosity	m^2/s	v_f	2.8203×10^{-6}	0.024
Gas velocity	m/s	u_{max}	5.89	0.01
Gas concentration	mol/m^3	C_0	103.44	10
Gas diffusion coefficient	m^2/s	D_f	1.96×10^{-5}	1/6
Saturated adsorption capacity	mol/m^3	N_m	-	/10.344
Solid diffusion coefficient	m^2/s	D_s	3.92×10^{-8}	1/3000
Constant adsorption rate	$m^3/mol\cdot s$	k_1	1.47×10^7	0.0161
Constant desorption rate	s^{-1}	k_{-1}	2.94×10^6	0.0010

4. Results and Discussion

4.1. Effect of Fluid Flow and Mass Transfer

The simulation conditions are 120 K temperature and 0.1 MPa pressure, while the left inlet velocity satisfies the Poiseuille flow with a maximum velocity of 0.01 (lattice units), and the inlet methane concentration is maintained at 10 (lattice units).

Figure 3 shows the change of methane concentration in the calculation domain at 4000-500000 lattice time steps. It can be found that when adsorption starts to occur, the methane in the pores of the Maxsorb III adsorption bed is rapidly adsorbed and the methane concentration in the pores drops to 0. The pore size of Maxsorb III is very small, and the fluid-solid boundary used in LBM simulation is a rebound boundary without slip, so the fluid flow rate in the pore is very small, close to 0, and the diffusion rate of methane is very small compared to the adsorption rate. Therefore, in the whole simulation process, the methane concentration is always 0 after the initial methane is adsorbed in the pore. Only when Maxsorb III particles outside the pore reach saturation adsorption, methane begins to diffuse into the pore.

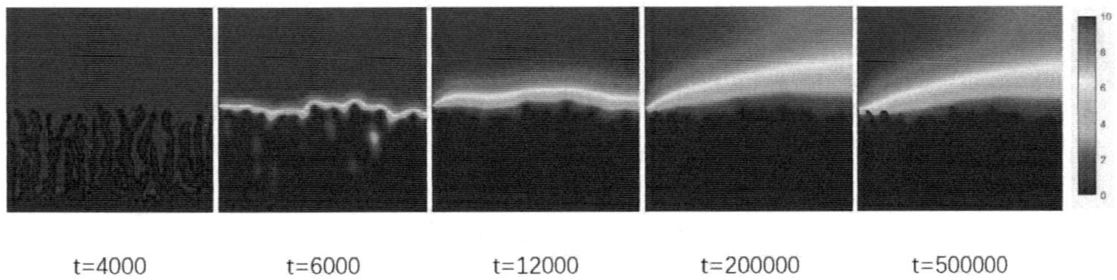

Figure 3. Figure with the change of methane concentration in the calculation domain at 4000-500000 lattice time steps.

4.2. Effect of Adsorption and Heat Transfer

The change of adsorption capacity of Maxsorb III adsorption bed and temperature change in computational domain with 4000-500000 lattice time steps are shown in figure 4. At the beginning, the methane adsorption rate is very fast, and the adsorption capacity on the surface of adsorption bed reaches 3312 (lattice unit) in 2000 lattice time steps. According to the content in the previous section, after methane concentration within the pores of the adsorbent bed drops to 0, the adsorption only occurs on the surface of the adsorption bed. After 500,000 lattice time steps, the diffusion of methane in Maxsorb III particles is only close to the adsorption amount on the surface of the adsorption bed in the area with a thickness of about 12 (lattice unit).

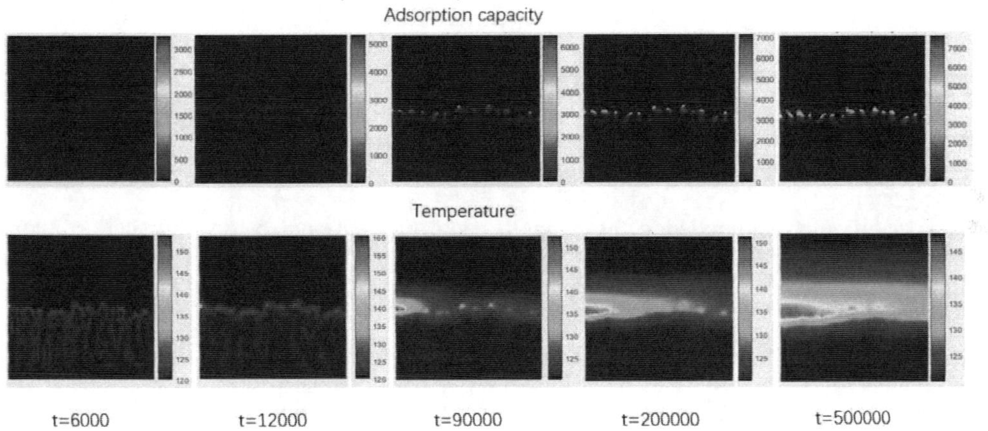

Figure 4. Figure with the change of adsorption capacity of Maxsorb III adsorption bed and temperature change in computational domain with 4000-500000 lattice time steps.

Adsorption is an exothermic process and the temperature change in the computational domain is directly influenced by the adsorption rate. The upper left region of the adsorption bed is near the inlet, where the methane concentration is high and the adsorption rate is high, so the temperature increase in this region is the largest. The variation of the adsorption capacity and temperature in the calculation domain is shown in figure 5. The temperature increases rapidly at the initial stage and reaches a peak value of 167K at 18000 lattice time step. Then the adsorption gradually reaches saturation, and the adsorption rate decreases and heat conducts to the fluid domain and the interior of the adsorption bed, so the temperature decreases gradually. As the temperature drops, the saturated adsorption capacity increases, the adsorption capacity increases slowly. Finally, the adsorption capacity and the temperature stabilize at 7650(lattice unit) and 148K, respectively, in the later simulation period.

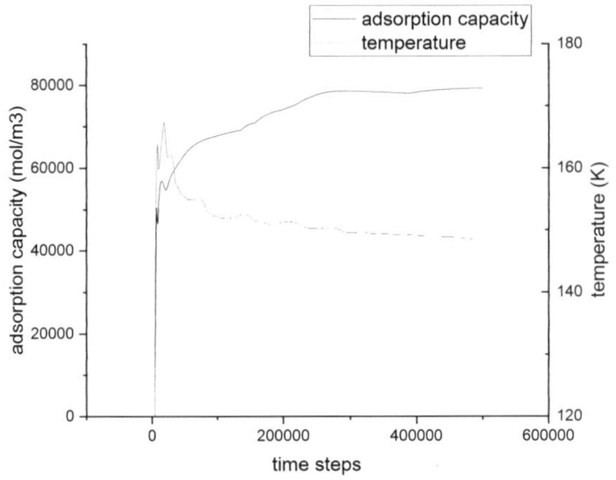

Figure 5. Figure with the variation of the maximum adsorption capacity and maximum temperature.

5. Conclusions

The adsorption heat transfer process at the pore-scale level of porous media was studied based on LBM. The experimental data of Maxsorb III particle saturated adsorption capacity were fitted by Tóth adsorption isotherm model. The methane concentration and adsorption rate on the surface of the adsorption bed were calculated according to Langmuir adsorption kinetics model. The methane adsorption capacity and temperature field in the calculation domain were calculated by the fourth-order Runge-Kutta finite difference method. It is concluded that the maximum temperature increase in the adsorption process reaches 47K, and the temperature change caused by the adsorption exotherm will have a significant impact on the saturated adsorption capacity.

Acknowledgments

The authors gratefully acknowledge the supports from the National Nature Science Foundation of China (Grant No. 52001207) and the China Postdoctoral Science Foundation (Grant No. 2019M651500).

Reference

[1] Wang H, Bai J Q, Qu Z G and Chang M 2019 Numerical study of heat transfer in process of gas adsorption in a Cu-benzene- 1, 3, 5-tricarboxylic acid particle adsorption bed *Int. J. Hydrog. Energy* **44** 11989–2002

[2] Zhou L, Qu Z G, Chen L and Tao W Q 2015 Lattice Boltzmann simulation of gas–solid adsorption processes at pore scale level *J. Comput. Phys.* **300** 800–13

[3] Guo Z and Shu C 2013 *Lattice Boltzmann Method and Its Applications in Engineering* ([Hackensack,] New Jersey: World Scientfic) pp 15-33

[4] Zhou L, Qu Z G, Ding T and Miao J Y 2016 Lattice Boltzmann simulation of the gas-solid adsorption process in reconstructed random porous media *Phys. Rev. E* **93** 043101

[5] Wang H, Qu Z G and Zhou L 2017 A combined GCMC and LBM simulation method for CH 4 capture in Cu-BTC particle adsorption bed *Int. Commun. Heat Mass Transf.* **88** 48–53

[6] Li Q, Kang Q J, Francois M M, He Y L and Luo K H 2015 Lattice Boltzmann modeling of boiling heat transfer: The boiling curve and the effects of wettability *Int. J. Heat Mass Transf.* **85** 787–96

[7] Line O Experimental and theoretical studies on adsorbed natural gas storage system using activated carbons 2 p 80

Focusing of Brownian Particles in a Fluid Induced by the Temperature Difference

Yuxiang Ying, Kaixuan Zheng and Deming Nie*

Institute of Fluid Mechanics, China Jiliang University, Hangzhou 310018, China
Email: nieinhz@cjlu.edu.cn

Abstract. In this work the Brownian motion of particles in a two-dimensional channel was numerically studied through a fluctuating-lattice Boltzmann method. The effects of the fluid temperature difference on the biased Brownian motion of particles were examined. It is shown that the particles are likely to move towards and stay at the low temperature zone. More importantly, the focusing of Brownian particles may be realized if the temperature difference is large enough. This paper also provides a quantitative analysis on this issue.
Keywords. Brownian motion, focusing, lattice Boltzmann method.

1. Introduction

Brownian motion, the random movement of submicro/nano particles suspended in a liquid or gas, is caused by collisions between these particles and the molecules of the liquid or gas. Brownian motion is very common in nature as well as in many engineering fields. The nanofluid may be one of the most famous examples. Experimental study [1, 2] demonstrated that the nanofluids exhibit enhanced thermal conductivity and convective heat transfer ability compared to those of base fluids, which is believed to be connected with the Brownian motion of nanoparticles (e.g., nano metals or metaloxides) [3].

Due to its significance both in engineering and academic fields, much effort has been dedicated to the study of particle Brownian motion. For example, the Brownian motion constrained between two walls was experimentally investigated by Lin et al. [4]. Similarly, Benesch and Yiacoumi [5] studied the wall effects through a numerical method. In particular, the short-time characteristics of the Brownian motion were checked by Iwashita and Yamamoto [6] who performed direct numerical simulations of a sphere driven by thermal fluctuations of a fluid. The hydrodynamic interactions between two Brownian particles were experimentally examined by Radiom et al. [7]. Recently, the motion of active Brownian particles (also known as microswimmer) has received much attention. In comparison with the passive Brownian particles, the microswimmers propel themselves with directed motion and exhibit out-of-equilibrium nature. Volpe et al. [8] proposed a mathematical model to simulate the motion of active Brownian particles in homogeneous and complex environments which is based on a set of stochastic differential equations. Ouyang et al. [9] presented a numerical study on the motion of microswimmers near a wall using an immersed boundary – lattice Boltzmann method. Three types of microswimmer (i.e., pusher, neutral swimmer and puller) were taken into in their study. Reinken et al. [10] developed a fourth-order continuum theory to account for the collective behaviours of a large number of microswimmers in a Newtonian fluid.

It is known that the fluid temperature is a key factor which largely determines the intensity of the Brownian motion of particles. In other words, the Brownian particles move randomly at a higher speed in a hot fluid than in a cold fluid because the flow fluctuations are directly linked to the fluid temperature. If the fluid temperature is the same everywhere, which is the common case, each Brownian particle moves in any direction with the same possibility. This results in the classical Brownian diffusion. However, things may be quite different if there is a temperature difference in the fluid. Our previous study [11] showed that the Brownian motion may become biased when the fluid temperature is not homogeneous. That is, most of Brownian particles were observed to go into the cold area from the hot area. This trend is getting more significant as the temperature difference is larger [11]. As a first step towards the study on the biased Brownian motion, it was assumed that there is no transition region from the hot region to cold region in our previous work [11]. In other words, a sharp interface was used to separate the cold fluid and hot fluid, which does not exist in the real world. To improve the energy equations along with a parameter of heat diffusion coefficient were introduced to solve the fluid temperature in this work. Similarly, the fluctuating lattice Boltzmann method [12] was employed in this work to study the effects of fluid temperature difference on the focusing of Brownian particles.

2. Method

The fluctuating-lattice Boltzmann method (FLBM) [12] was implemented to solved the motion of fluid. A brief description of the FLBM is provided here. The lattice Boltzmann equations are given by,

$$f_i(\boldsymbol{x}+\boldsymbol{e}_i\Delta t, t+\Delta t) - f_i(\boldsymbol{x},t) = -\frac{1}{\tau}\left[f_i(\boldsymbol{x},t) - f_i^{(eq)}(\boldsymbol{x},t)\right] + f_i^{(B)}(\boldsymbol{x},t) \tag{1}$$

where $f_i(\boldsymbol{x}, t)$ is the distribution function corresponding to the i-th discrete velocity $\boldsymbol{e}_i$, $f_i^{(eq)}(\boldsymbol{x}, t)$ is the equilibrium distribution function, Δt is the time step and τ is the relaxation time. In particular, a stochastic term, i.e. $f_i^{(B)}(\boldsymbol{x}, t)$, is introduced to represent the thermal fluctuations of fluid, which is related to the fluctuating stress in the Navier-Stokes equations [12].

The fluid density ρ and velocity $\boldsymbol{u}$ are calculated through the following formulations,

$$\rho = \sum_i f_i, \quad \rho\boldsymbol{u} = \sum_i f_i \boldsymbol{e}_i \tag{2}$$

The D2Q9 lattice model, which is very popular for two-dimensional lattice Boltzmann computations, can be formulated by its velocity vectors as follows,

$$\boldsymbol{e}_i = \begin{cases} (0,0), & \text{for } i=0, \\ (\pm1,0)c,\ (0,\pm1)c, & \text{for } i=1 \text{ to } 4, \\ (\pm1,\pm1)c, & \text{for } i=5 \text{ to } 8, \end{cases} \tag{3}$$

where $c=\Delta x/\Delta t$ and Δx is the lattice spacing. The equilibrium distribution functions are chosen as,

$$f_i^{(eq)}(\boldsymbol{x},t) = w_i\rho\left[1+\frac{3\boldsymbol{e}_i\cdot\boldsymbol{u}}{c^2}+\frac{4.5(\boldsymbol{e}_i\cdot\boldsymbol{u})^2}{c^4}-\frac{1.5\boldsymbol{u}\cdot\boldsymbol{u}}{c^2}\right] \tag{4}$$

where the weights are assumed to be $w_0=4/9$, $w_{1\text{-}4}=1/9$ and $w_{5\text{-}8}=1/36$.

As shown in [12], the relationship between the stochastic term and the fluctuating stress is formulated as,

$$\sigma_{\alpha\beta}^{(B)} = -\tau\sum_i f_i^{(B)} e_{i\alpha} e_{i\beta} \tag{5}$$

According to the fluctuation-dissipation theorem, the following equations need to be satisfied [12],

$$
\begin{aligned}
&\left\langle \sigma_{\alpha\beta}^{(B)} \right\rangle = 0 \\
&\left\langle \sigma_{\alpha\beta}^{(B)}(x,t)\,\sigma_{\gamma\delta}^{(B)}(x' t') \right\rangle = 2k_B T \mu \left(\delta_{\alpha\gamma}\delta_{\beta\delta} + \delta_{\alpha\delta}\delta_{\beta\gamma} - \frac{2}{3}\delta_{\alpha\beta}\delta_{\gamma\delta} \right) \delta_{xx'}\delta_{tt'}
\end{aligned}
\tag{6}
$$

where <> represents averaging over an ensemble, k_B is the Boltzmann constant, T and μ are the fluid temperature and dynamic viscosity, respectively. Note that each member of the fluctuating stress tensor is sampled from a Gaussian distribution with zero mean and a given variance of $2k_B T\mu$. By using a Chapman-Enskog expansion, the Navier-Stokes equations can be recovered from the lattice Boltzmann equations [12],

$$\partial_t \rho + \partial_\alpha (\rho u_\alpha) = 0 \tag{7a}$$

$$\partial_t (\rho u_\alpha) + \partial_\beta (\rho u_\alpha u_\beta) = -\partial_\alpha p + v\partial_\beta \left[\partial_\alpha (\rho u_\beta) + \partial_\beta (\rho u_\alpha) \right] + \partial_\beta \sigma_{\alpha\beta}^{(B)} \tag{7b}$$

The kinematic viscosity of fluid is determined through $n = c^2 \Delta t(\quad -0.5)/3$. According to equation (5), the stochastic term was assumed to have the following forms which guarantee the conservation of mass and momentum [11],

$$
\begin{aligned}
&f_0^{(B)} = 0 \\
&f_1^{(B)} = f_3^{(B)} = \frac{1}{2\tau}\sigma_{yy}^{(B)} \\
&f_2^{(B)} = f_4^{(B)} = \frac{1}{2\tau}\sigma_{xx}^{(B)} \\
&f_5^{(B)} = f_7^{(B)} = -\frac{1}{4\tau}\left[\sigma_{xx}^{(B)} + \sigma_{yy}^{(B)} + \sigma_{xy}^{(B)} \right] \\
&f_6^{(B)} = f_8^{(B)} = -\frac{1}{4\tau}\left[\sigma_{xx}^{(B)} + \sigma_{yy}^{(B)} - \sigma_{xy}^{(B)} \right]
\end{aligned}
\tag{8}
$$

In addition, the bounce-back scheme proposed by Ladd [13] was applied to deal with the curved surfaces of solid particles. The hydrodynamic force and torque of particles were computed through the momentum-exchange method. Then the Brownian motion of particles can be updated by solving the Newton's equations.

3. Problems

This work aims to present a numerical study on the focusing of Brownian particles caused by the fluid temperature difference. As illustrated in figure 1, multiple particles with density ρ_p and diameter d, confined in a two-dimensional channel with length L and height H, undergo Brownian motion due to the random collisions of fluid molecules. The density and viscosity of fluid are denoted as ρ_f and v, respectively. It is assumed that the channel walls have a higher temperature (T_w) than that of the fluid (T_f), i.e., $T_w \geq T_f$. The temperature distribution is then determined through solving the energy equations for which another set of lattice Boltzmann equations were adopted [14]. For more details, readers are referred to [14]. It should be noted here that the fluid temperature difference depends strongly on the heat diffusion coefficient of the fluid (α).

Journal of Physics: Conference Series **2097** (2021) 012025 doi:10.1088/1742-6596/2097/1/012025

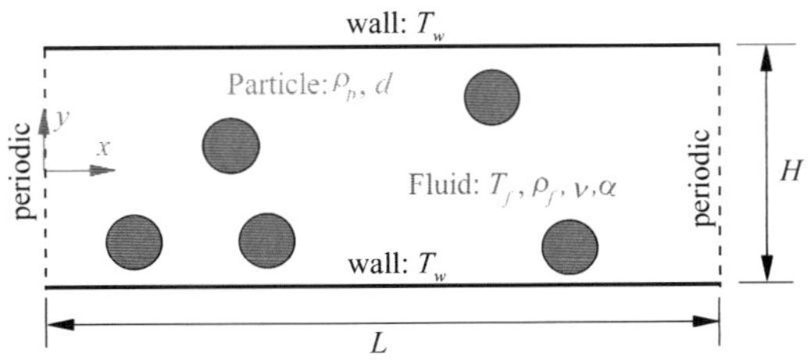

Figure 1. Schematic diagram and notations of the present problem.

For this work some parameters are fixed as follows: $\rho_f = 1$, $T_f = 10^{-4}$, $\nu = 1/100$, $\alpha = 1/600$, $\rho_p = 10$, $d = 13$. The channel height is fixed at $H = 10d$ and its length $L = 4H$. Note that all parameters are all in lattice units. A dimensionless parameter, i.e., $\lambda = T_w/T_f$, was introduced to represent the temperature ratio of the channel walls to the fluid.

In addition, no-slip boundary conditions were imposed on both walls of the channel, as shown in figure 1. Both lateral sides of the channel were assumed to be periodic, which indicates that one particle comes into the channel from the left side once it leaves from the right side, and vice versa.

4. Results

For the sake of saving space, the validation was not included in this paper since the effectivity of the fluctuating – lattice Boltzmann method has been examined in our previous work [11, 12]. For all cases in this work, an array of particles with 5 rows and 20 columns (i.e., 20×5) were initially placed in the channel with equal interparticle distance. It is obvious that the number of the particles is 100.

Figure 2 shows the instantaneous flow fluctuations at different times for $l = 1$, which corresponds to the case of $T_w = T_f$. The fluctuations are illustrated by showing the normalized magnitude of the fluid velocity, i.e., $|\boldsymbol{u}|/U_c$. The reference velocity, i.e., U_c, is defined as $U_c = (k_B T/M)^{0.5}$. Note that M is the mass of a particle. As shown in figure 2, the particles undergo random walks in the channel since there is no temperature difference. As a result, a random but homogeneous distribution of the particles is always seen due to the hydrodynamic interactions and collisions between them [figure 2(b, c)]. This is the classical Brownian diffusion.

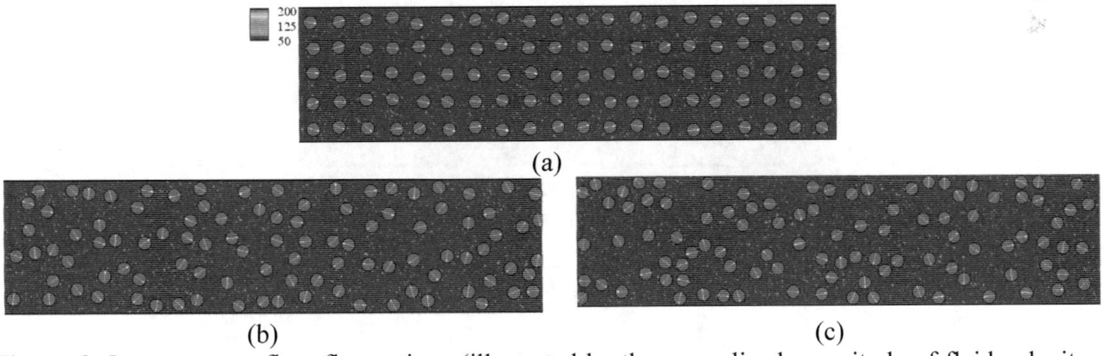

Figure 2. Instantaneous flow fluctuations (illustrated by the normalized magnitude of fluid velocity, i.e., $|\boldsymbol{u}|/U_c$) at different times for $\lambda = 1$ (i.e., $T_w = T_f$): (a) $t = 10^4$, (b) $t = 10^5$ and (c) $t = 3\times10^5$. Note that the reference velocity is defined as $U_c = (k_B T/M)^{0.5}$ with M being the mass of a particle. The same as below.

The situation becomes different when the channel walls are heated. That is, the temperature of the walls is higher than that of the fluid. Figure 3 shows the corresponding results for $\lambda = 5$, i.e., $T_w = 5T_f$. It is seen that the flow fluctuations adjacent to the walls are much stronger than those at the central area of the channel, reflecting the temperature difference in the fluid. It appears that most of the particles at the high temperature zone eventually move into the central area of the channel, as shown in figure 3. The particles exhibit a pattern of biased Brownian motion, which differs from the classical Brownian diffusion (figure 2). The mechanism behind this can be explained as follows. A particle undergoing Brownian motion experiences a larger random force when it is surrounded by a hot fluid than by a cold fluid. Therefore, the particles may be pushed back by the hot fluid when they are trying to move into the high temperature zone from the low temperature zone. If the temperature difference is large enough, the interface between the hot fluid and cold fluid may act as a barrier which prevents the particles from crossing. This is further illustrated by figure 4(a, b), which shows the final state of the particle distribution for $\lambda = 10$ and $\lambda = 15$, respectively.

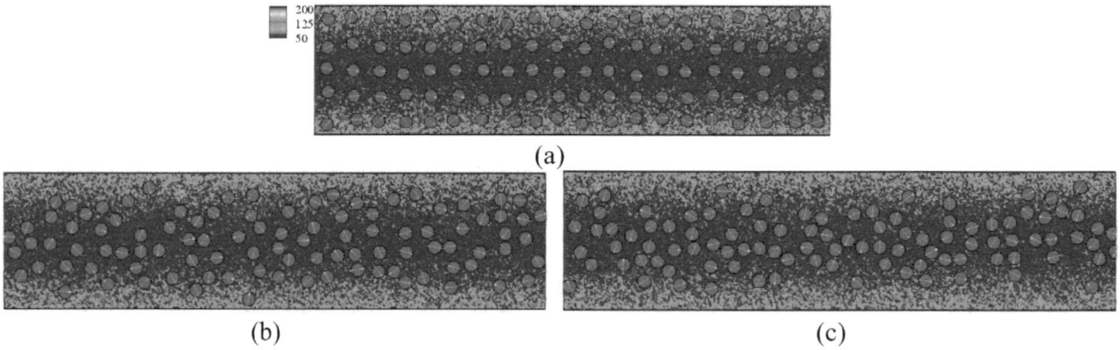

Figure 3. The same as figure 2 except $\lambda = 5$ (i.e., $T_w = 5T_f$).

It is seen from figure 4 that most of the particles are focusing at the central area of the channel due to very strong flow fluctuations near the walls. The larger the temperature difference is, the more significant the focusing of particles is [figure 3(c) and figure 4(a, b)]. This suggests that the biased Brownian motion induced by a temperature difference can be used to drive small particles (sub-micro and nano particles) to a certain region, e.g., the central area of a channel/tube or the corners of a container.

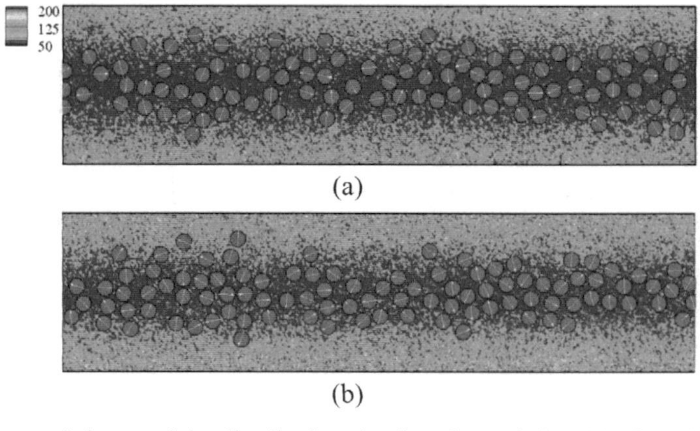

Figure 4. Final state of the particle distribution in the channel for (a) $\lambda = 10$ and (b) $\lambda = 15$, respectively.

To provide a quantitative analysis on the particle focusing, a dimensionless parameter was introduced,

$$Y(t) = \frac{2\sum_{i=1}^{N}\sqrt{y_i^2(t)}}{NH} \tag{9}$$

where $y_i(t)$ is the vertical position of the i-th particle and N denotes the number of the particles. This definition indicates that $Y(t)$ represents the averaged distance between all particles and the channel axis (see also figure 1). The smaller the value of $Y(t)$ is, the more prominent the focusing is. Figure 5 shows the time history of $Y(t)$ for all values of λ considered. It is seen that $Y(t) \sim 0.5$ for $\lambda = 1$ which corresponds to the case of homogeneous fluid temperature. This is in accord with the nature of the classical Brownian diffusion. However, for $\lambda \geq 5$ the value of $Y(t)$ is seen to decrease sharply with time at $t > 10^3$, suggesting that the particles are driven to move towards the central area of the channel. It is obvious that $Y(t)$ reaches a smaller minimum value for a larger λ, which is consistent with the observation made in figure 3 and figure 4.

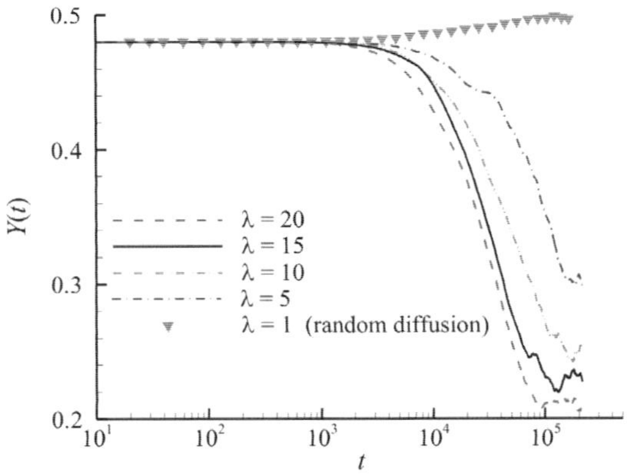

Figure 5. Time history of the averaged distance between the particles and the channel axis [$Y(t)$] for different temperature ratios (λ).

5. Conclusion

A two-dimensional fluctuating lattice Boltzmann method was used to simulate the Brownian motion of particles in a fluid which is confined in a channel. Much attention was paid to a pattern of biased Brownian motion induced by the fluid temperature difference. Results reveal that the Brownian particles are likely to move towards and eventually stay at the low temperature zones of the fluid. This may lead to the focusing of Brownian particles if the temperature difference is large enough. The effects of the fluid viscosity, heat diffusion coefficient and the particle inertia will be further examined in the future work.

Acknowledgments

This work was supported by the National Natural Science Foundation of China (Grant Nos. 11272302, 11632016 and 11972336).

References

[1] Tischer C, Altmann S, Hoerber J, Stelzer E and Florin E L 2001 *Appl. Phys. Lett.* **79** 3878.
[2] Jeney S, Stelzer E, Grubmüller H and Florin E L 2004 *Chem. Phys. Chem.* **5** 1150.
[3] Astumian R D 1997 *Science* **276** 917.
[4] Derényi I and Astumian R D 1998 *Phys. Rev. E* **58** 7781.
[5] Keblinskia P, Phillpot S R, Choi S U S and Eastman J A 2002 *Int. J. Heat Mass Transf.* **45** 855.
[6] Lee S P, Choi S U S and Li S P 1999 *J. Heat Transf.* **121** 280.
[7] Prasher R, Bhattacharya P and Phelan P E 2005 *Phys. Rev. Lett.* **94** 025901.

[8] Volpe G, Gigan S and Volpe G 2014 *American J. Phys.* **82** 659.
[9] Ouyang Z, Lin, J and Ku, X 2018 *Rheol. Acta* **57** 655.
[10] Reinken H, Klapp S, Bär M and Heidenreich S 2017 *Phys. Rev. E* **97** 022613.
[11] Nie D M and Wang C B 2020 *Therm. Sci.* **24** 1.
[12] Nie D M and Lin J Z 2009 *Particuology* **7** 501.
[13] Ladd A J C 1994 *J. Fluid Mech.* **271** *285.*
[14] Gong S and Cheng P 2012 *Int. J. Heat Mass Transf.* **55** 4923.

Study on the Calculation Method for Pumping Process in as Vacuum System

Si Huang[1*], Jiawei Guo[1], Tiankun Yi[1], Songfeng Li[2] and Taizhong Wu[2]

[1] School of Mechanical and Automotive Engineering, South China University of Technology, Guangzhou Guangdong, China
[2] Guangdong Kenflo Pump Co. Ltd., Foshan Guangdong, China
Email: 1158172998@qq.com

Abstract. The calculation methods for vacuum system pumping were usually based on some simple theoretical models, the corresponding results had significant deviations from actual situations. In this study, medium and low vacuum systems (including vacuum chambers, pipes and pumps) were taken as research objects. With a measured vacuum system, and relationship between pump's suction flowrate and inlet pressure, a new calculation method for vacuum pumping time was proposed, in which laminar or turbulent model was selected according to the pipeline's flow state. New and traditional laminar method were used to calculate the pumping process of the measured system, which found that in the middle and high pressure stage, the pipeline flow was in turbulent state and the resistance was non-negligible. If the influence of turbulence was ignored, the calculated pressure drop would be faster than actual situation. The calculation result was verified by actual measurement result, indicating that new method is practical for vacuum pumping time calculation.
Keywords. Vacuum pump, vacuum system, pumping process, calculation method.

1. Introduction

The vacuum system is mainly composed of vacuum pumps, vacuum chambers, pipelines, valves, and program controllers. It is widely used in chemical, electronic, medical, and aerospace technologies [1-5]. The characteristic relationship between the suction rate of the vacuum pump S_p and the suction pressure P_o is the main factor that determines the performance of the vacuum system [6, 7]. In recent years, scholars have carried out many research on vacuum-system pumping. Zhou [8] used mathematical modeling to check the performance of the compressor in the theoretical method, and verified the effect of the new testing-process through simulation tests; Xu [9] developed a calculation program for the constant-speed pumping, and obtained the required pumping time to the specific vacuum degree in the vacuum circulation-degassing process; Ma [10] used a CFD (Computational Fluid Dynamics) method to simulate the high-altitude environment simulation system with specific Mach number; Sui [11] studied the pumping process of a space-environment-simulator with laminar flow model, and obtained the relationship between the pressure in the simulator and the pumping time.

In summary, there were few calculation methods for system vacuuming, and simple theoretical models were generally used, thus the corresponding results deviated greatly from the actual situation. Therefore, in this paper, a quick and effective calculation method was proposed. This method was based on the characteristic relationship between the suction rate of the vacuum pump S_p and the suction pressure P_o, and laminar or turbulent model would be selected according to the flow state of the pipeline, which provided a theoretical basis for the design and engineering application of the system vacuum.

2. Calculation Method

2.1. Theoretical Model of Vacuum System Pumping

As shown in figure 1, the research object was a low-medium vacuum system, which could be simplified to be composed of a vacuum chamber, a main pumping pipeline and a vacuum pump [12]. The theoretical calculation was based on the following assumptions: (1) The fluid media was regarded as ideal gas, and the pumping process was approximately an isothermal process; (2) The gas was a continuous Newtonian fluid.

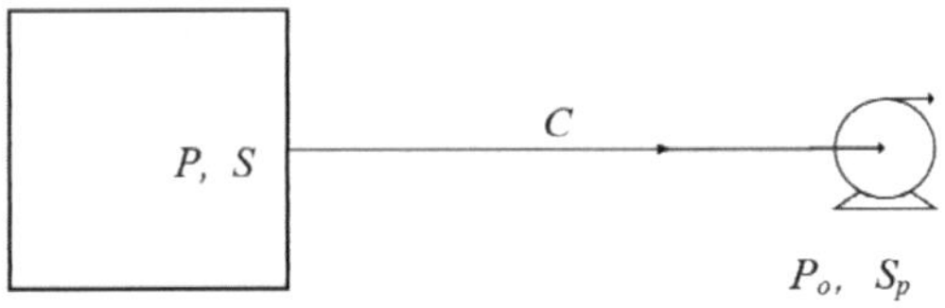

Figure 1. Simplified vacuum system model.

The gas passing through any cross section of the pipeline per unit time, satisfies the following continuous equation [13, 14]:

$$Q = Q_o - Q_l = -V_0 \frac{dP}{dt} \tag{1}$$

Q_o satisfies:

$$Q_o = S_p P_o = SP = C(P - P_o) \tag{2}$$

Equation (2) can be rewritten as:

$$S = S_p C / (S_p + C) \tag{3}$$

The conductance of the pipeline *C* is defined by the equation (2), as the gas flow divided by the change value of pressure of pipeline ports, that is, the volumetric flow rate through the pipeline per unit pressure. So the conductance of the pipeline can be used to indicate the passing capacity of air flow within the pipeline, and the value of conductance can be regarded as the resistance of the pipeline to air flow. So the physical meaning of equation (3) is that after the volumetric flow of pump at pipeline outlet Sp affected by the conductance of the pipeline *C*, the actual volumetric flow at the pipeline inlet can be calculated.

$$Q_l = S_l P_a = \frac{S_m}{\rho_a} P_a \tag{4}$$

From equations (1), (2), (3) and (4), we can get the pumping equation of the vacuum system:

$$V_0 \frac{dP}{dt} = \frac{S_m}{\rho_a} P_a - \frac{S_p C}{S_p + C} P \tag{5}$$

2.2. Conductance Equation of Pipeline

The flow state in the pipeline needs to be judged according to the Reynolds number. When less than 2000, it is laminar, otherwise it is turbulent. The expression of Reynolds number is [14]:

$$Re = \frac{Du\bar{\rho}}{\mu} \tag{6}$$

The ideal gas in isothermal state satisfies [14]:

$$\frac{P_a}{\rho_a} = \frac{\bar{p}}{\bar{\rho}} \tag{7}$$

the average gas pressure $\bar{p}$ in the pipeline was defined as:

$$\bar{p} = (P + P_o)/2 \tag{8}$$

From equations (6), (7), and the average gas flow speed u in the pipeline was calculated by the volumetric flow of the vacuum pump S_p, so there was:

$$Re = \frac{Du\bar{p}\rho_a}{\mu P_a} = \frac{4S_p\bar{p}\rho_a}{\pi D\mu P_a} \tag{9}$$

For a uniform circular cross-section pipe, the pressure difference between the two ports can be expressed by Fanning equation [14], which is:

$$P - P_o = \lambda \frac{L}{D}\frac{\bar{\rho}u^2}{2} \tag{10}$$

The resistance coefficient λ relates to flow state:

$$\lambda = \begin{cases} 64/Re, & \text{Laminar} \\ 0.3164/Re^{0.25}, & \text{Turbulence} \end{cases} \tag{11}$$

Thus, from equations (7), (9), (10) and (11), we could get the expression of gas flow in the pipeline Q_o:

$$Q_o = \bar{p}u\mathrm{A} = \begin{cases} a\bar{p}(P - P_o), Re < 2000 \\ b\bar{p}^{-\frac{5}{16}}(P - P_o)^{\frac{4}{7}}, Re \geq 2000 \end{cases} \tag{12}$$

Where $a = \dfrac{\pi \cdot D^4}{128 \cdot \mu \cdot L}$ and $b = 2.2526 \cdot D^{\frac{19}{7}} \cdot P_a^{\frac{3}{7}} \cdot \mu^{-\frac{1}{7}} \cdot L^{-\frac{4}{7}} \cdot \rho_a^{-\frac{3}{7}}$.

From equations (2) and (12), we could get the expression of the conductence of pipeline C:

$$C = \begin{cases} a\bar{p}, Re < 2000 \\ b\bar{p}^{-\frac{5}{16}}(P - P_o)^{-\frac{3}{7}}, Re \geq 2000 \end{cases} \tag{13}$$

3. Application of Calculation Methods

In order to solve equation (5), it could be considered that when the time step Δt was small enough, the volumetric flow of the vacuum pump S_p was approximately unchanged, and there was a linear relationship between Δt and ΔP, so equation (5) could be rewritten as:

$$\Delta t_i = -\mathrm{V}_0 \frac{\Delta P(S_{pi} + C_i)}{S_{pi}C_iP_i} \tag{14}$$

where $\Delta t_i = t_i - t_{i-1}$, $\Delta P = P_i - P_{i-1}$ and $\bar{p}_i = (P_i + P_{oi})/2$.

The specific calculation method was: dividing the initial pressure P_0 to the end pressure P_e into n segments. With the initial value $P_0 = P_{o1}$ and $\Delta P_1 = P_1 - P_0$, then calculating $\bar{p}_1$. According to P_{o1} and the S_p-P_o characteristic curve of the vacuum pump, to calculate the Reynolds number Re_1 and the conductence of pipeline C_1. Then the corresponding time step $\Delta t_1 = t_1 - t_0$ was calculated by equation (13), $t_0 = 0$. In next iterative calculation, making $P_{o2} = (P_1 + P_{o1})/2$, calculating time step like above, and finally got the total pumping time $\sum_{i=1}^{n} \Delta t_i$.

The tested and measured system of Reference [11] was selected as the calculation case: volume of the vacuum chamber V_0 was 740 m^3, the total length of the main pipeline L was 9 m, the diameter D was 0.32 m. The gas media in the vacuum chamber was 20 °C air, μ was 1.820×10^5 Pa·s, ρ_a was 1.205 kg/m^3, S_m was 0 kg/s, calculation initial pressure $P_0 = P_{o1}$ was 101325 Pa, calculation end pressure P_e was 10 Pa. At one port of the main pipeline, four same vacuum pumps were connected in parallel, and the equivalent pumping flow was the superposition of four pumps [15]. The actual S_p-P_o characteristic curve of single vacuum pump was shown in figure 2.

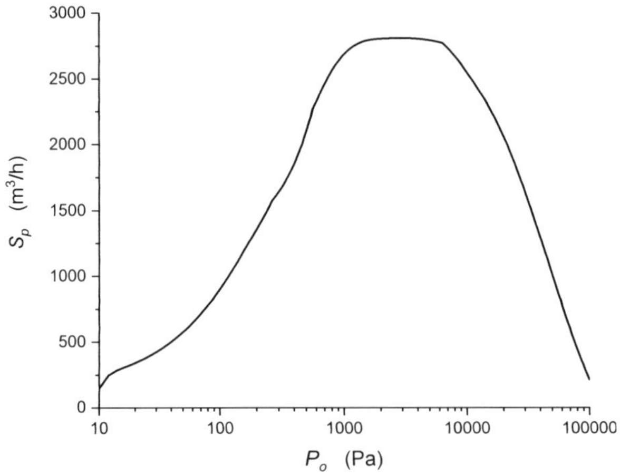

Figure 2. The actual characteristic S_p-P_o curve of single vacuum pump.

For ease of calculation, the actual S_p-P_o characteristic curve in figure 2 was fitted to obtain the following equations:

$$S_p = \begin{cases} -170348.07\cdot e^{(-P_o/1.36)} - 1961.72\cdot e^{(-P_o/212.56)} + 2127.71, \ P_o < 384 \\ -3793.80\cdot e^{(-P_o/289.70)} + 2810.98, \ 384 \le P_o < 6218, \\ 3434.75\cdot e^{(-P_o/48136.69)} - 218.53, \ P_o \ge 6218 \end{cases} \quad (15)$$

The entire calculation process was shown in figure 3, programming with Matlab to do numerical calculation, the pressure step ΔP_i was 1.0 Pa.

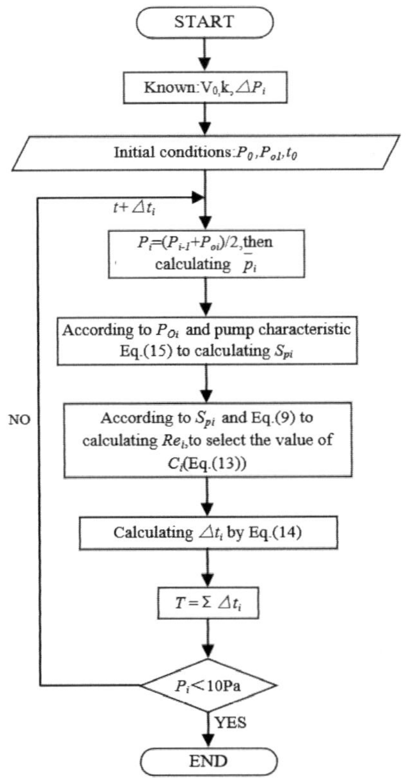

Figure 3. Calculation process.

4. Analysis and Verification of Calculation Results

Figure 4 shows the comparison between the theoretical calculation and Reference [11]'s measured results. The two curves are the results obtained by the conventional laminar method and the turbulent method considered in this paper. As shown in figure 4, turbulence has a significant effect on the calculation results of the pumping time. In the medium and high pressure stage, the *P-t* curve calculated by laminar method drops too fast, and the *P-t* curve calculated by turbulent method drops relatively slowly; in the low pressure stage, the trend of the *P-t* turbulent curve is basically close to the laminar curve.

In figure 4, the trend of turbulent *P-t* curve is consistent with the measured *P-t* curve, and the pressure dropping speed is faster till $P = 500\,\text{Pa}$, and after the inflection point $P = 500\,\text{Pa}$, the pressure dropping speed gradually slows down. The end time of the two curves above reaching $P = 10\,\text{Pa}$ are $t = 3.03(10913$ s) and $t = 3.19(11484$ s) respectively, the relative error is -4.97%. Judging from the calculation results, the method considering turbulence is obviously superior to the traditional method only considering laminar flow. When the vacuum degree gradually increases, the phenomenon of gas leakage and deflation in the vacuum system also increases accordingly. The measured pressure drop is slower when it is close to $P = 10\,\text{Pa}$, which is the main reason for the deviation from the calculation results. In addition, from the turbulent curve in figure 4, it can be seen that under the same pumping flow speed of the vacuum pump, even if the high pressure stage is affected by the turbulent flow, it has a faster pressure drop speed than the low pressure stage. The pumping time in low pressure stage is more sensitive to variation of pumping speed, so attention should be paid to the pumping speed of the vacuum pump in the low pressure stage to improve vacuum pumping efficiency.

Figure 5 is the *Re-t* curve calculated according to equation (9) during the entire pumping process. As shown in figure 5, Reynolds number of pipeline is high in the medium and high pressure stage, and the flow is in a turbulent state. Turbulence increases the resistance of the pipeline, which slows down the drop speed of *P-t* curve. When $t \geq 1.79$ (6450s) , that is the pressure vacuum chamber *P* is less than 384 Pa, the gas density in the pipeline decreases accordingly, and the gas flow in the pipeline changes from turbulent to laminar ($Re < 2000$). In other words, the airflow in the pipeline is in turbulent state for at least half of the entire pumping process, so the influence of turbulent flow should not be ignored in the calculation.

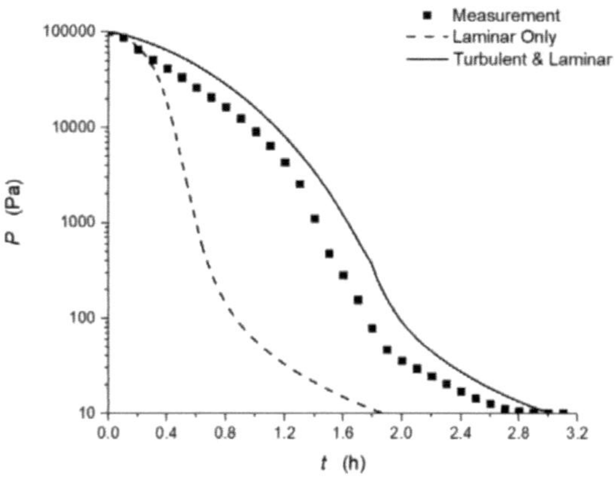

Figure 4. Comparison between theoretical calculation and measurement.

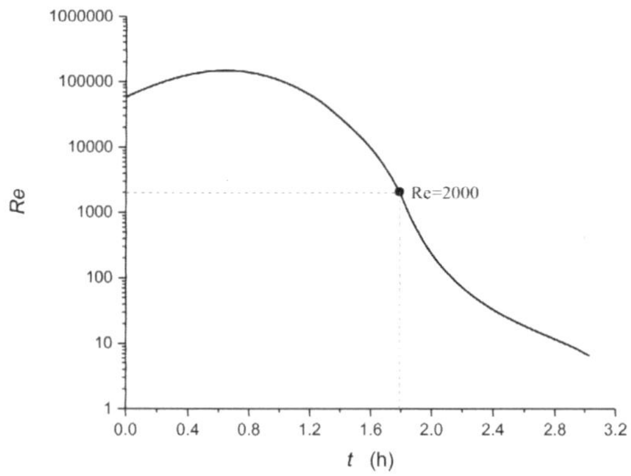

Figure 5. Curve about Reynolds number in the pipeline *Re* and time *t*.

5. Conclusions

This paper proposed a calculation method for vacuum system degassing, and the following conclusions were obtained by analysis and verification of calculation results:

(1) The influence of turbulence on the calculation results of pumping time is significant and notable, mainly in the middle and high pressure stage of pumping process. At this time, Reynolds number is large, the flow is in turbulent state and the resistance of the pipeline increases correspondingly, the *P-t* curve drop too fast and there is a big deviation from the actual situation if calculating by laminar method.

(2) During the pumping process, it takes longer time in the low pressure stage. The pumping time is more sensitive to the variation of the pumping speed in the low pressure stage. Therefore, attention should be paid to the pumping speed of the vacuum pump in the low pressure stage to improve vacuum pumping efficiency.

(3) The trend of *P-t* curve, which is calculated by the turbulent or laminar flow model selected by the airflow state, is close to the measured curve. The error is within the allowable range of actual project, indicating that this is a practical and feasible system method for vacuum pumping time calculation.

Appendices

A	Area of circular pipe of cross-section, m^2.
C	Conductance of main pipe, m^3/s.
C_i	Conductence of pipeline of certain period, m^3/s.
D	Diameter of pipeline, m.
L	Total length of pipeline, m.
P	Pressure in vacuum chamber, Pa.
P_0	Initial pressure of calculation.

P_a	Local atmospheric pressure, Pa.
P_o	Suction pressure of vacuum pump, Pa.
P_{oi}	Suction pressure of vacuum pump inlet of certain period, Pa.
P_e	End pressure of calculation.
Q	Gas flow into vacuum system, Pa·m^3/s.
Q_l	Leakage flow of vacuum chamber, Pa·m^3/s.
Q_o	Pumping flow of vacuum pump, Pa·m^3/s.
S_l	Leakage volumetric flow of vacuum chamber, m^3/s.
S_m	Leakage mass flow of vacuum chamber, kg/s.
S_p	Volumetric flow of vacuum pump inlet, m^3/s.
S_{pi}	Volumetric flow of vacuum pump inlet of certain period, m^3/s.
V_0	Volume of vacuum chamber, m^3.
$\overline{p}$	Average gas pressure in pipeline, Pa.
$\overline{p}_i$	Average pressure of pipeline at beginning of corresponding period, Pa.
u	Average gas velocity in pipeline, m/s.
λ	Resistance coefficient.
μ	Dynamic viscosity of gas, Pa·s.
$\overline{\rho}$	Average gas density in pipeline, kg/m^3.
ρ_a	Corresponding density of local atmospheric pressure, kg/m^3.
ΔP	Pressure step of certain period, Pa.
Δt	Time step of certain period, s.

References

[1] Akbari V M, Bidi M and Najafi A F 2018 Energy, Exergy analysis and optimization of solar thermal power plant with adding heat and water recovery system *Energy Conversion and Management* **171** 1639-50

[2] Sanyal K, Dhara S and Misra N L 2017 Improved approach for the determination of low-Z elements in uranium samples using a vacuum chamber TXRF spectrometer *X-Ray Spectrometry* **46**(5) 442-47

[3] Kikawada S, Uimonen J and Lahitnen J 2014 Energy Saving in Paper Machine Vacuum System *Japan Tappi Journal* **68** 519-24

[4] Liu L 2006 Development of process and equipment of RH vacuum refinery technology *Kang T'ieh/Iron and Steel (Peking)* **41** 1-11

[5] Schäfer K, Zimmermann H and Krühsel G 2003 *Altitude Simulation Bench for VINCI Engine* 39th AIAA/ASME/SAE/ASEE Joint Propulsion Conference

[6] Morris T A, Marciniak M A, Wollenweber G C and Turk J A 2006 Analysis of uncertainties in infrared camera measurements of a turbofan engine in an altitude test cell *Infrared Physics & Technology* **48**(2) 130-53

[7] Strozier J A, Sterling M, Schultz J A and Ignatiev A 2001 Wake vacuum measurement and analysis for the wake shield facility free flying platform *Vacuum* **64**(2) 119-44

[8] Zhou Q, Zhang L, Zhuang C L and Jia H L 2017 Study on Matching Calculation Method and Experimental Technology of Compressor Gas Source for ASTF *Chinese Journal of Turbomachinery* **59**(02) 48-55

[9] Xu H and Branch W S 2019 Design and Simulation Software for Fume Extraction Unit of RH Vacuum Refining Furnace *Chinese Journal of Vacuum Science and Technology* **39(03)** 28-35

[10] Ma P, Yan L and Li H 2017 Aerodynamic design and optimization of high altitude environment simulation system based on CFD *AIP Conference Proceedings* **1839**(1) 020128

[11] Sui J 2017 *Calculation of Vacuum Pumping Process in Space Environment Simulatior* (Liaoning: Northeastern University)

[12] Hucknall D J and Morris A 2003 *Vacuum Technology: Calculations in Chemistry* The Royal Society of Chemistry
[13] Weissler G L, Carlson R W and Hoffman R W 1980 *Vacuum Physics and Technology*
[14] Landau L and Lifschitz E 1987 *Fluid Mechanics* 2nd ed. vol 6 Pergamon Press
[15] Badr H and Ahmed W 2015 *Pumping Machinery Theory and Practice* John Wiley & Sons, Ltd. pp 1-371

Prediction Models of Droplet Characteristics Based on the Locally Convergent Configurations in PMMA Microchips

Zhongxin Liu[1], Zhiliang Wang[2], Chao Wang[1] and Jinsong Zhang[1*]

[1] School of Mechatronic Engineering and Automation, Shanghai University, Shanghai City, 200444, China
[2] School of Mechanics and Engineering Science, Shanghai University, Shanghai City, 200444, China
Email: zhangjs@shu.edu.cn

Abstract. This paper novel designed the local convergence configuration in the coaxial channels to study the two-phase flow (lubricating oil (continuous phase, flow rate Q_c) / deionized water (dispersed phase, flow rate Q_d)). Two geometric control variables, the relative position (x) and tapering characteristics (α), had the different effects on the droplet formation. The increase of relative position x caused the higher frequency and finer droplets, and the increase of convergence angle α, took the opposite effects. The results indicated that the equivalent dimensionless droplet length L_d/W_{out} and the flow rate ratio Q_d/Q_c had an exponential relationship of about 1/2. Similarly, it was found that the dispersed droplets generating frequency and the two-phase capillary number, $Ca_{TP} = u_{TP}\mu_c/\sigma$, had an exponential relationship. The advantage of the convergent configurations in micro-channel was the size and efficiency of droplet generation was very favorable to be controlled by α and x.
Keywords. Prediction model, Two-phase flows, Droplet generation, Convergent coaxial flow

1. Introduction

Lots of literatures have shown that there is a power function relationship between then droplet size and the two-phase flow rate ratio, $L_d/W_{out} \propto (Q_d/Q_c)^k$, called the 'flow-rate-controlled-flow' diagram [1–7] in certain flow parameter zones, where the range of k is $-0.5 \sim 1$. There are relatively less studies on droplet frequency f. It was proposed that there are many different relationships between the droplet frequency and other parameters $f \propto (Q_d/Q_c)^{2/3}$ [8], $f \propto u_d L_d/W_{out}$ [9], and $f \cdot t_{cap} = 0.79 Ca_{TP}^{1.35}$ [10, 11], where t_{cap} is the characteristic capillary time. It can be seen that the prediction models of droplet frequency are not consistent or physically conflicting, and need further in-depth studies.

Prior to those, we put forward a convergent geometry of coaxial micro-channel in our early work [12], which is characterized by the tapering cone-shaped outer tube and the inner needle in a round configuration. Recently, we took the experimental studies of micro-channel tapering on droplet forming acceleration in liquid paraffin/ethanol coaxial flows [13]. Two convergence angles ($\alpha = 0^\circ$ and $\alpha = 2.8^\circ$), limited to the constrains of making perfect conical inner wall by stretching glass tube while heating, were considered to show the big differences of flow charts and droplet characters on slight tapering in round co-axial digital micro-flows. Based on the previous theoretical and numerical results, in this paper we intend to take a more detailed discussion of the effects about tapering geometry on droplet forming through experiments. PMMA microchips of rectangle cross-section channels, instead of round millimeter tubes, are adopted, and serials of samples are observed and analyzed.

2. Materials and Micro-Channel

The materials used are 5W-20 lubricating oil (continuous phase, density ρ_c=826 *kg/ m*^{3}, viscosity μ_c=40.96 *mPa·s*, interfacial tension σ=20.08 *mN/m*) and deionized water (dispersed phase, density ρ_d=986.2 *kg/ m*^{3}, viscosity μ_d=1.23 *mPa·s*, interfacial tension σ=20.08 *mN/m*) separately.

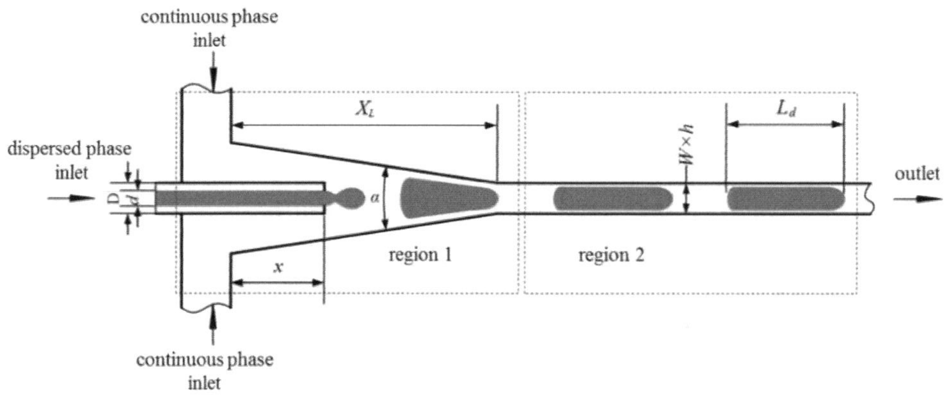

Figure 1. The co-flowing in the tapering micro-channel geometry.

Table 1. Two parameters to define tapering geometries of PMMA microchip samples and four typical flow patterns.

Samples	Convergence Angle α (°)	Nozzle Stretch x (μm)	Flow pattern
A3	5	1170	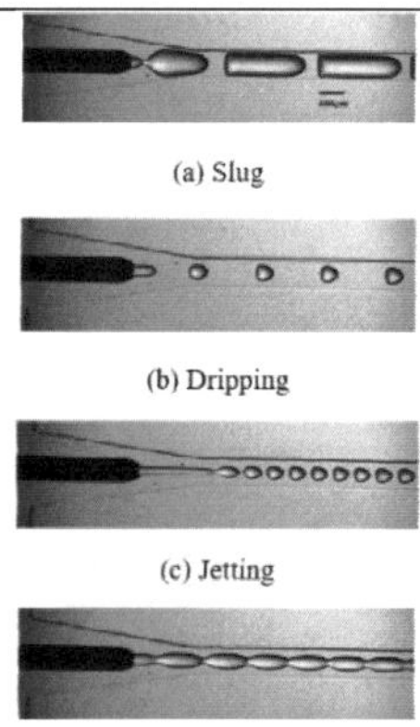
B3	11	1170	
C3	17	1170	(a) Slug
D3	29	1170	
C1	17	0	(b) Dripping
C2	17	620	
C4	17	1620	(c) Jetting
C5	17	2160	
E	0	-	(d) Sausages

Figure 1 shows the coaxial flow configuration and geometric parameters of convergent micro-channel on a PMMA microchip. The dispersed phase and continuous phase flow in from the left side, and flow out toward the right side. Both phases mix at the convergence area, whose length is X_L = 2680 μm and convergence angle α. The syringe needle, whose inner diameter is d = 200 μm and outer diameter D = 400 μm, of dispersed phase is inserted into this tapering area to form two-phase flow (region 1). The downstream micro-channel is a straight tunnel (region 2), with rectangular cross-section of width W_{out} = 400 μm and height h = 600 μm. L_d is the axial size (length) of the droplet generated.

Many samples of PMMA microchips are prepared. The combinations of the two geometrical parameters, convergence angle α and needle displacement x, are shown in table 1, labelled as sample A3, B3, C3, D3, C1, C2, C4, C5 and E.

3. Result and Discussion

3.1. Scaling up Droplet Length in Slug and Dripping Regimes

Figure 2(a) shows the variation of L_d/W_{out} corresponding to the increase of the flow rate ratio Q_d/Q_c at different needle displacements x. The dashed line is an empirical slug-dripping line at $L_d/W_{out} \approx 1.2$ to divide these two regimes. Above which, such as $L_d/W_{out} > 1.2$, is the slug regime for much bigger droplets. We find the linear correlation between L_d/W_{out} and Q_d/Q_c in double logarithm coordinates, which can be described as

$$L_d / W_{out} = A(Q_d / Q_c)^k, \tag{1}$$

where A represents fitting coefficients, and exponent k is 0.49 ± 0.05 under variation of the needle displacement x.

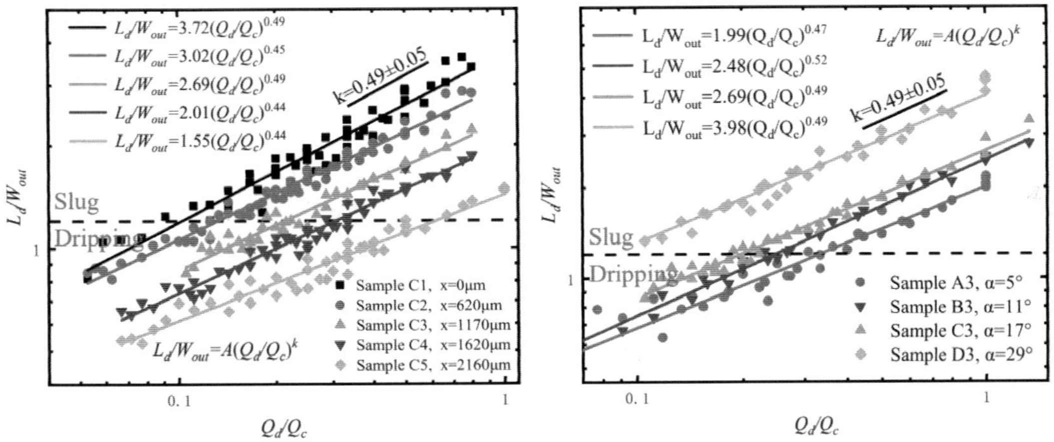

Figure 2. Droplet length L_d/W_{out} varies with Q_d/Q_c. (a) Samples C1–C5 for $\alpha = 17^\circ$ with x changing from 0, 620, 1170, 1620, to 2160 μm; (b) samples A3, B3, C3 and D3 for $x = 1170\ \mu m$ with α changing from 5°, 11°, 17° to 29°.

Similar to figure 2(a), figure 2(b) gives the variation of L_d/W_{out} corresponding to the increase of the flow rate ratio Q_d/Q_c at different convergence angles α. The same fitting formula equation (1) is derived even for the scaling exponent k.

These are very interesting results, and it is not only for the same scaling law of droplet length for both slug and dripping regimes. It shows that the increase of inner needle displacement x can produce finer droplets, which is in consistent with our previous results obtained in round tube experiments and simulations. While increasing the convergence angle α, contrarily, takes opposite effects. Hence, there must exist a intrinsic variable to reflect these phenomena. Analyzing the essential physical or geometrical characteristics, we find that in the tapering configuration, both the increase of needle displacement x and the decrease of convergence angle α cause that the outer flow (continuous oil phase) meets the inner flow at a relative small channel width, which means that the outer flow is accelerated. Therefore, the local micro-channel width right at the inner needle exit, W_{local}, can be the intrinsic variable, and is defined by

$$W_{local} = W_{out} + 2(X_L - x)\tan(\alpha / 2). \tag{2}$$

Hence, we rearrange the data in figure 2(a) and figure 2(b) by multivariate nonlinear regression analysis and replot, as shown in figure 3. Here, the formula can be derived as

$$L_d / W_{local} = 1.32(Q_d / Q_c)^{0.49}. \tag{3}$$

In equation (3), all the lines in figure 2(a) and figure 2(b) collapse to form a single line, and the exponent and coefficient are given by constants. Therefore, the influence of the needle displacement x and convergence angle α is unified.

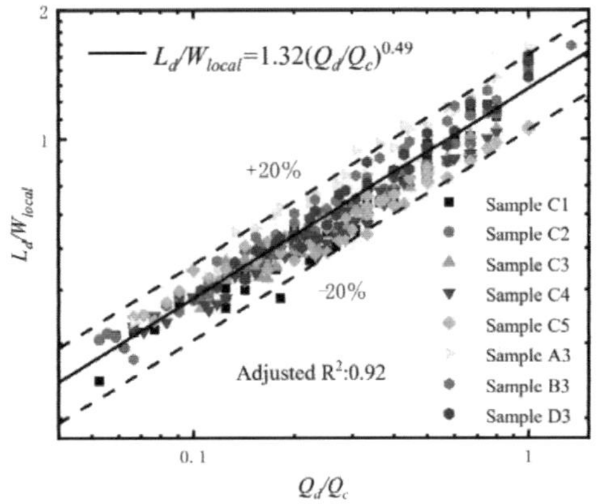

Figure 3. Droplet length L_d/W_{local} varies with Q_d/Q_c for samples C1–C5, A3, B3 and D3.

3.2. Scaling up Droplet Generating Frequency in Slug and Dripping Regimes

In the present tapering micro-channel, the influence of the needle displacement x and the convergence angle α on the droplet length L_d is exponentially related to the water/oil flow rate ratio Q_d/Q_c, while the droplet-generating frequency f is exponentially related to the total water and oil flow rates $Q_{total} = Q_d + Q_c$, as shown in figure 4(a) and figure 4(b).

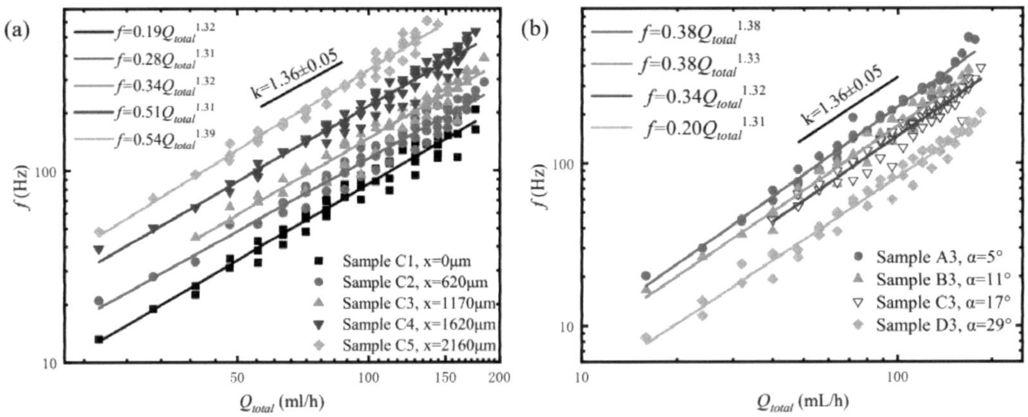

Figure 4. Droplet forming frequency f varies with Q_{total} . (a) Samples C1–C5 for $\alpha = 17^\circ$ with x changing from 0, 620, 1170, 1620, to 2160μm; (b) samples A3, B3, C3 and D3 for $x = 1170\mu$m with α changing from 5°, 11°, 17° to 29°.

When the needle displacement x and convergence angle α are fixed, the droplet frequency f shows an increasing trend with the increase of the two-phase flow rate Q_{total} . When the convergence angle α is constant, the droplet frequency f increases with the increase of the nozzle needle displacement x, as shown in figure 4(a). When the nozzle needle displacement x is constant, the droplet frequency f decreases as the convergence angle α increases, as shown in figure 4(b). In both figure 4(a) and figure 4(b), we gain relation for the droplet frequency f with Q_{total} formulated as

$$f = AQ_{total}^{\ k}, \tag{4}$$

where A represents experimental coefficients separately, and $k = 1.36 \pm 0.05$ is the corresponding exponent. For the same reasons of deriving equation (3), we can combine the influence of the needle displacement x and convergence angle α on the droplet frequency f into a degenerated form. By introduce the capillary time $t_{cap} = \sqrt{\rho d^3 / \sigma}$, we nondimensionalize equation (4), and the dimensionless droplet forming frequency $f \cdot t_{cap}$ can be expressed as

$$f \cdot t_{cap} = 1.82 Ca_{TP}^{1.36}, \tag{5}$$

where $Ca_{TP} = u_{TP}\mu_c/\sigma$ represents the two-phase capillary number defined on the downward velocity and oil viscosity, and $k = 1.36$ is the exponent. As seen in figure 5, the error between equation (5) and experimental data is less than 30%.

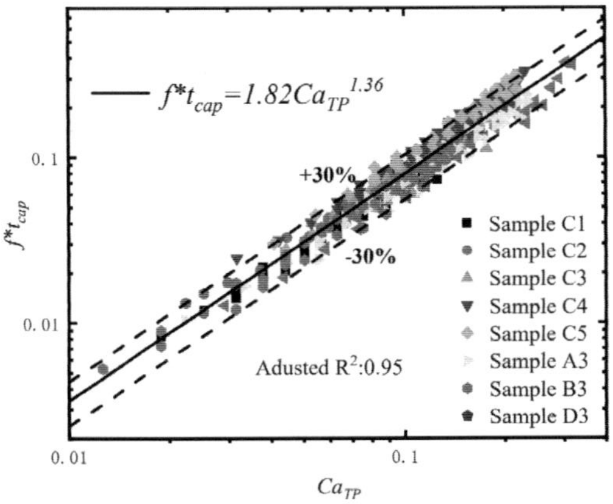

Figure 5. Dimensionless droplet forming frequency $f\, t_{cap}$ varies with Ca_{TP} for samples C1–C5, A3, B3 and D3.

3.3. Extend the Present Models to Non-Tapering Geometries, Test and Compare

Sample E, listed in table 1, is a configuration without tapering (convergence angle $\alpha = 0°$). We use equation (3) to predict the dimensionless droplet length L_d/W_{local}, compare the predicted values $(L_d/W_{local})_{pre}$ with their experiment values $(L_d/W_{local})_{exp}$ directly from sample E, and plot in figure 6(a). Also, We use equation (5) to predict the dimensionless droplet-generating frequency $f \cdot t_{cap}$, compare the predicted values $(f\, t_{cap})_{pre}$ with their experiment values $(f\, t_{cap})_{exp}$ from sample E, and plot in figure 6(b). It is found that the errors are both less than 20%. From figure 6(a) and figure 6(b), we can find that the predictive models are proved to fit usage of non-tapering geometries.

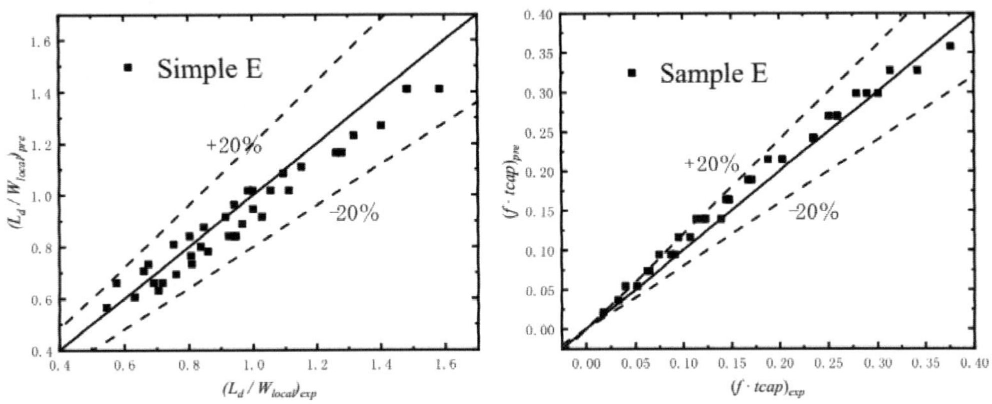

Figure 6. Scaling laws extend to predict for and compare with the experiment data from sample. E. (a) $(L_d /W_{local})_{pre}$ .vs. $(L_d /W_{local})_{exp}$, (b) $(f\, t_{cap})_{pre}$ .vs. $(f\, t_{cap})_{exp}$.

To end our discussion, we would like to compare our scaling models with those derived for straight (non-tapering) micro-channels [14–16], as listed in table 2. The original formula is translated into the normalized models, as in equation (6)–(9), for comparing on the same bases.
We use all these models as in table 2 on a non-tapering geometry sample E and on a tapering geometry sample C3 for comparing. Only the droplet length is considered here, owing to lack of literature work on droplet forming frequency. The correlation between the predictive model values $(L_d/W_{out})_{pre}$ and the experimental data $(L_d/W_{out})_{exp}$ is plotted in figure 7.

Table 2. Predictive model in this paper vs. those derived only for non-tapering structure.

Literature	Original predictive model	Normalized model	Equation
Present	$L_d/W_{local} = 1.32(Q_d/Q_c)^{0.49}$	$L_d/W_{out} = 1.32(Q_d/Q_c)^{0.49}(W_{local}/W_{out})$	(6)
Ref.[14]	$d/W = 1.277(Q_d/Q_c)^{0.031}Ca_c^{-0.18}$	$L_d/W_{out} = 1.277(Q_d/Q_c)^{0.031}Ca_c^{-0.18}$	(7)
Ref.[15]	$L/W = 1 + Q_g/Q_l$	$L_d/W_{out} = 1 + Q_d/Q_c$	(8)
Ref.[16]	$L_b/D_1 = 2.2(Q_g/Q_l)^{0.7}Ca^{0.1}$	$L_d/W_{out} = 2.2(Q_d/Q_c)^{0.7}Ca^{0.1}$	(9)

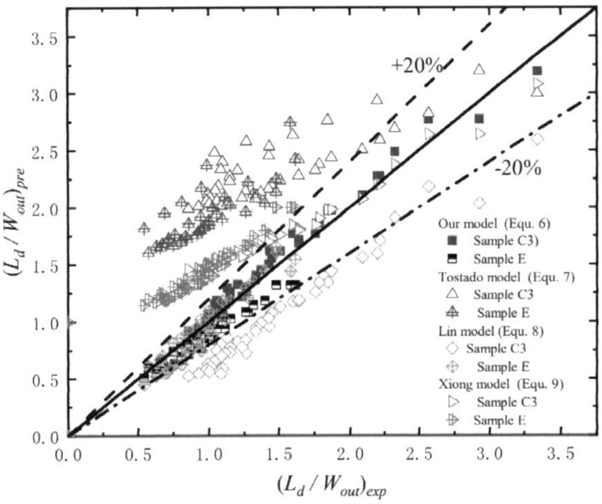

Figure 7. Comparison the four models, equation (6)–(9), predicted values on sample C3 and sample E with their corresponding experimental values, for showing consistency among these models both on non tapering and tapering geometries.

In figure 7, the diagonal means the perfect coincidence between the predicted values $(L_d/W_{out})_{pre}$ and the experimental values $(L_d/W_{out})_{exp}$ for the upper left $(L_d/W_{out})_{pre} > (L_d/W_{out})_{exp}$, or reversely, $(L_d/W_{out})_{pre} < (L_d/W_{out})_{exp}$. It is shown that the present model is solid for sample C3 and sample E within 20% error limit lines, which means its validation both for non-tapering and tapering geometries. Lin's model [16] shows well consistency for non-tapering sample E, but cannot extent its prediction to tapering sample C3, where its predicted value $(L_d/W_{out})_{pre}$ is lower than the corresponding experimental value $(L_d/W_{out})_{exp}$. Moreover, both Tostado's [14] and Xiong's [15] prediction models overestimate and give high deviation from the related experimental value both on non-tapering sample E and tapering sample C3.

4. Conclusion
We find unified scaling laws valid for both of slug and dripping regimes, not only on the droplet length, but also on their forming frequency. The droplet length is exponentially proportional to the water/oil flow rate ratio, and the droplet forming frequency is exponentially proportional to the two-phase capillary number. These laws, though derived from tapering geometries, can be extended to non-tapering cases. Therefore, we can simulate straight channel flow results in tapering ones and realize real-time and effective manipulations of droplet forming characteristics. The comparisons with the literatures on non-tapering cases also show the validation and universality of the present work.

Acknowledgements
This article is funded by the National Key Research and Development Program "8.5 Generation Printing OLED Display Industrial Demonstration Project" (2017YFB0404503).

References

[1] Tice J D, Song H, Lyon A D and Ismagilov R F 2003 Formation of droplets and mixing in multiphase microfluidics at low values of the reynolds and the capillary numbers *Langmuir* **19** 9127–33
[2] Ward T, Faivre M, Abkarian M and Stone H A 2005 Microfluidic flow focusing: Drop size and scaling in pressure versus flow-rate-driven pumping *Electrophoresis* **26** 3716–24.
[3] Xu J H, Luo G S, Li S W and Chen G G 2005 Shear force induced monodisperse droplet formation in a microfluidic device by controlling wetting properties *Lab on A Chip* **6** 131–136.
[4] Garstecki P, Fuerstman M J, Stone H A and Whitesides G M 2006 Formation of droplets and bubbles in a microfluidic t-junction scaling and mechanism of break-up *Lab on a Chip* **6** 437– 446
[5] Cubaud T and Mason T G 2008 Capillary threads and viscous droplets in square microchannels *Physics of Fluids* **20** 501–123.
[6] Xu J, Li S, Tan J and Luo G 2008 Correlations of droplet formation in t-junction microfluidic devices: from squeezing to dripping *Microfluidics and Nanofluidics* **5** 711–717.
[7] Sheu T S, Chen Y T, Lih F L and Miao J M 2010 Ferrofluid-in-oil two-phase flow patterns in a flow-focusing microchannel *Physics Procedia* **9** 147–151.
[8] Derzsi L, Kasprzyk M, Plog J P and Garstecki P 2013 Flow focusing with viscoelastic liquids *Physics of Fluids* **25** 368–141.
[9] Lian J, Luo X, Huang X, Wang Y, Xu Z and Ruan X 2019 Investigation of microfluidic co-flow effects on step emulsification: Interfacial tension and flow velocities *Colloids and Surfaces A: Physicochemical and Engineering Aspects* **568** 381–390.
[10] Liu C, Zhang Q, Zhu C, Fu T, Ma Y and Li H Z 2018 Formation of droplet and string of sausages for water-ionic liquid [bmim][pf6] two-phase flow in a flow focusing device *Chemical Engineering and Processing - Process Intensification* **125** 8–17.
[11] Qian J, Li X, Gao Z and Jin Z 2009 Mixing efficiency and pressure drop analysis of liquid-liquid two phases flow in serpentine microchannels *Journal of Flow Chemistry* 1–11.
[12] Wang Z L 2015 Speed up bubbling in a tapered co-flow geometry *Chemical Engineering Journal* **263** 346–355.
[13] Zhang J, Wang C, Liu X, Yi C and Wang Z L 2020 Experimental studies of microchannel tapering on droplet forming acceleration in liquid paraffin/ethanol coaxial flows *Materials* **13** 1–16.
[14] Tostado C, Xu J, Du A and Luo G 2012 Experimental study on dynamic interfacial tension with mixture of sdspeg as surfactants in a coflowing microfluidic device *Langmuir* **28** 3120–28.
[15] Xiong R, Bai M and Chung J N 2007 Formation of bubbles in a simple co-flowing micro-channel *Journal of Micromechanics & Microengineering* **17** 1002.
[16] Lin X, Bao F, Tu C, Yin Z, Gao X and Lin J 2019 Dynamics of bubble formation in highly viscous liquid in co-flowing microfluidic device *Microfluidics and Nanofluidics* **23**

Research of Pressure Fluctuation with Experiment and Numerical Simulation for Dredging Pump

Mingming Liu*, **Haifei Zhuang and Lei Cao**

CCCC National Engineering Research Center of Dredging Technology and Equipment Co. Ltd, Shanghai, 201208, China
Email: lmm611love@126.com

Abstract. In order to reveal the dredge pump flow instability characteristics, the cavitation and pressure fluctuation in experimental study are carried out, the pressure fluctuation frequency domain and time domain characteristics of three different position inside the volute are analyzed. The results showed that, before cavitation, the main frequency at different positions at different flow rates is 1 times the main frequency of the blade. The fluctuation amplitude near the volute tongue and diffusion section is slightly larger than that at other positions. Before cavitation, the fluctuation amplitude at the same position off design flow is slightly higher than that near the design flow. Cavitation has little influence on the main frequency of the pressure fluctuation. After cavitation, the pressure fluctuation amplitude in the low flow point and the position of the volute tongue under each condition has little change, but cavitation aggravates the pressure fluctuation in the other conditions. Besides, the comparison between simulation and experiment results shows the dredge pump performance curve is in good agreement with the simulation curve, and the simulation results of pressure amplitude at different positions are basically consistent with the experiment results, which verifies the reliability of the numerical simulation method.
Keywords. Dredge pump, pressure fluctuation, experiment, numerical simulation.

1. Introduction

Dredger pump is one of the key transport equipment of dredging engineering ships such as trailing suction dredger and cutter suction dredger. Its operation stability has a very important impact on the dredging efficiency of dredging ships. The dredge pump is mainly a centrifugal dredge pump. The dynamic and static interference between the impeller and the volute makes the pressure in the pump change with time and form pressure pulsation. The flow instability caused by cavitation of dredge pump and off design operation of dredge pump will aggravate the pressure fluctuation. The larger pressure fluctuation may further aggravate the generation of local cavitation, or even cause resonance [1-2], which has a direct impact on both the stability of the dredger's own structure and the dredging efficiency. Therefore, it is of great significance to study the pressure fluctuation characteristics of dredge pump for improving the operation stability and dredging efficiency of dredge pump.

At present, many numerical simulation and experimental studies have been carried out on the pressure fluctuation in centrifugal pump by domestic and foreign scholars. the pressure fluctuation in different positions of the volute is monitored through the test by Parrondo [3] , it shows that the dynamic and static interference between the impeller volute plays a key role in the pressure fluctuation; A centrifugal pump with low specific speed is used to study the vibration and pressure fluctuation by Gao [4], and the results indicate that the pressure fluctuation signal and the pump body vibration signal have strong coherence at each basic frequency, and Gao thinks the main reason of low frequency vibration of centrifugal pump is the interference of dynamic and static in centrifugal pump and unbalance of rotor; Yuan [5] analyzed the pressure fluctuation of impeller and volute under the design condition by numerical simulation, it is found that the main frequency of fluctuation is similar to that of blade, and the high frequency part mainly occurred in impeller channel, and no high frequency fluctuation is observed in the volute; Yang [6] takes double-walled dredge pump as the research object, he calculates and analyzes the pressure fluctuation characteristics under unsteady cavitation state, it is found that the main frequency of pressure fluctuation in the circumferential direction of the volute is blade frequency, the cross-section area of the volute has a great influence on the pressure fluctuation, and the pressure fluctuation at the tongue is affected by cavitation flow; Hui [7] monitor the pressure fluctuation in the impeller and volute channel by numerical calculation, it was found that the pressure fluctuation of centrifugal pump was higher under the condition of small flow. Besides, the pressure fluctuation at the volute tongue and outlet was obvious due to vortex and backflow under the condition of large flow.

Although domestic and foreign scholars have done a lot of research on the pressure fluctuation of centrifugal pump and achieved some results, there are still few experimental studies on the pressure fluctuation of centrifugal pump after cavitation, and there are some differences in the flow characteristics of different series of dredge pumps. Therefore, this paper conducted an experimental study on the pressure fluctuation of a dredged mud pump before and after cavitation. By monitoring the pressure fluctuation of the volute section, the time domain and frequency domain characteristics of the pressure at the test position were analyzed to reveal the variation law of the pressure fluctuation before and after cavitation. Considering the long test period and high cost, the numerical simulation of pressure fluctuation before cavitation was carried out, and the accuracy of the numerical method was verified by comparing the experiment results, so as to provide a certain basis for the subsequent analysis of the flow instability characteristics of different types of dredge pumps.

2. Parameters and Structure of Dredge Pump

The dredge pump parameters are as follows: impeller type is centrifugal impeller, suction diameter is 350mm, impeller diameter is 910mm, design flow is 1800m^3/h, design head is 45m, impeller speed is 600rpm, blade number is 4, discharge diameter is 250mm. The dredge pump structure includes impeller, volute, front and rear wearing liner, etc. in order to simplify the structure, the pump is not equipped with water sealing structure, as shown in figure 1.

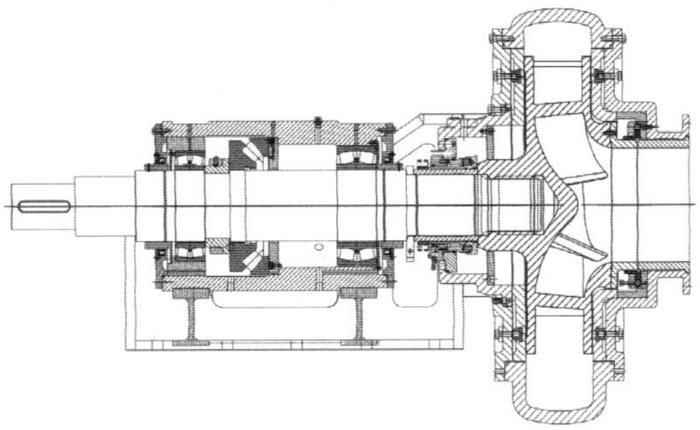

Figure 1. structure of dredge pump.

3. Experiment Method

The experiment accuracy reaches the level 1 accuracy requirements of GB / T 3216-2016. In this test, the motor speed is 1500 rpm, and the pump speed is 611-613 RPM through the reducer. The position of pressure fluctuation test point including M1, M2 and M3 is shown in figure 2.

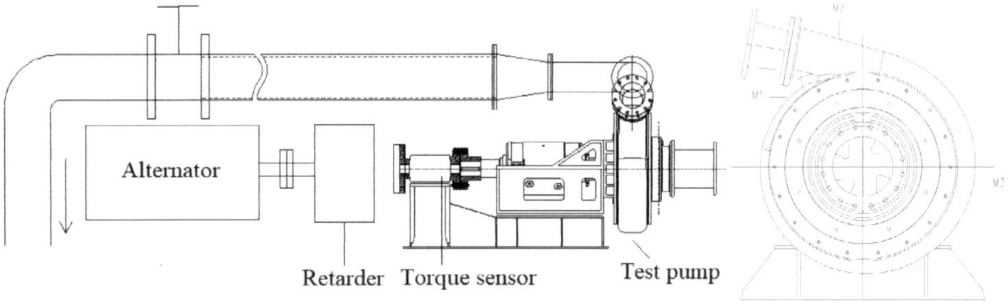

Figure 2. Experiment system and position of pressure fluctuation sensor.

The instruments used in the experiment, such as vacuum sensor, discharge pressure sensor and flowmeter, have been calibrated and qualified. The torque sensor model is JC3a-10000, equipped with secondary instrument which can output 4-20mA signal. The pressure fluctuation sensor is connected with a data collector. Its model is ws-5931u160216c16. The sampling frequency is 2000 Hz, and the sampling time of each test point is 10 s. The measured flow range is 600 ~ 2600 m^3/h. The pressure fluctuation and cavitation test points include five flow points: 600 m^3/h, 1000 m^3/h, 1400 m^3/h, 1800 m^3/h and 2200 m^3/h. In the cavitation experiment, the inlet pressure is continuously reduced by adjusting the inlet pipeline valve, and the head drop of 5% is taken as the evaluation standard of cavitation at the corresponding flow point. It should be noted that:

- The pipe resistance between pressure measuring points should be considered in head calculation, and the experiment data should be corrected in data analysis.
- The installation height of the vacuum sensor and the discharge pressure sensor is the same; The influence of different pipe diameters on the velocity head has been considered in the experiment system, so the velocity head will not be calculated in data processing.
- The torque sensor and the pump shaft are connected by a pair of elastic pin couplings, and the coupling efficiency should be considered in the calculation of pump power.
- All experiment data are converted to 600 rpm based on Similarity Laws.

4. Experiment Results

Taking the impeller rotating 2 times as the time axis, the amplitude is the difference between the instantaneous pressure and the average pressure. Figure 3 is the time domain diagram of the test pressure fluctuation before and after cavitation at the design flow rate (1800 m^3/h). It can be seen from the diagram that the pressure fluctuation changes periodically before cavitation. Before cavitation, the fluctuation amplitude at M1 is greater than at M2, and the fluctuation amplitude at M2 is greater than at M3, The reason is that M1 is affected by the interference between impeller and volute tongue, so the fluid flow is more complex. M2 is close to the diffusion section of volute, so the flow field is also unstable. While M3 is far away from the outlet of volute tongue and diffusion, so the flow at M3 is relatively stable; Compared with before cavitation, cavitation has little effect on the amplitude of pressure fluctuation at M1 and M3. But the maximum amplitude of pressure fluctuation at M2 after cavitation is about twice that before cavitation, which indicates that a large number of cavitation bubbles formed by cavitation flow to the volute outlet under the action of impeller rotation, which intensifies the pressure fluctuation near M2. Fourier transform the above experiment results to obtain the frequency domain diagram of pressure fluctuation, as shown in figure 4, where fn (10Hz) is the rotation frequency of the impeller, and the number of blades is 4. Therefore, the dominant frequency of blades (referred to as blade frequency) is 40 Hz. It can be seen that the pressure fluctuation frequency of each measuring point is mainly blade frequency, and the secondary main frequency is twice blade frequency; Before and after cavitation, the position of the main frequency of the pressure fluctuation remains unchanged, and the main frequency and the secondary frequency are still 1 and 2 times of the blade frequency respectively.

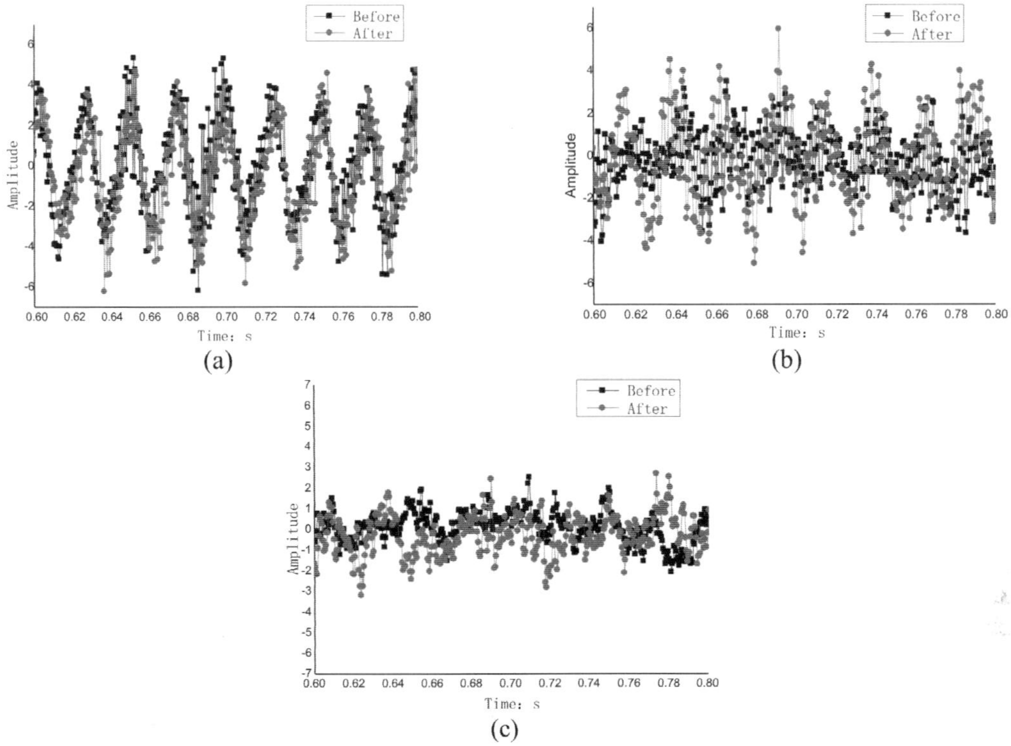

Figure 3. Time domain at design point before and after cavitation. (a), (b), (c) represent the position of M1, M2, M3 respectively.

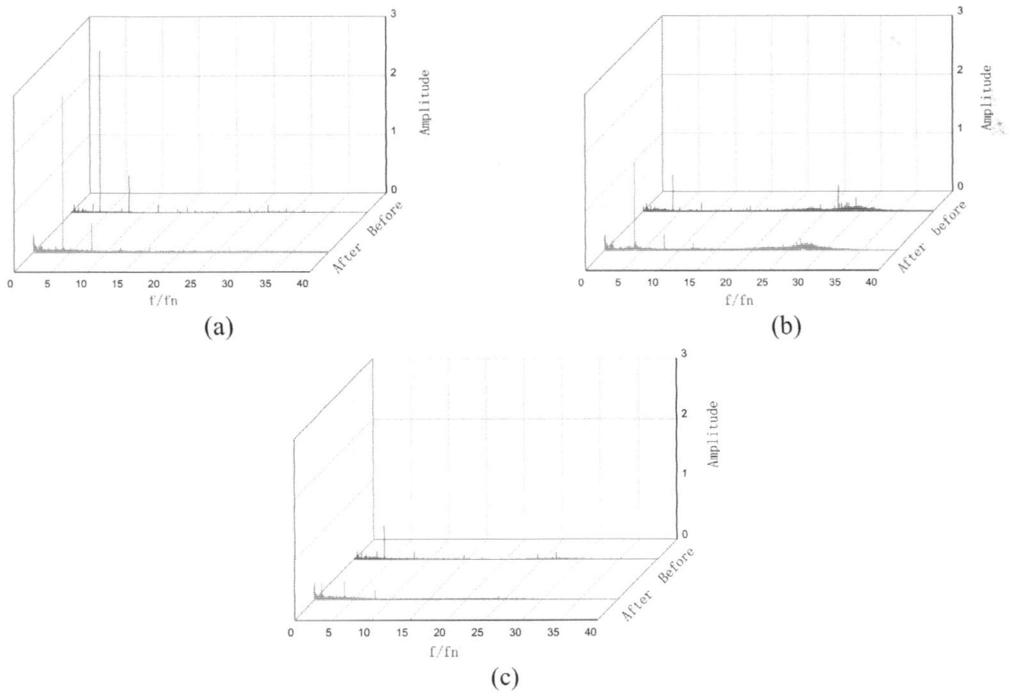

Figure 4. Frequency domain at design point before and after cavitation.

Figure 5 and figure 6 show the frequency domain experiment results of pressure fluctuation at M1, M2 and M3 before and after cavitation at five flow points of 600 m^3/h, 1000 m^3/h, 1400 m^3/h, 1800 m^3/h and 2200 m^3/h respectively. It can be seen that the main frequency of pressure fluctuation at all measuring flow points before and after cavitation is blade frequency; Before cavitation, the maximum amplitude of fluctuation is larger at the flow rate of 600 m^3/h, and there is little difference between different test positions, which indicates that the whole flow in the dredge pump is extremely unstable at low flow rate, while at other measuring flow points, the maximum amplitude of fluctuation is larger at M1 position with the same flow rate, which indicates that the dynamic and static interference of impeller and volute has the greatest impact on pulsation at this time The fluctuation amplitude at M2 and M3 is relatively low; Compared with that before cavitation, cavitation has little influence on the maximum amplitude of pulsation at each position when the flow rate is 600 m^3/h, which indicates that cavitation has little influence on the flow field in the pump, the main reason is that the flow points is far from the design flow rate, the flow in the pump is very complex, and the fluctuation amplitude is larger than other conditions, at this time, the cavitation is hardly to affect the flow field. In addition, compared with that before cavitation, M1 is the least affected by cavitation, and the distribution of fluctuation amplitude is the same as that before cavitation. M2 is the most affected by cavitation at 1800 m^3/h and 2200 m^3/h. Different from M2, M3 is far away from the volute outlet, and the fluctuation amplitude is weakened at the flow rate of 2200 m^3/h, This shows that the cavitation aggravates the backflow and flow instability in the diffusion section of the volute when the flow rate is large.

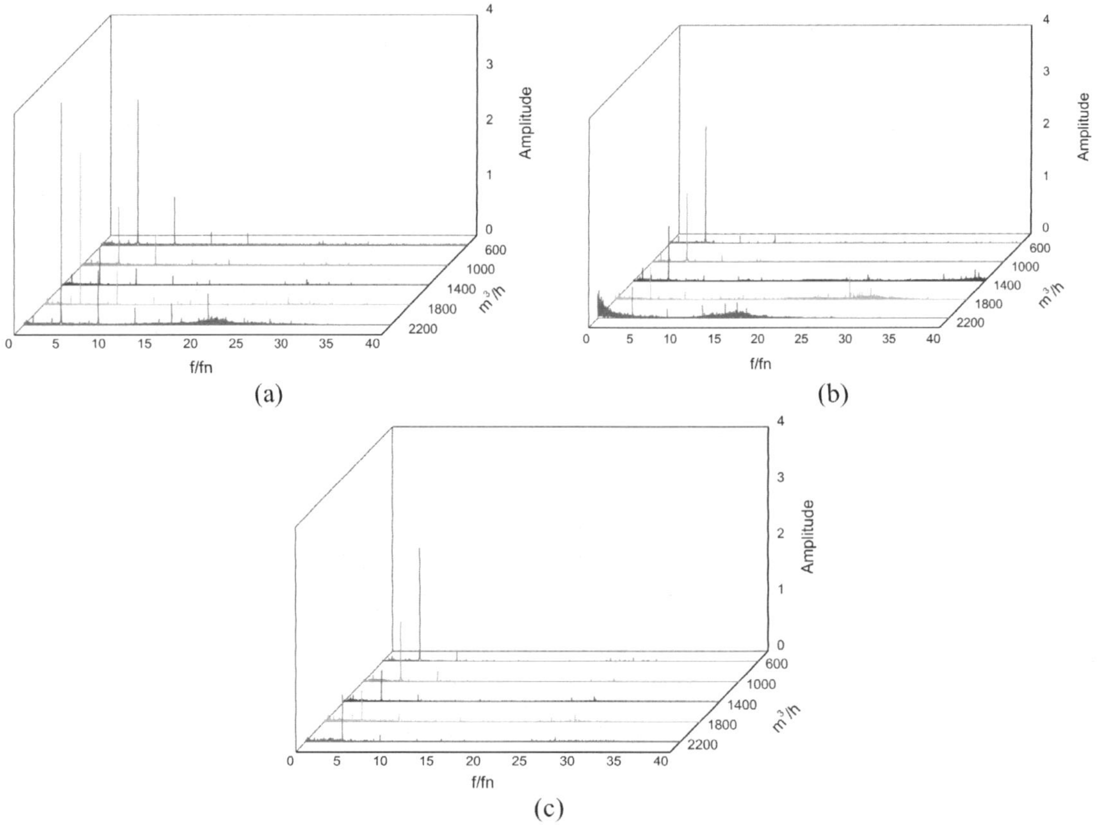

Figure 5. Frequency domain under different flow before cavitation.

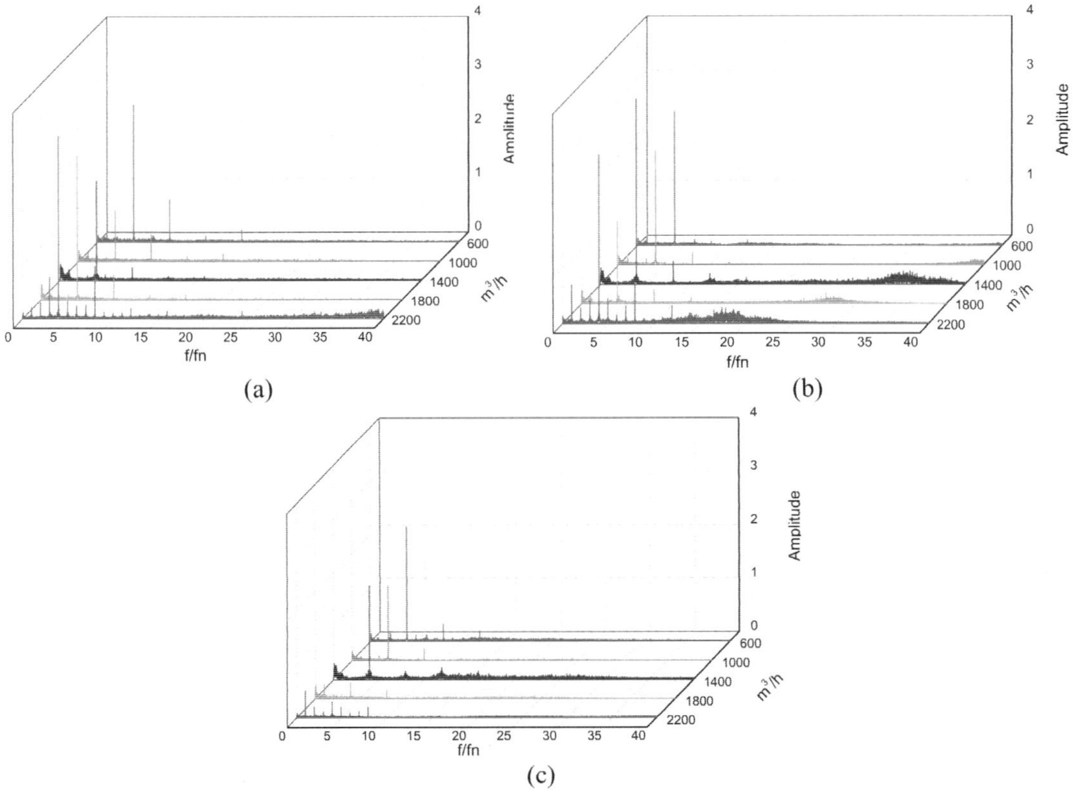

Figure 6. Frequency domain under different flow after cavitation.

5. Comparison between Simulation and Experiment

In dredging engineering, different models of dredgers are equipped with different dredge pumps, and the flow characteristics of different dredge pumps are quite different. Considering that the experiment itself requires a long period and the experiment conditions are more complex, therefore, the numerical simulation of pressure fluctuation before cavitation of dredge pump is further carried out, which can provide a certain basis for the subsequent analysis of the flow characteristics of different dredge pumps.

5.1. Numerical Simulation

The numerical simulation is carried out based on ANSYS CFX 19.2. The computation model includes impeller, volute and inlet and outlet extension, as shown in figure 7. The unstructured tetrahedral grid is used as the calculation grid. Through the grid independence at the design point, the final grid is 2 million. The normal speed and opening are taken as the boundary of inlet and outlet respectively. The rotating reference coordinate is used for the moving surface of impeller; Volute wall adopts no slip wall; The convergence accuracy is set to 1E-4; Considering that the SST model is more accurate for the prediction of flow separation [8], the SST model is selected as the turbulence model for numerical calculation.

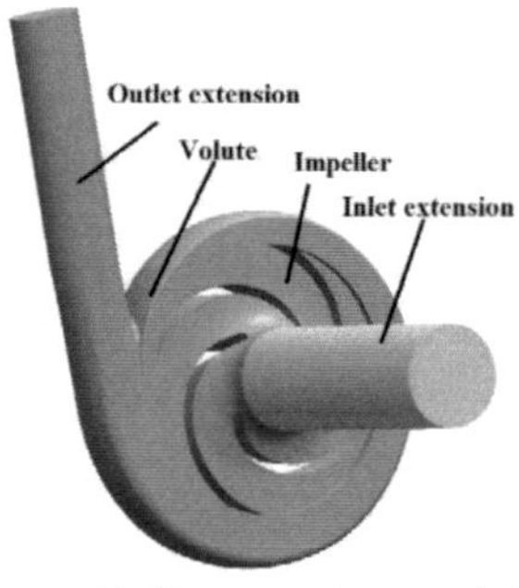

Figure 7. Computation model.

5.2. Energy Performance

Figure 8 shows the energy performance comparison between the simulation and the experiment results of the dredge pump. It can be seen that the performance curves of the simulation and the experiment are generally in good agreement. The efficiency curve of the dredge pump is flat and the high efficiency area is wide. The simulated head value and efficiency are slightly higher than the experimental value. At the design flow point of 1800 m^3/h, the simulation head is 46.3 m, and the simulation efficiency is 87%, which is 0.8 m higher than the experiment head and 3 percentage points higher than the experiment efficiency. The above analysis shows that the numerical model and experimental results can well reflect the energy performance of the dredge pump, which provides a strong support for further analysis of the pressure fluctuation.

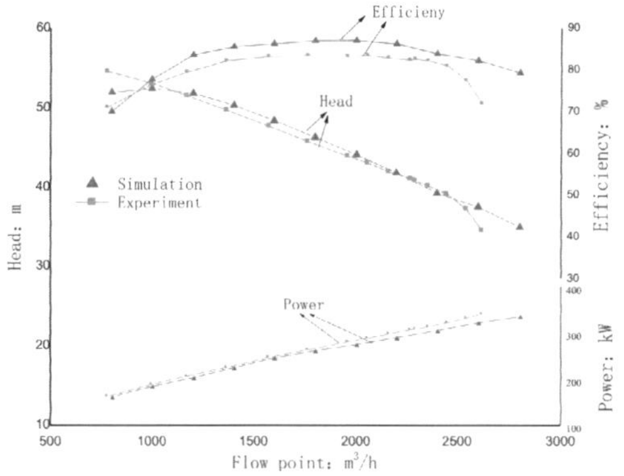

Figure 8. Comparison of energy performance between experiment and simulation for dredge pump

5.3. Pressure Fluctuation

Figure 9 is comparison of the time domain diagram of pressure fluctuation at M1, M2 and M3 before cavitation at design flow point between experiment and simulation. It can be seen from the picture that the simulation results of pressure fluctuation amplitude at different positions are basically consistent with the experiment results. Figure 10 shows the frequency domain diagram of pressure fluctuation at M1, M2 and M3 before cavitation at design flow rate. It can be seen that the simulation results are consistent with the experiment results. The pressure fluctuation frequency of each measuring flow point is mainly blade frequency, and the secondary main frequency is twice blade frequency; In the main frequency part, the simulated amplitudes of M1 and M3 are basically consistent with the experimental results, and the simulated fluctuation amplitudes of M2 are slightly larger than the experimental values, about 1.5 times of the experimental values; In addition, only the water part of the pump is calculated and analyzed, and the whole system of the pump is not calculated, so the high frequency part is not predicted in the simulation.

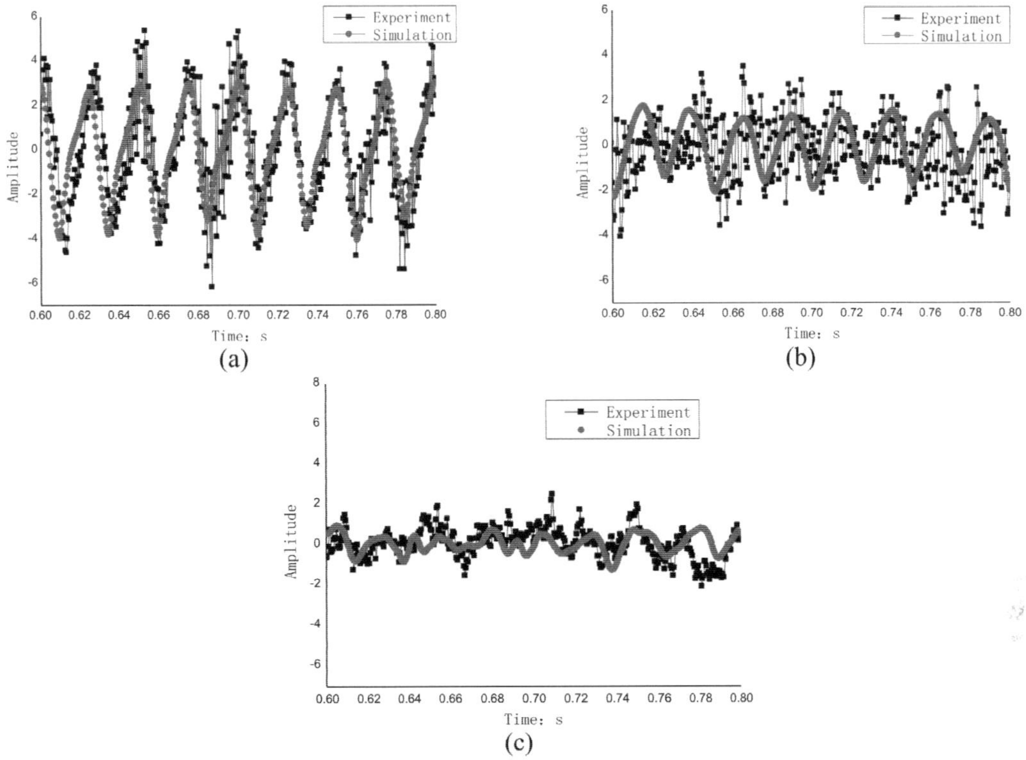

Figure 9. Pressure fluctuation time domain at design flow point.

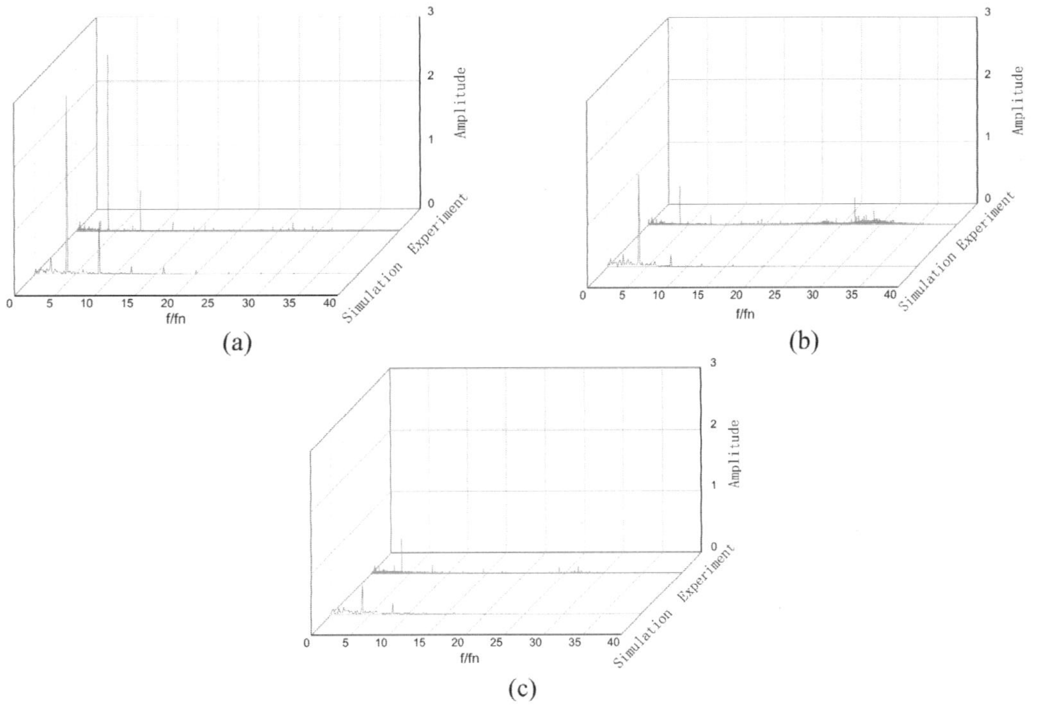

Figure 10. Pressure fluctuation frequency domain at design point.

6. Conclusion

Through the experiment and numerical simulation on the pressure fluctuation of a dredge pump, we can draw the following conclusions:

- Before cavitation, the pressure fluctuation amplitude near the volute tongue is larger than that of the volute diffusion section, and the pressure fluctuation amplitude far away from the tongue and outlet is the smallest; At the same position, there is no significant difference in the amplitude of pressure fluctuation under design flow point, and the amplitude of pressure fluctuation is larger under small flow and large flow.
- Before and after cavitation, the main frequency of pressure fluctuation remains unchanged, which is blade frequency; Under the design flow and large flow point, cavitation has a great influence on the maximum amplitude of pressure fluctuation at the diffusion section of the volute, which is about twice and four times of the maximum amplitude before cavitation respectively, but has little influence on other positions; Under the low flow point, there is no obvious change in the pressure fluctuation characteristics before and after cavitation, and different positions have little effect on the maximum amplitude of the pressure fluctuation.
- The simulation results of pressure fluctuation amplitude at different positions are basically consistent with the experiment results, and the main frequency position of pressure fluctuation at each measuring point is consistent, which verifies the reliability of the numerical method.

Reference

[1] Hu F F, Chen T, Wu D Z and Wang L 2013 Experiment study of cavitation induced vibration and noise of guide vane mixed flow pump *Journal of Drainage and Irrigation Machinery Engineering* **31**(12) 1021-24.

[2] Wang K, Liu H L, Zhou X H and Wang W B 2016 Experimental research on pressure fluctuation and vibration in a mixed flow pump *Journal of Mechanical Science and Technology* **30**(1) 179-84.

[3] Parrondo J L, González J and Fernández J 2002 The effect of the operating point on the pressure fluctuations at the blade passage frequency in the volute of a centrifugal pump *Journal of Fluids Engineering* **124**(3) 784 -90.

[4] Gao B, Yang L, Ni D, Zhang N, and Gu G 2018 Identification of relationship between the pressure fluctuation and vibration in a centrifugal pump based on coherence analysis *Journal of Engineering Thermophysics* **39**(01) 89-93.

[5] Yuan S Q, Ni Y Y, Pan Z Y and Yuan J P 2009 Unsteady Turbulent Simulation and Pressure Fluctuation Analysis for Centrifugal Pumps *Chinese Journal of Mechanical Engineering* **22**(01) 64-69.

[6] Yang J J, Li X J, Pan Z Y and He S 2020 Effects of cavitating flow on induced pressure fluctuation near cutwaters in double-volute centrifugal pump *Journal of Drainage and Irrigation Machinery Engineering* **38**(05) 445-50.

[7] Hui B and Xu H 2018 Characteristic analysis and test of pressure pulsation of centrifugal pump *Electronics World* **12** 25-26.

[8] Zhu C, Wang Z Y and Mao G H 2006 Study on flow field in an elbow draft tube by 3D numerical simulation *Journal of Hydroelectric Engineering* **25**(01) 76-80.

AUTHOR INDEX

Institute of Physics
Dirac House, Temple Back
Bristol BS1 6BE UK

ISSN: 1742-6588
ISBN 978-1-7138-5068-7